D1224331

Structure and Dynamics of Nucleic Acids, Proteins, and Membranes

Structure and Dynamics of Nucleic Acids, Proteins, and Membranes

Edited by

E. Clementi

and

S. Chin

International Business Machines Corporation
Kingston, New York

PLENUM PRESS • NEW YORK AND LONDON

Library of Congress Cataloging in Publication Data

International Symposium on Structure and Dynamics of Nucleic Acids, Proteins, and
    Membranes (1986: Riva, Italy)
    Structure and dynamics of nucleic acids, proteins, and membranes.

    "Proceedings of the International Symposium on Structure and Dyanmics of Nucleic
Acids, Proteins, and Membranes, held August 31–September 5, 1986, in Riva del Garda, Italy, under the sponsorship of the National Foundation for Cancer Research and
the International Business Machines Corporation"—T.p. verso.
    Includes bibliographies and index.
    1. Molecular dynamics—Congresses. 2. Molecular structure—Congresses. 3. Pro-
teins—Congresses. 4. Nucleic acids—Congresses. 5. Membranes (Biology)—Congress-
es. I. Clementi, Enrico. II. Chin, S. (Steven), 1957-   . III. National Foundation for
Cancer Research. IV. International Business Machines Corporation. V. Title. [DNLM:
1. Membrane Proteins—congresses. 2. Nucleic Acid Conformation—congresses. 3.
Protein Conformation—congresses. 4. Structure–Activity Relationship—congresses.
QU 55 I6738s 1986]
QP517.M65I58   1986   574.87'328   87-2366
ISBN 0-306-42553-X

Proceedings of the International Symposium on Structure and Dynamics
of Nucleic Acids, Proteins, and Membranes, held August 31–September 5,
1986, in Riva del Garda, Italy, under the sponsorship of the National
Foundation for Cancer Research and the International Business
Machines Corporation

Preface

This volume collects a number of the invited lectures and a few selected contributions presented at the *International Symposium on Structure and Dynamics of Nucleic Acids, Proteins and Membranes* held August 31st through September 5th, 1986, in Riva del Garda, Italy. The title of the conference as well as a number of the topics covered represent a continuation of two previous conferences, the first held in 1982 at the University of California in San Diego, and the second in 1984 in Rome at the Accademia dei Lincei. These two earlier conferences have been documented in *Structure and Dynamics: Nucleic Acids and Proteins*, edited by E. Clementi and R.H. Sarma, Adenine Press, New York, 1983, and *Structure and Motion: Membranes, Nucleic Acids and Proteins*, edited by E. Clementi, G. Corongiu, M.H. Sarma and R.H. Sarma, Adenine Press, New York, 1985.

At this conference in Riva del Garda we were very hesitant to keep the name of the conference the same as the two previous ones. Indeed, a number of topics discussed in this conference were not included in the previous ones and even the emphasis of this gathering is only partly reflected in the conference title. An alternative title would have been *Structure and Dynamics of Nucleic Acids, Proteins, and Higher Functions*, or, possibly, "higher components" rather than "higher functions." However, since these titles would have been somewhat ambiguous, in the end we decided to keep the original conference title — in deference to tradition — and to amend the ambiguities in this introduction.

The International Symposium in Riva del Garda, as well as this volume, are structured into several subfields ranging from relatively simple, to less simple, to relatively complex, to even more complex. To be correct, the so-called "relatively simple" is, however, quite complicated and it deals with structure of macromolecules. The chapters by Basosi et al., DeSantis et al., Middendorf et al., Poltev et al., Scheraga, and Wuthrich report about studies mainly on the structure of proteins and DNA, the grand foundation of macromolecules. Indeed, it is well known, especially to chemists, that structure is the first and necessary characterization of macromolecules. With the help of conformational analyses and simulations, DeSantis et al. attempted to explain DNA's supercoiling; Scheraga ventured as far as to predict — even if preliminarily — the structure of an interferon; while Wuthrich's mastery in obtaining NMR data opens the door to structural studies of proteins in solutions and offers detailed suggestions to molecular dynamicists. We then move in towards the dynamics of macromolecules with the papers by Alpert, Brunori et al., Dobson and Evans, Gratton et al., Karplus and Parak et al. It is well known that protein dynamics range from femto-seconds to hours, thus covering many time-scales. The many orders of time-scales represent a most difficult challenge for molecular dynamics computer simulations; in this context the contribution by Gratton et al. becomes critically important and appears to call for either a reinterpretation or an extension of the time-scales for the motion of some residues previously obtained from theory. Dynamics, of course, is a most fundamental and vital representation which holds the key to our understanding of the *function* of macromolecules. In this context we recall also the contribution by Sordo et al. dealing with *ab initio* interaction potentials, an alternative starting point for simulations of structural and dynamical studies, and the very interesting and appealing approach by Frauenfelder attempting to explain, with a relatively simple physical model — *a spinglass* — a very complex biological system — *a protein*.

But these macromolecules do not exist in a vacuum and interact with each other in the biological environment in which they exist; this observation brings out the importance of the study of the solvation of macromolecules, which is covered by the contributions of Careri and Rupley, Clementi et al., Lindsay, and Swamy and Clementi.

We should point out that Careri's presentation at Riva del Garda was notably broader than the one published here; by stressing "processes" and "correlations of events," rather than "structure and dynamics," one likely can interpret more deeply molecular and cellular biology.

Macromolecules — their structural interpretation and their dynamical characterization, and the environmental effects — are certainly three important and classical chapters of molecular biology. In this conference, however, we wanted to extend our learning and so we also included contributions from histology and modeling of cellular membranes; the chapters by Conti and Sackmann et al. deal with the latter.

It is self-evident that neurons are cells with very specific functions and are built up as a very complex organization of macromolecules; thus studies on the relationships between all the above chemical systems are an accepted viewpoint in science today.

However, in Riva del Garda we had sessions on artificial vision, pattern and voice recognition, artificial intelligence, present-day supercomputers, and even *sixth generation computers*. The chapters of Clementi et al., Mingolla and Grossberg, Reeke, and Schulten et al. are representative of these topics, possibly unrelated to biology, biophysics and physiology, would we accept a rather old fashioned — but still alive — vision of natural sciences. Our motivation for the inclusion of these modern topics of information sciences lies in the *determination to go much further* at this meeting and to link the most complex manmade machine with one of the most complex structures emerged from the evolutionary process; therefore, we asked what the components of the *computer* might have in common with the components of the *brain*.

The panel discussion did represent a very important moment of the symposium (it is a great pity that considerable portions of this discussion are not reported because of technical difficulties with the recording device). Berni Alder was a magnificent "moderator." Now, one might ask why we want to ask so many complicated and challenging questions at a single meeting. The answer is very simple: we have reached a stage in our civilization where technology is being pushed to the very limit, where we are taking more and more risks because of our ignorance and greed, but also enthusiasm and creative curiousity. We are living in an era which, on one side, is filled with enormous excitement, new learning and discoveries, but, on the other side, might be sowing the seeds of great dangers for the immediate and/or long-range future. *Thus, we need to understand not only at a deeper level, but also at a broader level.* Interdisciplinarity is *not* a luxury — it is a need and, possibly, a *necessity*: to think about more and more powerful engines for computations and simulations, to attempt to realize pragmatically such machines, to hope to learn new ways of interconnecting machine components from physiological research on neuronal networks, is clearly only a hope, but a *rational* one. There might be an element of naivete in our goal; maybe a wise person might characterize our efforts as somewhat *childish*. Yet, we are only at the beginning in an existential need for understanding, discussing and building what might tomorrow become a truly thinking machine. When confronted with novel challenges, namely at every "beginning," we — man — will always be "adolescents" and exhibit "childish" trends, but hopefully on higher and higher evolutionary levels.

In Riva del Garda we have placed "seeds" for the sixth generation computers; let us gather once more in 1988 and reconsider where we shall be.

Steven Chin
IBM Corporation
Kingston, New York

Enrico Clementi
IBM Corporation
Kingston, New York

Contents

Part II. DYNAMICS OF MACROMOLECULES

Part III. SOLVATION OF MACROMOLECULES

Conformational Analysis of Polypeptides and Proteins for the Study of Protein Folding, Molecular Recognition, and Molecular Design[*]

Harold A. Scheraga
Baker Laboratory of Chemistry
Cornell University
Ithaca, New York 14853-1301, USA

Abstract

Conformational energy calculations provide an understanding as to how inter-atomic interactions lead to the three-dimensional structures of polypeptides and proteins, and how these molecules interact with other molecules. Illustrative results of such calculations pertain to model systems (α-helices and β-sheets, and inter-actions between them), to various open-chain and cyclic peptides, to fibrous pro-teins, to globular proteins, and to enzyme-substrate complexes. In most cases, the validity of the computations is established by experimental tests of the predicted structures.

Key words: Internal interactions, ribonuclease, protein folding, conformational energy calculations, structural elements of proteins, oligopeptides, fibrous proteins, homologous proteins, globular proteins, enzyme-substrate complexes

1. Introduction

Twenty-five years ago, we began to develop computational methods[1] for the con-formational analysis of polypeptides and proteins to study protein folding, protein structure, and interactions between proteins and other molecules (molecular recog-nition), and to suggest loci for site-specific mutations in natural proteins to modify their structure and reactivity (molecular design). The motivation for this develop-ment derived from our early experimental work on the structure of bovine pancreatic ribonuclease.[2] Before the X-ray structure of ribonuclease was known, we used solution physical chemical methods to identify three non-covalent tyrosyl . . . aspartate interactions,[2] whose locations were confirmed by the subsequently-determined X-ray structure[3] (Fig. 1). These, together with the known locations of

---

[*]   This paper first appeared in the *Israel Journal of Chemistry*, Vol. 27, 1986.

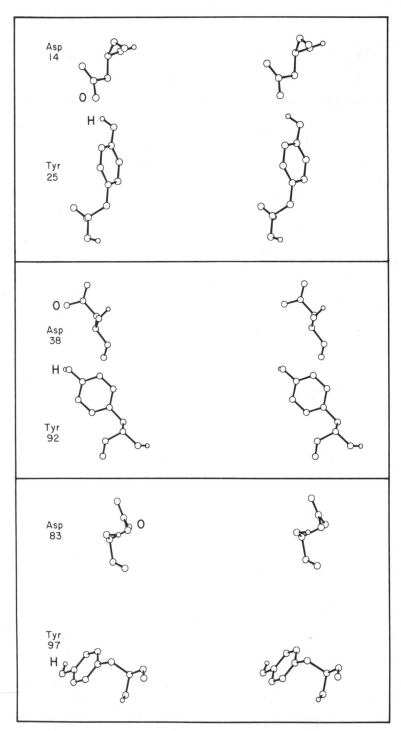

Figure 1. Three Tyr . . . Asp interactions deduced from solution physical chemical experiments.[2] The drawing is based on the subsequently-determined X-ray coordinates of Wlodawer et al.[3]

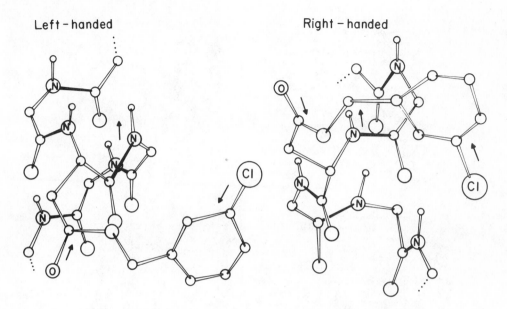

Figure 2. Orientation of the side chains of the lowest-energy left- and right-handed α-helices of poly (m-Cl-benzyl-L-aspartate).[22] The arrows represent the directions of the C-Cl, ester, and amide dipoles, respectively. The dipole-dipole interactions are more favorable in the left-handed form.

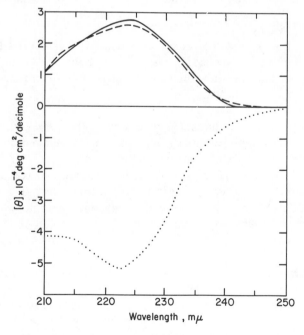

Figure 3. Circular dichroism spectra[32] of poly(β-benzyl-L-aspartates) in dioxane at 25°C. The symbols (——), (− − −), and (. . .) correspond to o-, m-, and p-Cl-benzyl derivatives, respectively.

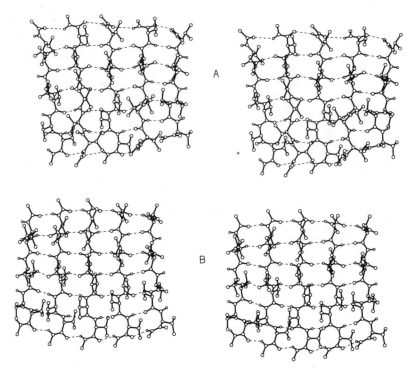

Figure 4. Stereo drawings of the minimum-energy β-sheets with five CH$_3$CO-(L-Val)$_6$-NHCH$_3$ chains.[24] (A) Antiparallel structure. (B) Parallel structure.

the four disulfide bonds, and the proximity of His 12, Lys 41 and His 119 in the active site of this enzyme, provided distance information that could serve as useful constraints to determine the three-dimensional structures of proteins by means of conformational energy calculations.

Initially,[1,4,5] only hard-sphere potentials were used in the computations, but subsequently more complete empirical energy functions were developed both in our laboratory[6-10] and elsewhere.[11-14] Our current main program is ECEPP/2, Empirical Conformational Energy Program for Peptides.[9,10] Procedures have also been introduced to take hydration and entropy effects into account, and to carry out energy minimization, Monte Carlo and molecular dynamics computations in a multi-dimensional space; these procedures have been reviewed on numerous occasions.[8,15-21]

While improvements in potential functions can be expected, the present ones are sufficiently accurate to provide agreement between various computational and experimental results, as will be shown here. The main obstacle to further progress is the multiple-minima problem, and much effort is being devoted to surmount this difficulty. In fact, this problem has already been solved for small open-chain and cyclic peptides and for fibrous proteins such as collagen, and progress is being made in the area of globular proteins.

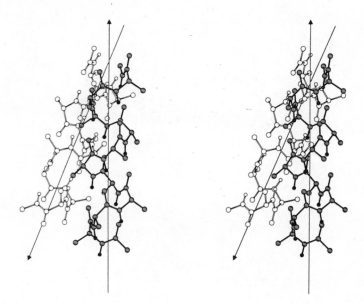

Figure 5. Stereo drawing of two $CH_3CO\text{-}(L\text{-}Ala)_{10}\text{-}NHCH_3$ α-helices in the lowest-energy packing arrangement.[27] The helix axes are indicated by arrows, with the head of the arrow pointing in the direction of the C-terminals of each helix.

This article will describe some of the conformational problems that have been eluci-dated by this computational methodology, and will summarize some of the current efforts being made to overcome the multi-minima problem for globular proteins. We shall describe first the results of computations on model systems of increasing complexity, and then consider the conformations of small open-chain and cyclic

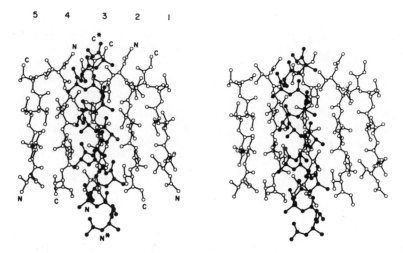

Figure 6. Stereo drawing of the lowest-energy packing arrangement of a $CH_3CO\text{-}(L\text{-}Ala)_{16}\text{-}NHCH_3$ α-helix and an antiparallel $CH_3CO\text{-}(L\text{-}Val)_6\text{-}NHCH_3$ β-sheet.[29]

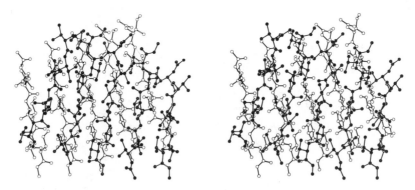

Figure 7. Stereo drawing of the lowest-energy packing arrangement of a CH$_3$CO-(L-Ile)$_6$-NHCH$_3$ parallel-chain β-sheet (open atoms) and a CH$_3$CO-(L-Val)$_6$-NHCH$_3$ antiparallel-chain β-sheet (filled atoms).[30]

peptides, fibrous proteins, and globular proteins. Further details can be found in several recent reviews.[18-21]

2. Model Systems

Conformational energy calculations have demonstrated how interatomic interactions lead to the preferred twists of α-helices[7,22] and β-sheets,[23-26] and to the preferred modes of packing of α-helices with α-helices,[25,27,28] α-helices with β-sheets,[29] β-sheets with β-sheets,[30] and pairs of triple helices of collagen with each other.[31] Some examples are provided in Figs. 2–8. Figure 2 illustrates how side chain-backbone dipole-dipole interactions play a dominant role in leading to a left-handed helix in poly(m-Cl-benzyl-L-aspartate).[22] The predicted[22] left-handedness of

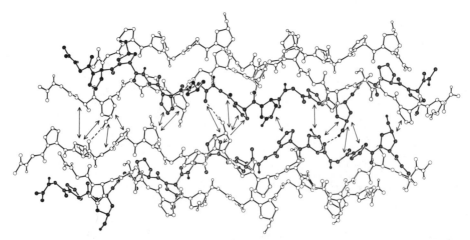

Figure 8. Computed lowest-energy packing arrangement[31] of two [CH$_3$CO-(Gly-Pro-Hyp)$_5$-NHCH$_3$]$_3$ triple helices, showing the near-parallel alignment of the two triple helices and O-H . . . O=C hydrogen bonds (dashed lines) between the triple helices. The arrows indicate residues that are in contact between the two triple helices.

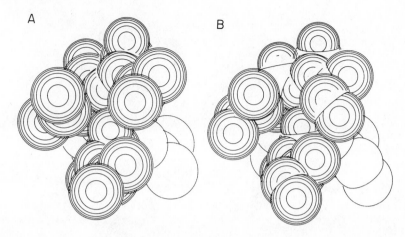

Figure 9. Space-filling models[36] of α-helical poly(L-valine) (A) and poly(L-isoleucine) (B).

the o- and m-derivatives, and the right-handedness of the p-derivative, were subsequently verified by circular dichroism measurements,[32] as indicated in Fig. 3. Figure 4 shows the computed right-handed twist of poly-L-valine β-sheets.[24] The lowest-energy packing arrangements for α . . . α, α . . . β, β . . . β pairs, and pairs of triple-helical collagen structures are illustrated in Figs. 5 – 8, respectively.

Conformational transitions have also been treated by this methodology, e.g. the interconversion of the cis and trans forms of poly(L-proline),[33] of the α- and

Table I. Comparison of Lowest-energy Structures Obtained by Two Different Methods[a]

	SMAPPS	Build-up plus Energy Minimization
	Backbone Dihedral Angles (DEG)	
Tyr ϕ	−86	−87
ψ	154	155
Gly ϕ	−155	−157
ψ	95	96
Gly ϕ	71	71
ψ	−90	−91
Phe ϕ	−90	−91
ψ	−40	−38
Met ϕ	−165	−165
ψ	−50	−48
Energy (kcal/mol)	−10.64	−10.66

[a]In this computation, the side chains were constrained to have the conformations of Fig. 12. More recently, this constraint has been eliminated (G.H. Paine and H.A. Scheraga, unpublished results).

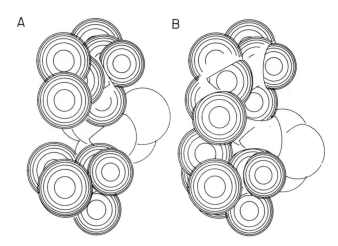

Figure 10. Space-filling models[36] of the β-conformation (the dominant species in the coil form) of poly(L-valine) (A) and poly(L-isoleucine) (B).

ω-helical forms of a crystalline homopolyamino acid,[34] and the α-helical and coils forms of various homopolyamino acids.[35-37] For example, the helix-coil transition curves for poly(L-valine) and poly(L-isoleucine) in water differ markedly from each other, primarily because of the different degrees of hydration of the side chains in the helical and coil forms of the two polymers;[36] the different effects of hydration arise from the difference in side chain-side chain separation, because of the extra methyl group in isoleucine, as illustrated in Figs. 9 and 10.

A model poly(L-alanine) chain has recently been used to test a new procedure for overcoming the multiple-minima problem.[38] This procedure is based on the assumption that each residue must have optimal electrostatic energy; i.e., the dipole moment of each residue must be optimally aligned in the electrostatic field created by the *whole* molecule. If it is not, the orientation of the dipole moment (of each residue, in turn) is changed to improve the electrostatic energy. Since this involves a *local* movement (in the field of the whole molecule), it is computationally very fast. Then the energy of the whole molecule [taking all (ECEPP) interactions, not only electrostatic, into account] is minimized, and the whole procedure is repeated iteratively.

Thus far, this procedure has been tested on a 19-residue poly(L-alanine) chain with acetyl- and N-methyl amide terminal blocking groups. The global minimum of this structure (in the absence of water) is an α-helix. The starting conformations were *very far* from the helical conformation and, in trivially short computation time, the global minimum was achieved.[38] The top stereo diagram of Fig. 11 illustrates one of the starting conformations (optimized by conventional energy minimization), which is very far from a helical one; i.e. the minimization procedure was trapped in this high-energy local minimum. After application of the electrostatic-optimization

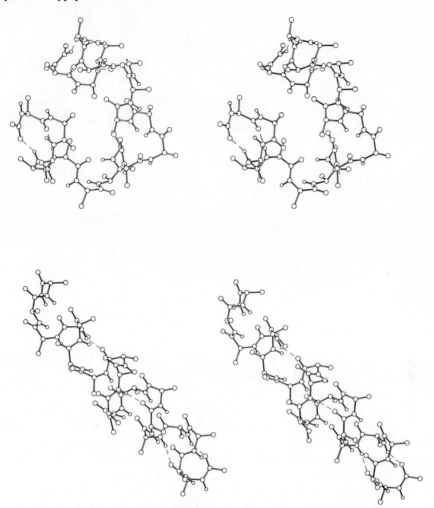

Figure 11. Test of procedure to surmount the multiple-minima problem.[38] Stereo-diagrams illustrating (top) a compact conformation of $CH_3CO\text{-}(Ala)_{19}\text{-}NHCH_3$ after complete ECEPP energy minimization to reach this particular local minimum, and (bottom) the global-minimum (α-helix) structure attained by first optimizing the local electrostatic interactions and then carrying out a complete (ECEPP) energy minimization.

procedure (and subsequent energy minimization with the complete ECEPP function), the global-minimum (α-helix) structure at the bottom of Fig. 11 was obtained. Unlike the usual minimization procedures, which make *small* changes in the dihedral angles, this new procedure can make very large changes (even $100° - 200°$) in these independent variables, as illustrated in Fig. 11. Whereas conventional energy minimization gave the structure at the top of Fig. 11, the new procedure surmounted local barriers and attained the global minimum. This method is now being tested on other structures for which the global minimum is not an α-helix.

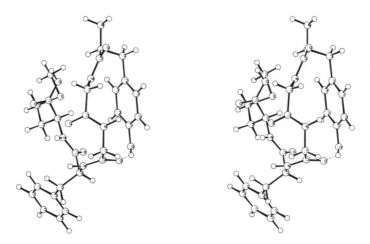

Figure 12. Stereo drawing of lowest-energy conformation of Met-enkephalin.[39]

3. Open-Chain and Cyclic Peptides

A "build-up" procedure[18,19] has been used to surmount the multiple-minima problem for small open-chain and cyclic peptides. In this method, longer peptides are built up from all possible combinations of low-energy conformations of shorter-peptides. This assures an adequate coverage of conformational space. Energy minimization is carried out at each stage of the build-up procedure. The problem of storage of numerous conformations is alleviated by (legitimately) eliminating inconsistent conformations. For example, in building up the pentapeptide A-B-C-D-E from the tetrapeptides A-B-C-D and B-C-D-E, all tetrapeptides in which the conformations of the common tripeptide fragment B-C-D are not similar can be eliminated. When this procedure was applied to the pentapeptide Met-enkephalin,[39] with the sequence Tyr-Gly-Gly-Phe-Met, the hairpin-bend structure of Fig. 12 was obtained; a more recent improvement of the build-up procedure[40] led to a slightly altered conformation with a somwhat lower energy. Simultaneously, a new procedure, based on statistical mechanics and designed to surmount the multiple-minima problem by an adaptive importance sampling Monte Carlo method,[41] was applied to this same pentapeptide. This procedure (SMAPPS, Statistical mechanical algorithm for predicting protein structure) gave identical results to those obtained by the build-up method, as can be seen in Table I.

The results of Table I obtained by two very different computational procedures, using the same potential functions, demonstrate the validity of both procedures. Work is currently in progress to pack this molecule into a crystal to determine the possible influence of intermolecular interactions on the conformation of this short linear peptide. (Parenthetically, it may be pointed out that SMAPPS evaluates the partition function directly. Hence, it may possibly provide a solution to the problem of computing free energies directly by a Monte Carlo procedure; tests of this possibility are in progress.)

A larger oligopeptide, the 20-residue membrane-bound portion of melittin has also yielded to the build-up procedure. The lowest-energy, largely α-helical structure[42] is shown in Fig. 13. X-ray[43] and NMR[44] structural information is available for melittin as a tetramer in a crystal or as a monomer bound to micelles, respectively; considering possible environmental effects in either of these forms, the general agreement with experiment is satisfactory.

Gramicidin S, a cyclic decapeptide, serves as an example of the application of the computational (build-up) methodology to a cyclic molecule. The requirement to close the ring constitutes an additional constraint during the energy minimization. Figure 14a shows the computed structure,[45,46] and Fig. 14b provides a diagram of the subsequently determined X-ray structure[47] which agrees[46,48] with the predicted one. When the structure was computed[45] in 1975, it was stated that "the ornithine side chain may be free to occupy more than one rotational state . . . (and there is) a hydrogen bond between a δ-NH$_2$ proton of Orn and the backbone CO of Phe. There is no experimental evidence indicating the existence of this interaction." When the X-ray structure was reported[47] in 1978, it was stated that "there is an intra-molecular hydrogen bond . . . between . . . the ornithine side-chain nitrogen atom and the D-phenylalanine carbonyl oxygen atom which has not been predicted." Figure 14a, which has a computed rotational variant[46] of the previously computed[45] side-chain conformation of ornithine, clearly shows the predicted hydrogen bond (the original computed rotational variant[45] had the Orn and Phe partners interchanged). The distortion of the X-ray structure in the lower right-hand corner of Fig. 14b (not seen in the symmetrical computed one of Fig. 14a) probably arises from the presence of a urea molecule in the crystal.

4. Fibrous Proteins

Collagen is an example of a fibrous protein which involves interchain association to form a triple-stranded coiled-coil structure. Conformational energy calculations have been carried out on several synthetic poly(tripeptide) analogs, poly(Gly-X-Y), of collagen.[49-52] Because of the regularity conditions imposed on each tripeptide in the computations, the number of degrees of freedom was small, so that, again, the multiple-minima problem was surmounted by an adequate coverage of conformational space, using the build-up procedure. The computations indicated that poly(Gly-Pro-Pro), poly(Gly-Pro-Hyp), and poly(Gly-Pro-Ala) form stable triple-stranded coiled-coil collagen-like structures, whereas poly(Gly-Ala-Pro) does not, all in agreement with experiment. Figure 8 illustrates two calculated low-energy (collagen-like) structures in favorable contact with each other.

The computations also provided an explanation for the association of the chains [in contrast to the single-chain structures of α-helical forms of, e.g., poly(γ-benzyl-L-glutamate)]; viz., the resulting interchain interactions among the

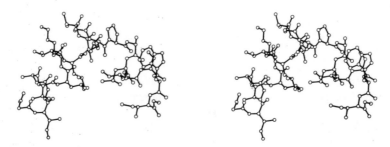

Figure 13. Stereo view of the lowest-energy structure calculated for residues 1-20 of melittin.[42]

three chains of collagen lower the energy of a tripeptide unit below that in the non-associated single chain. Furthermore, the inter-chain interactions induce a slight conformational change in going from the low-energy form of the single chain to the lower-energy form of the triple-stranded complex.

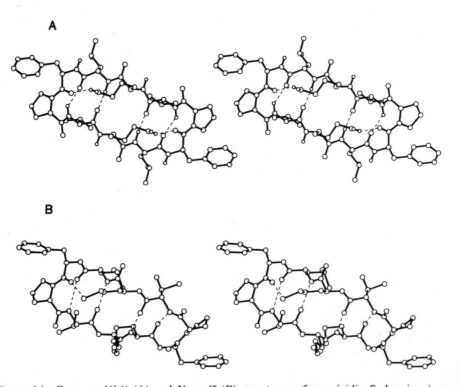

Figure 14. Computed[45,46] (A) and X-ray[47] (B) structures of gramicidin S showing (among other things) a hydrogen bond between the ornithine side chain and the phenylalanine backbone carbonyl group.

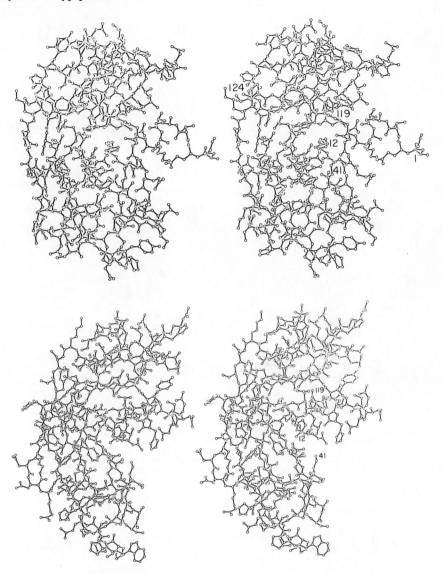

Figure 15. Stereo ORTEP diagrams of the full (heavy)-atom structures of ribonuclease (top) and angiogenin (bottom).[63] The active-site residues His-12, Lys-41, and His-119 (Bovine pancreatic ribonuclease numbering) are labeled in each structure.

After completion of the calculations[49] on poly(Gly-Pro-Pro), it was learned that Okuyama et al[53] had carried out a single-crystal X-ray structure analysis of (Pro-Pro-Gly)$_{10}$. Our structure is in agreement with theirs, with an rms deviation of 0.3Å for all (non-hydrogen) atoms, based on a comparison between the X-ray coordinates (kindly provided to us by Professor M. Kakudo) and our computed ones.

Similar methodology, already referred to in Section 2, is being used to investigate the formation of fibrils in collagen. The packing of two triple-stranded structures

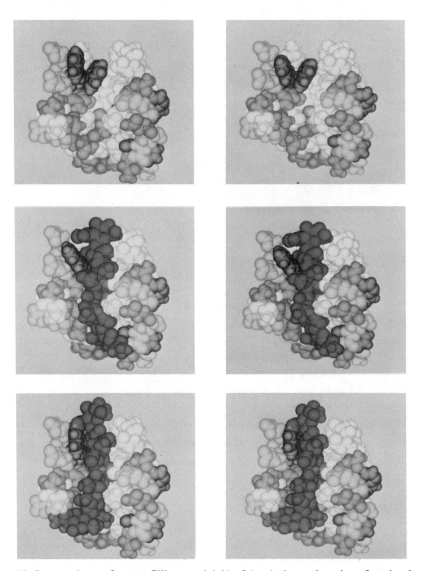

Figure 16. Stereo views of space-filling models[76] of (top) the active site of native hen egg white lysozyme; (middle) energy-minimized model-built hexamer [(GlcNAc)$_6$] bound to the active site (right-sided mode); and (bottom) lowest-energy hexamer bound to the active site (left-sided mode).

is shown in Fig. 8. Computations with assemblies of more than two triple-stranded structures are in progress.

5. Globular Proteins

As an introduction to the applicability of our computational methodology to globular proteins, we consider first two types of problems, refinement of X-ray structures of proteins and calculation of structures of homologous proteins, where the multiple-minima problem is not so serious, because one starts the computations with structures that are close to the global-minimum ones.

A. Refinement of X-ray Structures of Proteins

We have used energy minimization to refine X-ray data by eliminating steric overlaps and providing low-energy structures. This procedure has been applied to several proteins, as summarized in Ref. 20.

We have also combined energy minimization with re-computation of electron density maps at each stage of refinement to make optimal use of low-resolution X-ray data.[54,55] The procedure was applied[55] to 2.5 Å-resolution data[56] for bovine pancreatic trypsin inhibitor to obtain a structure that was a satisfactory approximation to the one that had been obtained by use of 1.5 Å-resolution data.[57]

B. Calculation of Structures of Homologous Proteins

If the X-ray structure of a protein is known, the structure of a homologous protein can be obtained by "inserting" the amino acid sequence of the protein of unknown structure into the known structure of the other homologous protein. The energy of the structure obtained in this manner will be very high, especially because of overlaps in regions where the sequences of the homologous proteins differ. However, these overlaps can be relieved, and the energy reduced, by minimization.

We have applied this procedure to obtain a structure of α-lactalbumin from that of lysozyme[58] and those of several snake venom inhibitors from that of bovine pancreatic trypsin inhibitor.[59] Subsequently, Berliner and Kaptein[60] obtained evidence that is compatible with our predicted conclusions about tryptophan residues of α-lactalbumin, and Gerkin[61] reported a high pK_a for the N-terminal amino group, which "supports its interaction in an ion pair as proposed by Warme et al." Just recently, Koga and Berliner,[62] in further NMR studies, reported that "the experimental results were consistent with a putative three-dimensional α-lactalbumin model,[58] which predicted the close proximity of Ile-95, Tyr-103, Trp-60, and Trp-104." We recently applied this procedure to obtain a preliminary structure of angiogenin[63] (a protein that induces in vivo formation of blood vessels) by taking advantage of its homology to pancreatic ribonuclease A (see Fig. 15).

C. Strategies to Overcome Multiple-Minima Problem

In the absence of the structure of a homologous protein, other strategies must be used to obtain a starting conformation that can reach the global minimum by direct energy minimization. These strategies include the build-up and electrostatic-optimization procedures (mentioned in Sections 3 and 2, respectively), the use of distance constraints,[64,65] empirical data on short-, medium-, and long-range inter-actions.[64-68] and a factor analysis* of amino acid properties,[69,70] to limit the area of conformational space searched, and other methods of searching, e.g. SMAPPS[41] and relaxation of dimensionality.[71] These procedures are used separately, and in various combinations with each other, to locate the approximate native conforma-tion of a globular protein. They are all intended as the *initial* approaches in the computations. In the *final* stages, the results from all of these procedures are col-lated into an approximate three-dimensional structure whose energy should lie in the potential well containing the global minimum (i.e. this structure should be a good approximation of the native structure). Then, the conformational energy of this structure is minimized, taking all, pair interactions (over the whole molecule) into account. These methods are being tested on a protein of known structure (bovine pancreatic trypsin inhibitor) and are currently being applied to one of unknown structure (human leukocyte interferon[72]).

This same methodology can also be used in protein design. By substituting one or more amino acids in specific parts of the sequence, and then minimizing the energy of the resulting structure, it is possible to assess the effect of the mutation(s) on the stability of the protein. One can therefore determine the effect of removing or adding a disulfide bond, or, say, substituting an alanine for a proline residue that might be postulated to play a rate-limiting role in protein folding.

6. Enzyme-Substrate Complexes

The same computational methodology is also applicable to enzyme-substrate com-plexes, thereby providing information about the structures and energetics involved in molecular recognition and specificity.[73] Molecular details of such enzyme-substrate complexes are required in order to understand the mechanism of enzyme action. The computation pertains essentially to a docking process in which a flex-ible substrate approaches a flexible enzyme as the energy is minimized. Such calcu-lations have been carried out for complexes of α-chymotrypsin with oligopeptides[74,75] and of hen egg white lysozyme with oligosaccharides.[75,76] Figure 16 illustrates two computed binding modes[76] for hexasaccharide substrates of hen

---

*   Using a factor analysis, it was possible to express the physical properties of the 20 naturally occur-ring amino acids in terms of 10 orthogonal factors.[69] These factors are then used[70] to search the protein data bank for amino acid sequences that have similar *properties* as the corresponding sequences of an unknown protein. Presumably, such sequences would then have similar three-dimensional structures. However, we have shown[70] that the sequences (whose *properties* are being compared) must be rather long ($\geq$ 15 residues) before the comparisons become meaningful.

egg white lysozyme, a "left-sided" and a "right-sided" binding mode. The former was predicted to predominate for $(GlcNAc)_6$. This prediction was subsequently verified[77] by experiments involving competition between oligosaccharides and monoclonal antibodies for binding to hen egg white lysozyme and by measurements of the Michaelis-Menten constant K_M for hen egg white lysozyme and for a homologous lysozyme from ringed neck pheasant, the latter having several different amino acids in the "right-sided" binding site.

7. Conclusions

The methodology underlying the results described herein has provided an understanding of the behavior of model systems, i.e. as to how interatomic interactions lead to the experimental observations. Similar understanding has been gained about the conformational properties of small open-chain and cyclic structures and fibrous proteins. Several strategies are currently being explored to try to surmount the multiple-minima problem for globular proteins. If successful, they should lead to predictable protein structures and to predictable effects of site-specific mutagenesis. They have already led to testable predictions about enzyme-substrate interactions, and offer the hope of understanding the fundamental processes in molecular recognition.

Acknowledgments

This work was supported by the National Science Foundation (DMB84-01811), the National Institutes of Health (GM-14312), and the National Foundation for Cancer Research.

References

1. G. Némethy and H.A. Scheraga, Biopolymers, **3**, 155 (1965).
2. H.A. Scheraga, Fed. Proc., **26**, 1380 (1967).
3. A. Wlodawer, R. Bott and L. Sjolin, J. Biol. Chem., **257**, 1325 (1982).
4. G.N. Ramachandran, C. Ramakrishnan and V. Sasisekharan, J. Mol. Biol., **7**, 95 (1963).
5. H.A. Scheraga, S.J. Leach, R.A. Scott and G. Némethy, Disc. Faraday Soc., **40**, 268 (1965).
6. R.A. Scott and H.A. Scheraga, J. Chem. Phys., **45**, 2091 (1966).
7. T. Ooi, R.A. Scott, G. Vanderkooi and H.A. Scheraga, J. Chem. Phys., **46**, 4410 (1967).
8. H.A. Scheraga, Adv. Phys. Org. Chem., **6**, 103 (1968).
9. F.A. Momany, R.F. McGuire, A.W. Burgess and H.A. Scheraga, J. Phys. Chem., **79**, 2361 (1975).
10. G. Némethy, M.S. Pottle and H.A. Scheraga, J. Phys. Chem., **87**, 1883 (1983).
11. P. DeSantis, E. Giglio, A.M. Liquori and A. Ripamonti, Nature, **206**, 456 (1965).

12. D.A. Brant and P.J. Flory, J. Am. Chem. Soc., **87**, 2791 (1965).
13. A.T. Hagler, E. Huler and S. Lifson, J. Am. Chem. Soc., **96**, 5319 (1974).
14. J. Hermans, D.R. Ferro, J.E. McQueen and S.C. Wei, Jerusalem Symp. Quantum Chem. Biochem., **8**, 459 (1975).
15. H.A. Scheraga, Chem. Revs., **71**, 195 (1971).
16. C.B. Anfinsen and H.A. Scheraga, Adv. Protein Chem., **29**, 205 (1975).
17. G. Némethy and H.A. Scheraga, Quart. Rev. Biophys., **10**, 239 (1977).
18. H.A. Scheraga, Biopolymers, **20**, 1877 (1981).
19. H.A. Scheraga, Biopolymers, **22**, 1 (1983).
20. H.A. Scheraga, Carlsberg Research Commun., **49**, 1 (1984).
21. H.A. Scheraga, Annals N.Y. Acad. Sci., **439**, 170 (1985).
22. J.F. Yan, F.A. Momany and H.A. Scheraga, J. Am. Chem. Soc., **92**, 1109 (1970).
23. K.C. Chou, M. Pottle, G. Némethy, Y. Ueda and H.A. Scheraga, J. Mol. Biol., **162**, 89 (1982).
24. K.C. Chou and H.A. Scheraga, Proc. Natl. Acad. Sci., U.S., **79**, 7047 (1982).
25. H.A. Scheraga, K.C. Chou and G. Némethy, in "Conformation in Biology ," Ed. R. Srinivasan and R.H. Sarma, Adenine Press, pp. 1 – 10 (1983).
26. K.C. Chou, G. Némethy, M.S. Pottle and H.A. Scheraga, Biochemistry. **24**, 7948 (1985).
27. K.C. Chou, G. Némethy and H.A. Scheraga, J. Phys. Chem., **87**, 2869 (1983). Erratum: *ibid.*, **87**, 4772 (1983).
28. K.C. Chou, G. Némethy and H.A. Scheraga, J. Am. Chem. Soc., **106**, 3161 (1984). Erratum: *ibid.*, **107**, 2199 (1985).
29. K.C. Chou, G. Némethy, S. Rumsey, R.W. Tuttle and H.A. Scheraga, J. Mol. Biol., **186**, 591 (1985).
30. K.C. Chou, G. Némethy, S. Rumsey, R.W. Tuttle and H.A. Scheraga, J. Mol. Biol., **188**, 641 (1986).
31. G. Némethy and H.A. Scheraga, Biochemistry, **25**, 3184 (1986).
32. E.H. Erenrich, R.H. Andreatta and H.A. Scheraga, J. Am. Chem. Soc., **92** , 1116 (1970).
33. S. Tanaka and H.A. Scheraga, Macromolecules, **8**, 516 (1975).
34. Y.C. Fu, R.F. McGuire and H.A. Scheraga, Macromolecules, 7, 468 (1974).
35. N. Gō, M. Gō and H.A. Scheraga, Proc. Natl. Acad. Sci., U.S., **59**, 1030 (1968).
36. M. Gō and H.A. Scheraga, Biopolymers, **23**, 1961 (1984).
37. F.T. Hesselink, T. Ooi and H.A. Scheraga, Macromolecules, **6**, 541 (1973).
38. L. Peila and H.A. Scheraga, Biopolymers, in press.
39. Y. Isogai, G. Némethy and H.A. Scheraga, Proc. Natl. Acad. Sci., U.S., **74**, 414 (1977).
40. M. Vásquez and H.A. Scheraga, Biopolymers, **24**, 1437 (1985).
41. G.H. Paine and H.A. Scheraga, Biopolymers, **24**, 1391 (1985).
42. M.R. Pincus, R.D. Klausner and H.A. Scheraga, Proc. Natl. Acad. Sci., U.S., **79**, 5107 (1982).
43. T.C. Terwilliger, L. Weissman and D. Eisenberg, Biophys. J., **37**, 353 (1982).
44. L.R. Brown, W. Braun, A. Kumar and K. Wüthrich, Biophys. J., **37**, 319 (1982).
45. M. Dygert, N. Gō and H.A. Scheraga, Macromolecules, **8**, 750 (1975).

46. G. Némethy and H.A. Scheraga, Biochem. Biophys. Res. Commun., **118**, 643 (1984).
47. S.E. Hull, R. Karlsson, P. Main, M.M. Woolfson and E.J. Dodson, Nature, **275**, 206 (1978).
48. S. Rackovsky and H.A. Scheraga, Proc. Natl. Acad. Sci., U.S., **77**, 6965 (1980).
49. M.H. Miller and H.A. Scheraga, J. Polymer Sci.: Polymer Symposia, No. 54, pp. 171−200 (1976).
50. M.H. Miller, G. Némethy and H.A. Scheraga, Macromolecules, **13**, 470 (1980).
51. M.H. Miller, G. Némethy and H.A. Scheraga, Macromolecules, **13**, 910 (1980).
52. G. Némethy, M.H. Miller and H.A. Scheraga, Macromolecules, **13**, 914 (1980).
53. K. Okuyama, N. Tanaka, T. Ashida and M. Kakudo, Bull. Chem. Soc. Japan, **49**, 1805 (1976).
54. S. Fitzwater and H.A. Scheraga, Acta Cryst., **A36**, 211 (1980).
55. S. Fitzwater and H.A. Scheraga, Proc. Natl. Acad. Sci., U.S., **79**, 2133 (1982).
56. R. Huber, D. Kukla, A. Rühlmann, O. Epp and H. Formanek, Naturwissenschaften, **57**, 389 (1970).
57. J. Deisenhofer and W. Steigemann, Acta Cryst., **B31**, 238 (1975).
58. P.K. Warme, F.A. Momany, S.V. Rumball, R.W. Tuttle and H.A. Scheraga, Biochemistry, **13**, 768 (1974).
59. M.K. Swenson, A.W. Burgess and H.A. Scheraga, in "Frontiers in Physicochemical Biology," ed. B. Pullman, Academic Press, pp. 115−142 (1978).
60. L.J. Berliner and R. Kaptein, Biochemistry, **20**, 799 (1981).
61. T.A. Gerken, Fed. Proc., **42**, 2001 (1983).
62. K. Koga and L.J. Berliner, Biochemistry, **24**, 7257 (1985).
63. K.A. Palmer, H.A. Scheraga, J.F. Riordan and B.L. Vallee, Proc. Natl. Acad. Sci., U.S., **83**, 1965 (1986).
64. H. Wako and H.A. Scheraga, J. Protein Chem., **1**, 5 (1982).
65. H. Wako and H.A. Scheraga, J. Protein Chem., **1**, 85 (1982).
66. H. Meirovitch and H.A. Scheraga, Proc. Natl. Acad. Sci., U.S., **78**, 6584 (1981).
67. T. Kikuchi, G. Némethy and H.A. Scheraga, J. Comput. Chem., **7**, 67 (1986).
68. H. Wako, N. Saitô and H.A. Scheraga, J. Protein Chem., **2**, 221 (1983).
69. A. Kidera, Y. Konishi, M. Oka, T. Ooi and H.A. Scheraga, J. Protein Chem., **4**, 23 (1985).
70. A. Kidera, Y. Konishi, T. Ooi and H.A. Scheraga, J. Protein Chem., **4**, 265 (1985).
71. E.O. Purisima and H.A. Scheraga, Proc. Natl. Acad. Sci., U.S., **83**, 2782 (1986).
72. K.D. Gibson, S. Chin, M.R. Pincus, E. Clementi and H.A. Scheraga, Symp. on "Supercomputer Simulations in Chemistry," Montreal, Aug. 27, 1985, in press.
73. H.A. Scheraga, Pont. Acad. Sci. Scr. Var., **55**, 21 (1984). Also, in Proceedings of the XVIIIth International Solvay Conference on Chemistry, in press.
74. K.E.B. Platzer, F.A. Momany and H.A. Scheraga, Intl. J. Peptide and Protein Research, **4**, 201 (1972).

75. H.A. Scheraga, M.R. Pincus and K.E. Burke, in "Structure of Complexes Between Biopolymers and Low Molecular Weight Molecules," eds. W. Bartmann and G. Snatzke, John Wiley, Chichester, pp. 53 – 76 (1982).
76. M.R. Pincus and H.A. Scheraga, Macromolecules, **12**, 633 (1979).
77. S.J. Smith-Gill, J.A. Rupley, M.R. Pincus, R.P. Carty and H.A. Scheraga, Biochemistry, **23**, 993 (1984).

Conformation of Non-Crystalline Proteins Viewed by NMR

Kurt Wüthrich

Institut für Molekularbiologie und Biophysik
Eidgenössische Technische Hochschule-Hönggerberg
CH-8093 Zürich, Switzerland

Abstract

During the period 1979−83 methods were introduced for efficient, sequence-specific assignment of the ^{1}H NMR spectra of proteins. This now provides the basis for determination of three-dimensional protein structures in solution. In addition, sequence-specific resonance assignments are a key for detailed studies of the molecular dynamics and for investigations of intermolecular interactions.

1. Introduction

This article presents a survey of recent progress in NMR investigations of proteins. With the use of two-dimensional (2D) experiments, NMR has become the first technique applicable for determination of three-dimensional polypeptide structures in solution and other non-crystalline states. Comparable data have so far been obtained exclusively by X-ray diffraction in single crystals. The results obtained by NMR represent an important complementation and extension of the crystal structures, and NMR can of course also be applied for compounds where no suitable crystals for X-ray studies are available. Beyond determination of the spatial structures, NMR data bear also on the molecular dynamics and on intermolecular interactions. For all these potential applications it is an important asset that NMR experiments can be performed under conditions which may be closely related to the physiological milieu. On the other hand, structure determination by NMR in solution is limited to relatively small systems, for example for proteins (or domains excised from larger natural proteins) with molecular weights up to ca. 20,000.

NMR studies of macromolecular structures rely primarily on three recent developments. First, *two-dimensional spectroscopy*[1-3] improves the resolution of highly complex NMR spectra and efficiently delineates networks of scalar couplings and dipole-dipole couplings, which may extend over the entire macromolecule.[4] Second, new strategies for studies of proteins enabled the assignment of the individual NMR lines to distinct nuclear spins in the polypeptide chains.[4-7] Third, new concepts were introduced for determination of the spatial structure of polymer chains from data on scalar and dipolar couplings between nuclear spins, notably the implementation of *distance geometry algorithms* for this purpose.[4,5,8-12] In all of this, *sequence-specific resonance assignments have a pivotal role* and the reliability of each NMR investigation, for example the accuracy of a structure determination, will depend critically on the extent to which the NMR spectrum was assigned.

2. 2D ^1H NMR Spectra of Proteins

Figure 1 shows a 2D correlated (COSY) spectrum of the protons in a small protein. The lines on the diagonal from the upper right to the lower left *(diagonal peaks)* correspond to the resonances of individual ones of the approximately 500 hydrogen atoms in this molecule. Off diagonal *cross peaks* manifest interactions between different nuclear spins. This is indicated in the contour plot of a 2D nuclear Overhauser enhancement (NOESY) spectrum in Fig. 2A by solid lines connecting the two diagonal peaks i and j via the cross peak k. In COSY spectra[1-4] the cross peaks are caused by scalar, *through-bond* interactions (spin-spin couplings) between protons located in the same chemical structure and separated by three or less covalent bonds. COSY connectivities therefore allow to divide the resonance lines into groups arising from chemical entities, for example the individual amino acid residues in a protein. In NOESY spectra,[2-4] which have a similar appearance as COSY (Fig. 2), the cross peaks are a consequence of dipole-dipole, *through-space* couplings between protons which are located near each other in the folded molecule (closer than ca. 5.0 Å). They thus allow observation of parameters which are directly related with the three-dimensional molecular structure. The origin of the NOESY cross peaks lies in the fact that the populations of the spin states are affected by the dipolar interactions. (In conventional one-dimensional experiments, the NOE's are observed as variations in the intensity of one NMR line when the resonance of another, nearby spin is selectively irradiated with a radio-frequency field.)

For high resolution NMR studies the samples are placed in a highly homogeneous polarizing magnetic field of typically 11 Tesla, which is produced by a superconducting coil. A 2D NMR experiment is recorded in a two-dimensional time space. In the simplest case, COSY, the following experimental scheme is used: $90°$-t_1-$90°$-t_2, where $90°$ indicates a non-selective radio-frequency pulse, t_1 is the *evolution period*, and t_2 is the *observation period*. In COSY, the polarized nuclear spins are thus perturbed by a succession of two radio-frequency pulses. After the second pulse, the return of the magnetization to equilibrium is recorded versus time *(t_2-axis)*, and a second time dimension is obtained by repetition of the same experiment with different time separation of the two radio-frequency pulses *(t_1-axis;* typically 200 to 2000 increments of t_1 are recorded). For biomacromolecules the data matrix $s(t_1,t_2)$ consists usually of between 4 and 64 million data points. A two-dimensional Fourier transformation produces the desired frequency spectrum $S(\omega_1,\omega_2)$ (Figs. 1 and 2A). The NOESY experimental scheme consists of three radio-frequency pulses: $90°$-t_1-$90°$ -τ_m-$90°$-t_2, where the *mixing time,* τ_m, is kept constant throughout the measurements needed to record the data matrix $s(t_1,t_2)$. Advantages of 2D NMR for work with macromolecules include that scalar or dipolar interactions between distinct, individual spins are observed with the use of non-selective radio-frequency pulses, and that a complete set of all spin-spin inter-actions of a given kind in the entire molecule is manifested in a single experiment.

3. Sequence-specific Resonance Assignments

As the result of the assignment procedure, diagonal peaks are attributed to spins in distinct locations of the polypeptide chain (Fig. 2B). Once these sequence-specific assignments are available, the individual cross peaks can then be correlated with interactions between distinct spins along the chain. The example of Fig. 2C illustrates the information contained in a NOESY cross peak between two assigned protons, i.e. evidence for close approach between the locations of the two spins, which are in this case near the chain ends. (Without resonance assignments the same cross peak would merely indicate that any two protons of the polypeptide chain are in close proximity of each other, and this would not be sufficient for use as input for a structure determination; see below.)

Obtaining sequence-specific resonance assignments is not straightforward because in general a protein contains multiple copies of the different amino acids. Therefore, identification of the groups of scalar-coupled spins belonging to individual residues is in general not sufficient for determination of a unique sequence location. The problem can be solved either by isotope labeling of specified positions in the molecular structure, or with the use of NMR experiments capable of identifying the NMR lines of sequentially neighboring amino acids.[4-7] Prior to the introduction of

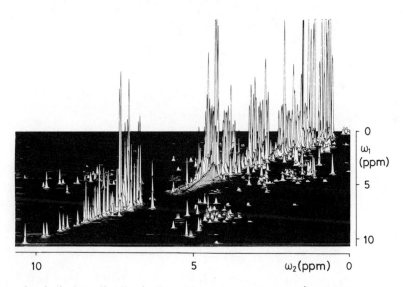

Figure 1. Stacked plot affording a three-dimensional view of a ^{1}H COSY spectrum of the basic pancreatic trypsin inhibitor (BPTI), a small protein with molecular weight 6500, in H_2O solution. Larmor frequency 500 MHz. The two frequency axes, ω_1 and ω_2, indicate difference frequencies in ppm of the Larmor frequency (chemical shifts). (*Reproduced from Ref. 7.*)

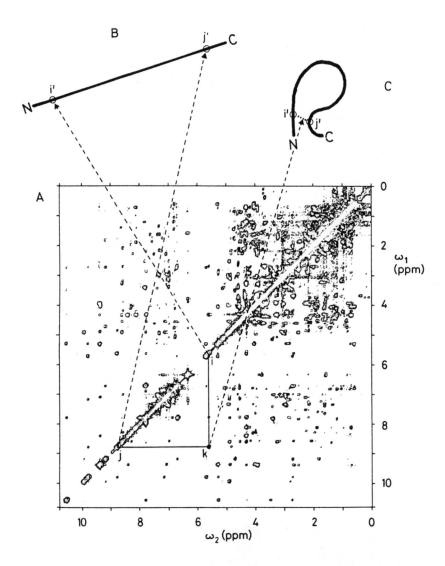

Figure 2. A. Contour plot of a ^{1}H NOESY spectrum of BPTI in D$_2$O solution. Larmor frequency 500 MHz. i and j identify two diagonal resonance peaks. These are connected by a horizontal line and a vertical line with a cross peak k manifesting a NOE between i and j.

B. The straight line represents a polypeptide chain. The circles indicate two protons i′ and j′ along the chain, which give rise to the diagonal peaks i and j in A, as indicated by the broken arrows.

C. Same as B, with formation of a loop manifested by the NOESY cross peak k between the two protons i′ and j′ (see text).

this latter approach during the period 1979 – 83, lack of resonance assignments was a major stumbling block opposing detailed analysis of the NMR spectra of proteins.

4. Spatial Structure Determination

The major source of information on the spatial molecular structures are the ^{1}H NOESY spectra. Typically, a large number of NOESY cross peaks are observed in globular proteins (Fig. 2A), indicating that these molecular structures contain numerous pairs of closely spaced hydrogen atoms. Combined with sequence-specific resonance assignments these *distance constraints* define the formation of loops by chain segments of variable lengths, as is illustrated for a single NOESY cross peak in Fig. 2C. With model building[13] or the use of suitable mathematical procedures, notably distance geometry,[8-12] the conformation space can then be searched for spatial arrangements of the polymer chain which are compatible with the experimental distance constraints.

In practice so far, we interpreted the NOESY cross peaks in terms of upper bounds on the distances between hydrogen atoms, rather than as exact distances (e.g. observation of a *strong* NOESY cross peak between two protons would correspond to an *upper distance limit* of ≤ 2.5 Å, or a *weak* NOESY cross peak would be interpreted as a distance-bound of ≤ 4.0 Å. A lower limit on the distance of ≥ 2.0 Å is in all cases imposed by the sum of the van der Waals radii of the two hydrogen atoms). Since in the linear polypeptide chain the protons in question may be separated by as much as 100 Å, or more, such distance constraints constitute nonetheless a drastic confinement of the accessible conformation space. However, the result of a structure determination based on such qualitative distance constraints is not a unique structure, but consists of a group of related conformers which are compatible with the experiments. In Fig. 3 this is illustrated with a segment of micelle-bound glucagon. Five conformers have been superimposed for minimal root mean square distances, which were obtained in five independent distance geometry calculations using the same experimental distance constraints as the input.

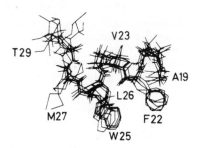

Figure 3. Superposition of five conformers of the segment 19-29 of micelle-bound glucagon (see text). (*Reproduced from Ref. 10.*)

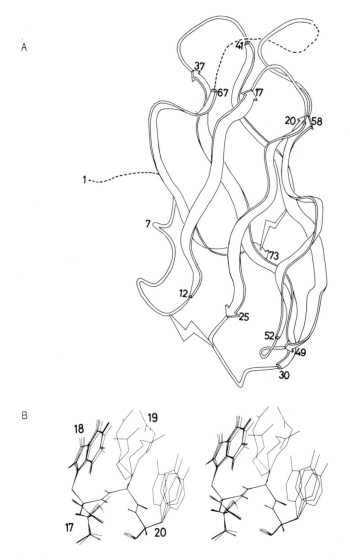

Figure 4. Conformation of Tendamistat in aqueous solution determined by NMR.

A. Backbone topology. The ribbon drawing was generated from the atom coordinates by the CONFOR program,[13] and some modifications were added by hand to improve the three-dimensional effect. The arrowed ribbons indicate the position and direction of the β-sheet strands. The rope structure was used for the loop areas, with the broken lines indicating those places where the data did not give a well confined solution. The two disulfide bridges are indicated by lightning bolts.

B. Stereo view of the complete structure of the active site consisting of the residues Ser 17, Trp 18, Arg 19, and Tyr 20. Similar to Fig. 3, four structures computed from the same experimental data were superimposed so as to minimize the root mean square distance of the backbone atoms. For the backbone and the indole ring all atoms are shown, for the rest of the structure the hydrogen atoms have been omitted. *(Reproduced from Ref. 15.)*

5. Initial Implications on Proteins in Solution

Structure determinations by NMR are of interest because the biological macro-molecules can be studied in environments which may closely resemble the physio-logical milieu. Glucagon (Fig. 3) was studied in the lipid-water interphase near the surface of perdeuterated dodecylphosphocholine micelles,[9, 10] which was chosen to mimic the environment of the polypeptide hormone on the surface of a biological membrane. More commonly so far, proteins were studied in aqueous solution. An example is the α-amylase inhibitor Tendamistat, which is a protein with 74 amino acid residues (Fig. 4). The architecture of the protein is characterized by the pres-ence of two antiparallel β-sheets.[14] In Fig. 4A one sheet consisting of the strands 12-17, 20-25 and 52-58 is in the foreground, and the other with the strands 30-37, 41-49 and 67-73 behind. The two β-sheets form a quite regular greek key barrel.[15] The presumed active site of this protein consists of the residues Trp 18-Arg 19-Tyr 20, for which a stereo view of the structure is displayed in Fig. 4B. The side chain of Arg 19 is sandwiched between the aromatic rings of Trp 18 and Tyr 20, which was also clearly manifested by high field shifts of the arginine ^{1}H NMR lines.[15]

While the use of NMR techniques for determination of three-dimensional protein structures is of very recent origin, NMR has long been employed for studies of dynamic processes in proteins (for reviews see, for example, Refs. 4 and 16). Such investigations rely on the observation of spectral properties in distinct NMR lines, which can be correlated with intramolecular motional processes. Once the NMR lines used for this purpose have independently been assigned, the intramolecular motions can be attributed to specified locations in the molecular structure, thus providing a map of the internal motility across the molecule.[17] NMR characteriza-tion of the solution conformation of proteins thus bears on both static and dynamic features. Even when a crystal structure of the same protein is available, this makes a complementation with independently analyzed NMR data highly interesting.

For four small proteins for which a structure determination by NMR was com-pleted in our laboratory, crystal structures are also available. This provides some initial indications of the relations between corresponding protein conformations in crystals and in solution. Overall, the implication is that different behaviour can be expected for "typical globular proteins" and for other polypeptide structures.

In the globular proteins BPTI (to be published) and Tendamistat[15,18] the same global architecture was observed in crystals and in solution. Differences between crystal and solution structures for these proteins are indicated on and near the molecular surface. More precise statements must await further refinement of the solution conformations. The intriguing implication with regard to functional inter-pretations of the solution structures is that two limiting principles of polypeptide-segments locked in unique spatial arrangements in the protein core and adopting local equilibrium situations with multiple populated conformers near the surface are both represented in the same molecule. Eventually, qualification of a structure-determination in solution might therefore include specification of the accuracy of the atom-positions in the molecular core, as well as a statistical description of the molecular surface.

For the non-globular polypeptide hormone glucagon different structures were found in single crystals[19] and in the non-crystalline environments provided by an aqueous solution[20] and by the lipid-water interphase in micelles (Fig. 3; Ref. 10). In metallothionein, where the core of the molecular structure consists of metal clusters involving coordinative bonds with cysteinyl residues of the polypeptide chain, important differences in the metal-cysteine coordination between the structures in crystals[21] and in solution[22] are indicated. From these two examples it appears that special care must be exercised in the functional interpretation of crystal structures of polypeptides deviating from the typical features of globular proteins.

Acknowledgments

The research from the author's laboratory described in this article was supported by the Schweizerischer Nationalfonds. I thank Mrs. E. Huber for the careful preparation of the manuscript and all the colleagues mentioned in the references for their enthusiastic collaboration on this project, without which none of it could have been realized.

References

1. W.P. Aue, E. Bartholdi, and R.R. Ernst, J. Chem. Phys., **64**, 2229 (1976).
2. G. Wider, S. Macura, Anil-Kumar, R.R. Ernst, and K. Wüthrich, J. Magn. Reson., **56**, 207 (1984).
3. R.R. Ernst, G. Bodenhausen, and A. Wokaun, Principles of nuclear magnetic resonance in one and two dimensions, Oxford University Press, Oxford (1986).
4. K. Wüthrich, NMR of proteins and nucleic acids, Wiley, New York (1986).
5. K. Wüthrich, G. Wider, G. Wagner, and W. Braun, J. Mol. Biol., **155**, 311 (1982).
6. M. Billeter, W. Braun, and K. Wüthrich, J. Mol. Biol., **155**, 321 (1982).
7. G. Wagner and K. Wüthrich, J. Mol. Biol., **155**, 347 (1982).
8. G.M. Crippen, J. Comp. Physics, **24**, 96 (1977).
9. W. Braun, Ch. Bösch, L.R. Brown, N. Gō, and K. Wüthrich, Biochim. Biophys. Acta, **667**, 377 (1981).
10. W. Braun, G. Wider, K.H. Lee, and K. Wüthrich, J. Mol. Biol., **169**, 921 (1983).
11. T.F. Havel and K. Wüthrich, Bull. Math. Biol., **46**, 673 (1984).
12. T.F. Havel and K. Wüthrich, J. Mol. Biol., **182**, 281 (1985).
13. M. Billeter, M. Engeli, and K. Wüthrich, J. Mol. Graphics, **3**, 79 (1985).
14. A.D. Kline and K. Wüthrich, J. Mol. Biol., **183**, 503 (1985).
15. A.D. Kline, W. Braun, and K. Wüthrich, J. Mol. Biol., in press.
16. G. Wagner, Quart. Rev. Biophys., **16**, 1 (1983).
17. G. Wagner and K. Wüthrich, J. Mol. Biol., **160**, 343 (1982).
18. J. Pflugrath, E. Wiegand, R. Huber, and L.Vértesy, J. Mol. Biol., in press.
19. K. Sasaki, S. Dockevill, D.A. Achmiak, I.J. Tickle, and T.L. Blundell, Nature, **257**, 751 (1975).

20. Ch. Bösch, A. Bundi, M. Oppliger, and K. Wüthrich, Eur. J. Biochem., **91**, 209 (1978).
21. W.F. Furey, A.H. Robbins, L.L. Clancy, D.R. Winge, B.C. Wang, and C.D. Stout, Science, **231**, 704 (1986).
22. W. Braun, G. Wagner, E. Wörgötter, M. Vasák, J.H.R. Kägi, and K. Wüthrich, J. Mol. Biol., **187**, 125 (1986).

Structures and Superstructures in Periodical Polynucleotides

P. DeSantis, S. Morosetti, A. Palleschi and M. Savino*
Dipartimento di Chimica
Università di Roma "La Sapienza"
00185 Roma, Italy

Introduction

A renewed attention has been focused on the structural aspects of DNA in the last few years. Such revival of interest starting from the elucidation of the structure of some synthetic oligonucleotides is going beyond toward the understanding of the basic molecular mechanisms of control of genetic information, which are strictly correlated to the question how DNA structure depends on the sequence.[1]

In the not so distant past, it was generally accepted that DNA could exist only in the traditional allomorphs A, B, C independent of its composition and sequence. In fact, X-ray fiber diffraction studies of various natural DNAs showed that the secondary structures were independent of the nature of the DNA and the transitions between the three forms A, B, C found in fibers were mainly due to the differential water and counterions activity as for a simple polyelectrolyte.[2] Synthetic DNA duplexes with two or three fold periodic sequence provided the best opportunity for settling the issue of whether and to what extent the sequence can influence DNA secondary structure. The most striking polymorphism was observed with the self-complementary poly (dC-dG), which, as first suspected on the basis of circular dichroism experiments,[3] can assume beside the canonical A and B forms a left-handed helical conformation, the Z form, in contrast to all previously studied DNAs. The major merit for these findings is, however, due to X-ray diffraction studies on single crystals of self-complementary oligonucleotides initiated by Rich et al.[4] These studies led to discover the Z form of DNA and revealed in the B form, significant fine features dependent on the sequence of bases. The possibility of inversion of the helical sense in correspondence of GC sequences together with the awareness reached in the same last few years of the importance of topological factors have determined the evolution of the concept of DNA from that of a structurally degenerate sequence of bases inert repository of the encoded genetic information, to that of a polymorphic system capable to direct, transmit, amplify at

---

* Dipartimento di Genetica e Biologia Molecolare and Centro per lo Studio degli Acidi Nucleici del CNR, Università di Roma "La Sapienza."

supermolecular level and integrate local molecular events, namely, to manage its information content.

More recently the discovery of periodical fluctuations of base pair distribution along the sequence, revealed a new fundamental aspect of B DNA in its ability to integrate the stereochemical differences of nucleotides where they are periodically distributed giving rise to potential or actual bends of the helical axis.[5]

The present paper reports some theoretical results obtained in our laboratory on the most representative classes of periodical DNAs, characterized by a regular alternance of purine and pyrimidine or by sequences with a ten-fold periodicity. To the first class belongs Z DNA and to the other one some periodical synthetic polynucleotides very recently investigated by Crothers et al. and Hagerman[6-8] as well as DNAs with a significant periodical sequence fluctuation. This paper aims to contribute to clarify some still pending questions on the nature of forces determining the Z structure and on the sequence specificity of this conformation, as well as on the mechanism of bending of natural and synthetic DNAs.

Two-fold Periodical DNAs: Conformational Analysis of DNA with Alternating Purine-Pyrimidine Sequence

In previous paper[9,10] mathematical methods were introduced to approach the problem of the conformational stability of the DNA under certain geometrical and symmetry constraints. We have shown that if a sugar pucker is chosen the six dimensional conformational space of uniform polynucleotide chains, is reduced to a four dimensional space under the constraints provided by the Watson-Crick base pairing. The two pairs of families characterized by base pairs stacking with right and left handedness and both the chain polarities were selected. Further constraints, as fixed monomeric translation along and rotation around the helical axis, reduced the problem to a two dimensional conformational space. Thus the B family of DNA was defined in terms of two independent angles of rotation and using potential energy functions, the most stable B DNA structure was predicted in good agreement with the X-ray structure analysis.

These methods, however, apply only to an ideal uniform double helical structure where the two strands are reported by two sets of pseudodyad axes both perpendicular to the helical axis: one lying on the average plane and the other between the average planes of base pairs. This corresponds to consider as determinant of the structure the common features of base pairs, neglecting their differences. These are, however, considered a posteriori as perturbations which become important when distributed in phase. On the contrary, in the case of regular alternance of purine and pyrimidine, the conformational coupling of nearest neighbor nucleotides is not negligible also in account of the exclusive ability of purine to exist in either the anti or syn conformation around the glycosidic bond. Therefore the conformational equivalence of all the nucleotide units cannot be assumed a priori since the repeating unit corresponds to a dinucleotide residue. As a consequence, the angles of rotation and the sugar pucker of a nucleotide residue are expected to be different

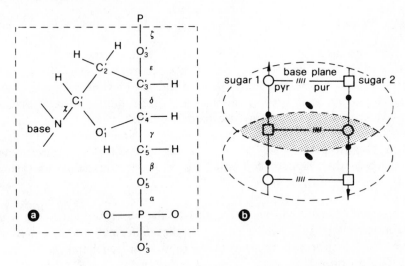

Figure 1. (a) Internal parameters of the polynucleotide chain. They are defined with zero at cis position and counterclockwise rotation of the nearest atom. The glycosil angle is similarly defined in terms of atoms $O'_1 - C'_1 - N_1 - C_2$ for pyrimidines and $O'_1 - C'_1 - N_9 - C_4$ for purines. **(b)** Schematic representation of the repeating unit of double helix alternating polynucleotides; $\bigcirc$ and $\square$ represent conformationally not equivalent sugars; $\blacklozenge$ represents the dyad axis; $\bullet$ represents the phosphorus atom.

from those of the adjacent residues. In this regular alternating structures only the set of the true dyad axes between the base planes are still present, while the pseudodyad axes lying on the base planes are absent. As an example this symmetry characterizes the Z form of poly (dC-dG).

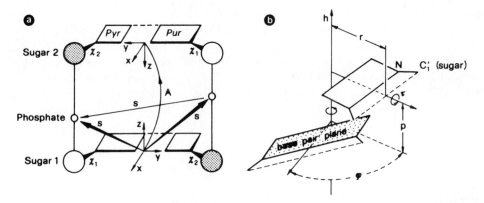

Figure 2. (a) Representation of a double-helix fragment; the geometrical transformation, the vectors and the local coordinate systems used in the mathematical procedure are indicated. **(b)** Cylindrical parameters.

Methods

Whilst the problem of selecting the most stable conformations of double helical alternating polynucleotides depends on 10 internal rotation angles (Fig. 1a) since the pair of sugar puckers has been fixed a priori (as previously shown), the practical lack of interactions between nonadjacent nucleotide units reduces the conformational problem to that of a base paired sequence of phosphate dinucleoside double helix having the same sugar puckers and the same angles of rotations χ_1, χ_2 around the two glycosidic bonds of the common base pair, when the repeating unit is considered as the union of two ensembles with an intersection, as sketched in Fig. 1(b).

Using this strategy the problem is reduced to five dimensions, namely the independent angles of rotation which define all the conformations of a pair of self-complementary dinucleotides connected by rigid hydrogen bonds, with the sugar puckers fixed a priori.

In fact, the chemical equivalence of the dinucleotide residues in alternating self-complementary strands allows us to assume their conformational equivalence; this results in the presence of a dyad axis perpendicular to the local helical axis which transforms a base pair into the adjacent one; this dyad axis will be generally not perpendicular to the true helical axis of the global structure, as, on the contrary, it is in the uniform double helices.

Let $\underline{A}$ be the transformation matrix which relates the cartesian systems suitably fixed on the base pair planes at half way between the glycosidic bonds so that these are related by a rotation of π around the axis as shown in Fig. 2(a). $\underline{A}$ can be defined in terms of the torsional angles along the nucleotide chain.

The conformational equivalence of the two strands requires that $\underline{AA} = \underline{E}$ the unitary matrix; thus $\underline{A} = \tilde{\underline{A}}$ is a symmetric matrix with trace -1 because equivalent to the operation of a dyad axis $\underline{d}$ which relates the two equivalent dinucleotide fragments; $\underline{d}$ is a vector perpendicular to the vector $\underline{h}$ which allows the superposition of the base pair planes through a simple rotation φ around it. In fact, $\underline{A\Pi}$ (where $\underline{\Pi}$ is the rotation matrix around x of 180°) represents the transformation of the system at the origin into the equivalent system fixed at the next base pair along the strand. It is a proper matrix which is equivalent to a pure rotation matrix $\underline{\Phi}$ around a suitable axis $\underline{h}$ through a similarity transformation $\tilde{\underline{T}}\underline{A}\ \underline{\Pi T} = \underline{\Phi}$. Thus trace $\underline{A\Pi} = a_{11} - a_{22} - a_{33} =$ trace $\underline{\Phi} = 2\cos\varphi + 1$ where φ is the angle of rotation around $\underline{h}$ (a_{ij} are elements of $\underline{A}$); this axis can be derived in the original system as the normalized eigenvector of the eigenvalue 1

$$\underline{A\Pi}\ \ \underline{h} = 1\underline{h}$$

so that

$$\underline{h} = \begin{vmatrix} a_{23} - a_{32} \\ - a_{13} - a_{31} \\ a_{21} + a_{12} \end{vmatrix} (2 \sin \varphi)^{-1}$$

$\underline{d}$ can be easily obtained as

$$\underline{d} = \begin{vmatrix} a_{11} + 1 \\ a_{21} \\ a_{31} \end{vmatrix} \left[(a_{11} + 1)^2 + a_{21}^2 + a_{31}^2 \right]^{-1/2}$$

and the tilting angle (τ) of the base plane with respect to local helical axis $\underline{h}$ is given by

$$\cos \tau = (a_{21} + a_{12}) (2 \sin \varphi)^{-1}$$

Also we can define the base pair repeat p along $\underline{h}$ through the scalar product with the displacement vector relating the two systems. Assuming bond angles and distances as well as the two sugar puckers fixed a priori, the possible conformations of the base paired dinucleotide unit can be given in terms of only two angles of rotation for each triplet of external parameters φ, τ and p (see Fig. 2b). For the purpose of this paper we have chosen these angles as the torsional angles around the glycosidic bonds of the purine χ_1 and the pyrimidine χ_2 in order to obtain the basic set of modular dinucleotide units.

In fact the presence of the dyad axis $\underline{d}$ allows the following equations to be written

$$\underline{A}s = - \underline{s} \tag{1}$$

where $\underline{s}$ is a vector connecting two equivalent atoms of the two strands (we have chosen the phorphorous atoms).

Thus $\underline{s} = \underline{s}_1 - \underline{s}_2$ where $\underline{s}_1$ and $\underline{s}_2$ define the position of the phosphorus atoms in the origin system (see Fig. 2a).

Equation 1 can be given in an explicit form for the pair of rotation angles α and ζ around the O-P-O phosphodiester bonds as

$$\underline{B} \, \underline{A} \, \Theta \underline{Z} \, \underline{C} \, \underline{s} = - \underline{s} \tag{2}$$

where $\underline{B}$ and $\underline{C}$ represent the transformation matrices as functions of the remnant conformational parameters, $\underline{A}$ and $\underline{Z}$ the pure rotation matrices around the x axes of the angles of rotation α and ζ; Θ represents the pure rotation matrix around the z axis of the bond angle at P atom. Equation 2 can be arranged to give the pair of equations

$$\underline{Z}\,\underline{C}\,\underline{s} = - \tilde{\Theta}\,\tilde{\underline{A}}\,\tilde{\underline{B}}\,\underline{s}$$
$$\underline{\Theta}\,\underline{Z}\,\underline{C}\,\underline{s} = - \tilde{\underline{A}}\,\tilde{\underline{B}}\underline{s} \tag{3}$$

The first elements of these vectorial equations are independent on ζ and α respectively owing to the form of the corresponding matrices.

Thus, the following scalar equations are obtained:

$$(\underline{C}\,\underline{s})_1 = - (\tilde{\Theta}\,\tilde{\underline{A}}\,\tilde{\underline{B}}\,\underline{s})_1$$
$$(\underline{\Theta}\,\underline{Z}\,\underline{C}\,\underline{s})_1 = (\tilde{\underline{B}}\,\underline{s})_1 \tag{4}$$

Since $\underline{s}$ is independent on both α and ζ, the equations in Eq. 4 allow the determination of the values of α and ζ and they also restrict the range of variability of the other angles of rotation.

In fact the equations in Eq. 4 are of the general form

$$a \cos \psi + b \sin \varphi = c$$

or the equivalent one

$$\cos(\psi - \psi_o) = \frac{c}{(a^2 + b^2)^{1/2}} \quad \text{where} \quad \psi_o = \text{tg}^{-1}(\frac{b}{a})$$

ψ standing for α or ζ with a, b, c, functions of the remnant angles of rotation $\chi_1, \chi_2, \varepsilon, \gamma, \beta$. This equation has in general two solutions for ψ but only when $a^2 + b^2 > c^2$.

Thus, the dyad symmetry allows the determination of α and ζ in terms of the remnant conformational parameters.

It is possible to operate a further reduction if constraints are introduced for the cylindrical parameters of the base pairs as the helical parameters φ, τ and p already defined.

In fact Eq. 1 can be transformed as

$$\tilde{\underline{T}}\,\underline{\Phi}\,\underline{T}\,\underline{\Pi}\,\underline{s} = -\underline{s} \tag{5}$$

which can be obtained in the form

$$a \cos \psi + b \sin \psi = c \tag{6}$$

ψ standing for one of the remnant angles of rotation $\chi_1, \chi_2, \varepsilon, \gamma, \beta$.

A further equation of the same form is obtained introducing explicitly the bond angle at the P atoms (ϑ).

In fact, calling $\underline{l}_1$ and $\underline{l}_2$ the normalized vectors along the OP and PO phosphodiester bonds their dependence on the torsional angles is

$$\underline{l}_1 = \underline{l}_1 \ (\chi_1, \ \varepsilon)$$
$$\underline{l}_2 = \underline{l}_2 \ (\chi_2, \ \gamma, \ \beta)$$

The existence of the dyad axis allows the scalar equations to be written

$$\tilde{\underline{l}}_1 \ [\tilde{T} \ \Phi \ T \ \Pi] \ \underline{l}_2 = \cos \ \vartheta \tag{7}$$

which is of the general form

$$a' \cos \psi + b' \sin \psi = c' \tag{8}$$

Equations 6 and 8, where ψ is the same angle of rotation, for example β, allow the unequivocal determination of $\sin \beta$ and $\cos \beta$ in terms of $\chi_1, \chi_2, \varepsilon, \gamma$ for a given pair φ and τ.

Finally the trigonometric identity $\sin^2 \beta + \cos^2 \beta = 1$ can be used in order to determine γ (or ε) in terms of the remnant three angles of rotation because a, b, c, and a', b', c' are functions of χ_1, χ_2 and γ.

Thus, the conformations of base paired nucleotide residues of alternating polynucleotides with given φ and τ are a function of three angles of rotation if the two sugar puckers are fixed a priori.

If the further constraint of the base pair stacking is introduced namely, if the conformations with an average distance about 3.4 Å between base pair planes are selected, the geometrically possible conformations of base paired nucleotide residues, characterized by base pairs stacking and fixed values of φ and τ are represented in the restricted two dimensional conformational space: χ_1, χ_2.

Incidentally, it is important to note that the intercalating conformations of the base paired dinucleotide unit can be selected in the same two dimensional conformational space if an average base pairs stacking distance of 6.8 Å is assumed.

The set of two dimensional conformational diagrams were tested for their conformational energy on the basis of conformational energy calculations using semiempirical van der Waals functions and fixed charge distributions.

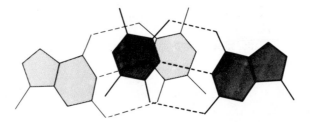

Figure 3. Theoretical stacking for the pyrimidine purine fragment in Z structure.

The energy conformational diagrams so obtained in terms of χ_1, and χ_2 represent the basic data for building up any kind of helical structures, either the uniform (B, A, C...), or the alternating (e.g. Z DNA) helices by assembling the local structures represented in the energy diagrams by conjugate pairs of angle of rotations: χ_1 and χ_2 and χ_2 and χ_1 respectively for the glycosidic bonds of pyrimidine purine and pyrimidine purine dinucleotides. A uniform helix will be represented by the points on the diagonals $\chi_1 = \chi_2$.

The set of χ_1, χ_2 diagrams with $\tau \sim 0°$ is particularly significant because it represents for all the values of φ the structures with the best stacking of the base pairs.

In this case the helical axis of the global structure coincides with the local azimuthal axes $\underline{h}$; as a consequence the dyad axes $\underline{d}$ are perpendicular to it. For $\tau \neq 0°$ this is not generally true except for uniform helices where $\chi_1 = \chi_2$.

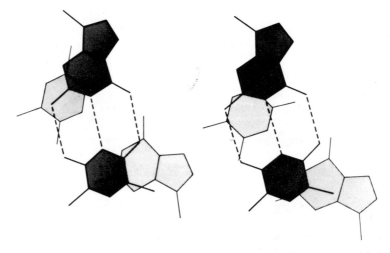

Figure 4. The two alternative theoretical stacking for pyrimidine purine fragments in Z structure.

Table 1. Energy and geometrical parameters of the theoretical Z DNA

	δ	χ_{pur}	γ	β	α	ζ	ε	δ	χ_{pyr}	$\varphi°$	pÅ	Evdw	Kcal/mole
	Conformational angles (°)									Helical parameters		Energy values	
Pyrimidine Purine fragments	83	62	160	180	77	74	−93	148	−146	−6.2	3.60	TA CG m^{5}CG	−10.92 − 9.01 −10.95
Purine-Pyrimidine fragments	83	55	50	−120	−135	−82	−93	148	−140	−51.5	3.70	AT GC m^{5}GC	−12.35 − 6.80 − 9.52
	83	55	80	−120	−157	−109	−78	148	−146	−60.2	3.55	AT GC m^{5}CC	9.14 −8.59 10.17

The Role of van der Waals Interactions in Determining the Structure of Z DNA

The energy allowed base paired dinucleoside monophosphate fragments of a double helix characterized by alternating syn − anti conformations around the glycosidic bonds, were selected for C $'_2$ endo − C $'_2$ endo and C $'_2$ endo − C $'_3$ endo puckers. In the case of C $'_3$ endo − C $'_3$ endo puckers no sterically allowed conformations were found; the average puckers were adopted as found by X-ray analyses,[11] for the sugar conformations of the C $'_2$ endo and C $'_3$ endo families.

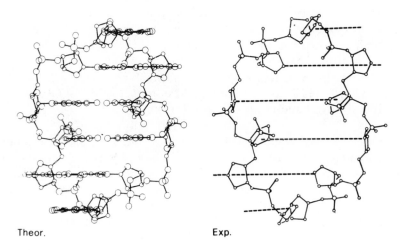

Theor.   Exp.

Figure 5. Comparison between theoretical and experimental Z structures; projection perpendicular to the helical axis.

The possible conformations for the pyrimidine purine fragments

$$\uparrow \begin{matrix} A-T \\ T-A \end{matrix} \downarrow \quad , \quad \uparrow \begin{matrix} G-C \\ C-G \end{matrix} \downarrow \quad \text{and} \quad \uparrow \begin{matrix} G-m^5C \\ m^5C-G \end{matrix} \downarrow$$

as well as the related one with 5-methyl cytosine which is known to stabilize Z DNA,[12] were found to be restricted to only one type of structure as shown in Fig. 3 with comparable conformational energy values as reported in Table I.

Two basic structures (shown in Fig. 4) were however found for the purine pyrimidine fragment

$$\uparrow \begin{matrix} T-A \\ A-T \end{matrix} \downarrow \quad \text{and} \quad \uparrow \begin{matrix} C-G \\ G-C \end{matrix} \downarrow$$

whose conformational parameters and energy are also reported in Table I.

By assembling the two blocks, two structures of comparable energy were obtained. These represent the repeating unit of Z helices which, on the projection perpendicular to the helical axis, appear practically identical to Z DNA[13] (see Fig. 5).

The main difference lies on the stacking of the base pairs of the purine pyrimidine fragment as shown in Fig. 6(a) and 6(b) where the comparison of the base pair stacking and the positions of phosphorus averaged over the two families, are illustrated. It is significant that the two stacking schemes appear very similar to those found for the Z_I and Z_{II} forms of Z DNA.[13] These structures however possess practically identical stacking of the pyrimidine purine block, as can be evidenced by comparing Fig. 6(a) with 6(b).

It is interesting that very similar base pair stackings were obtained from significantly different chain conformations, indicating that the base pairs and base pair-sugar interactions are the main contributions to the stability of the Z form of DNA. This is further supported by the strict similarity of purine pyrimidine stacking at the end bases of adjacent molecules, which obviously lack the phosphodiester bridges, as found in the crystal structure of $d(CG)_2$,[14-16] $d(CG)_3$,[13] the m^5C homologous derivative[17] and $d(m^5CGTAm^5CG)$.[18]

The theoretical torsional angles which characterize the Z families are reported in Table II and compared with Z_I and Z_{II} experimental Z DNA as well as with the angles of rotation found in the X-ray analyses of relevant compounds. These torsional angles are plotted in order to illustrate their common trend (see Fig. 7).

The agreement appears satisfactory whilst the theoretical results were obtained with a fixed pair of sugar puckers and that the experimental parameters are affected to some degree by indetermination connected with the low resolution of X-ray data. In particular, the conformation of the pyrimidine-purine block appears conforma-

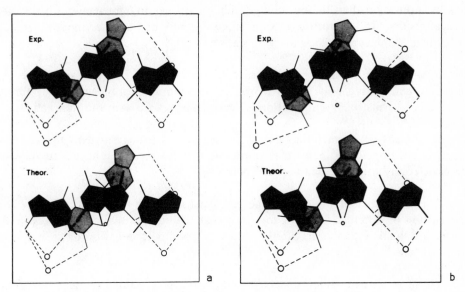

Figure 6. Comparison between theoretical and experimental Z stacking; projection along the helical axis; ○ represents the position of the phosphorus atom; o represents the position of the helical axis; **(a)** Z_I structure, **(b)** Z_{II} structure.

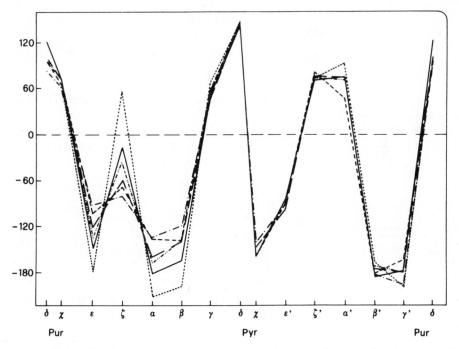

Figure 7. Diagram of the internal parameters along the repeating unit for the theoretical and experimental Z structures; — · — · — theoretical values,   – – – – and · · · · ·,[14] — —,[18] – – – —.[17]

tionally more rigid than the purine pyrimidine sequence in agreement with the theoretical results. In general the theoretical conformations fit better the idealized Z_I structure as it appears looking at Figs. 6(a) and 6(b).

It is important to note that the fine structural features of Z DNA, have been determined by considering only van der Waals interactions as previously made for A and B DNA, whilst their stabilization is dependent on the water and ions activities.

This fact seems to characterize also the structures of proteins where the van der Waals interactions determine for example the fine features of α-helices in spite of the basic contribution of the hydrogen bonds.

Thus, the fundamental biological macromolecules appear to be stabilized by the cooperation of different kinds of forces so that the energy minimization restricted to van der Waals interactions often results in a good determination of their molecular structure.

The role of water as counterions activities appears to be related to the equilibrium constants between definite conformations (at least at local structural level) corresponding to van der Waals energy minima.

In apparent contrast with the up-to-date experimental evidence, the Z structure seems to be suitable for alternating double helical poly (dA − dT), because the presence of methyl group stabilizes the Z form of DNA as for m^5C derivatives. Actually, Z form has been found in the crystal structure of d(m⁵CGTAm⁵CG) pyr-pur alternating hexamer,[18] where the stacking involving GT, TA and Am⁵C base pairs are practically equal to that found in the d(CG)₃ and d(m⁵CG)₃ X-ray structures.

Thus, the reluctance of poly (dA − dT) to assume the Z structure should be a consequence of the stabilization of the B-form by water molecules, as shown for the dodecamer d(CGCGAATTCGCG)[19] where a water spine of hydration of the position AATT freezes the B form; this explains also the well known resistance of AT rich DNA to transform in the A form lowering the relative humidity.[20]

Table II. Torsional angles of Z family

	δ	χ_pur	ε	ζ	α	β	γ	δ	χ_pyr	ε'	ζ'	α'	β'	γ'
Z_I theor	83	59	−93	−82	−135	−120	50	148	−143	−93	74	77	180	160
Z_{II} theor	83	59	−78	−109	−157	−120	80	148	−146	−93	74	77	180	160
Z_I(13)	99	68	−104	−69	−137	−139	56	138	−159	−94	80	47	179	−165
Z_{II}(13)	94	62	−179	55	146	164	66	147	−148	−100	74	92	−167	157
Z'(14)	122	70	−150	−17	177	−166	51	141	−160	−85	71	75	−175	−179
d(m⁵CGTAm⁵CG)(18)	100	69	−123	−61	−160	−140	46	143	−159	−97	75	72	−175	−179
d(m⁵CG)₃(17)	94	69	−139	−34	−168	−138	54	141	−157	−98	76	71	−173	−179

Ten-Fold Periodical DNAs

The possibility that the B form of DNA could be systematically bent or bendable because of periodicity in its sequence was first suggested by Trifonov et al.[5] They calculated the autocorrelation functions of a variety of eukariotic and viral DNAs and showed that some dinucleotides have a clear tendency to be periodically repeated along the chromatin DNA sequences. This was interpreted as a plausible cause of a preferential bendability of DNA in one direction, due to a specific wedge-like structure of dinucleotides in the sequence with implications in the recognition mechanisms of proteins. They proposed a mapping procedure to locate nucleosomes along a DNA sequence in the regions characterized by a significant potential bend of DNA axis, using a matrix of bendability derived from the analysis of chromatin DNA sequences.

A different algorithm of locating the potential or actual bend of DNAs was later proposed by Zurkin[21] based on the predicted distortions of PurPyr and PyrPur dinucleotide pairs structure according to the Calladine-Dickerson rule.[22]

The original idea about sequence dependent deformational anisotropy of DNA, obtained recently a direct experimental evidence by electric dichroism studies,[6,23] gel electrophoresis and gel filtration experiments.[7,24] These investigations demonstrate the existence of stable bends in certain natural, and synthetic DNAs intrinsic to their sequences. Such bends are localized in regions of DNA distringuished by recurring runs of adenine (and thymine in the complementary strand). Drew and Travers,[25] using a careful analysis of cleavage frequencies by DNase I on a circularized 169 bp tract of bacterium DNA, demonstrate that the minor groove of A and T bases are preferentially faced in, while the G and C region minor grooves face outward the circle. Such situation is reproduced in the corresponding nucleosome reconstituted with the same DNA.

Investigations on bent helical structure in DNA are at present a current subject of interest and their significance extends far beyond such structural property of DNA, whilst the mechanism of bending is still debated. In fact the experiments on periodical DNAs are extremely severe for the proposed previsional methods based on the wedge as well as the Calladine-Dickerson clash models. Crothers et al. and Hagerman[7,8] independently suggested the existence of cooperative effects because the simple considerations of dinucleotides properties appeared not to be sufficient to consistently explain the experimental data. Starting from conformational energy calculations of dinucleotide fragments of DNA, we have identified, in agreement with the results of Zurkin, the principal directions of bendability in both the minor and the major grooves, namely in the roll directions, while wedge deformations are energetically unfavorable. In particular, on account of both electrostatic and van der Waals interactions, AA and TT dinucleotides tend to reduce the narrow groove as well as AT, while TA, CA, CG and TG have the opposite behavior in partial agreement, with the Calladine-Dickerson rule. We have in a first approximation quantified such tendency in a bendability matrix which takes into account also the direction

i\j	A	T	G	C
T	−2.0	1.2	−1.6	0.2
A	1.2	1.9	−0.2	0.9
C	−1.6	−0.2	−1.3	−0.1
G	0.2	0.9	−0.1	0.5

$$\cdot \; e^{2\pi i \left(\frac{\eta}{v} + \frac{1}{4}\right)}$$

of deformation in B DNA: n is the sequence number and v is the period of DNA set equal 10.3.

The bendability of a DNA tract between the position n1 and n2 along the sequence is represented by a complex quantity equal to

$$B = \sum_{n_1}^{n_2} b_{ij}(n)$$

where $b_{ij}(n)$ is the element of matrix corresponding to the dinucleotide ij. The modulus of B represents the maximum bendability in the direction, defined by its phase angle, where contributions by the turns of DNA are in phase. A smooth folding of DNA is generated by assigning to each dinucleotide residue roll angles proportional to its contributions on the direction of maximum bendability. The existence of a linear trend limited to small angles between the sum of roll angles and the angle of curvature per turn of the double helix suggests a proportionality between curvature and bendability B. It is worth noting that practically the same folding is obtained, using whichever distribution of roll distortions of the single dinucleotides, providing that it is a periodical function with the DNA period. Thus practically the same curvature is obtained with the kink or minikink model, when the whole distortion is concentrated in one or two opposite in phase dinucleotides, respectively. The conformational versatility of the phosphodiester chains should assure a posteriori the suitable bridges between the base pairs.

We have first used this algorithm to predict the relative bend of synthetic periodical DNAs investigated very recently by Crothers et al. and Hagerman. Figure 8 shows the correlation between the retardation of the electrophoretic mobilities represented by the ratio R between the apparent number of bases and the true number of base pairs, against the modulus of bendability B per turn of the double helix calculated for the periodical DNAs reported in Table III. The correlation is quite satisfactory and explains consistently the dramatically anomalous behavior of such periodical synthetic DNA. It is noteworthy that in the algorithm used are not represented cooperative effects and only dinucleotide properties are taken into account pro-

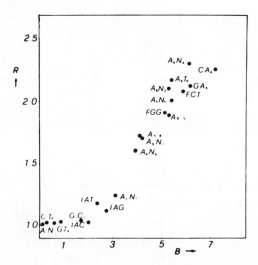

Figure 8. Correlation between the retardation of the electrophoretic mobility R and the modulus of bendability B of the periodical synthetic DNAs reported in Table III for a constant degree of polymerization N = 150 bp.

viding that they are associated to the correct direction and that no attempt has been made to obtain the "best fit" for the matrix of bendability.

Using the same method it is possible to account correctly for the direction of the curvature of the Dickerson dodecamer CGCGAATTCGCG which results in the narrowing of the small groove about the TT-AA sequence.

Table III. Monomeric sequences of periodical DNAs examined[7,8]

A_3N_7	GGGCCAAACCG	FCT	GGCAAAAATG
A_4N_6	GGCCAAAACG	G_5N_5	TCGTGGGGGC
A_5N_5	GGCAAAAACG	GA_4	GAAAATTTTC
A_6N_4	GGCAAAAAAC	CA_4	CAAAATTTTG
A_8N_2	CCAAAAAAAA	GT_4	GTTTTAAAAC
A_9N_1	CAAAAAAAAA	CT_4	CTTTTAAAAG
IAC	GGCAACAACG	C_3G_3	GGGTCGACCC
IAG	GGCAAGAACG	A_5N_{10}	CCGGCAAAAACGGGC
IAT	GGCAATAACG	A_{5-8}	CCAAAAACGGGCAAAAAAAA
FGG	CCGAAAAAGG	A_{8-5}	CCAAAAAAAAACGGGCAAAAA
A_6T_6	AAAACGGGTTTTTTGGGCAA		

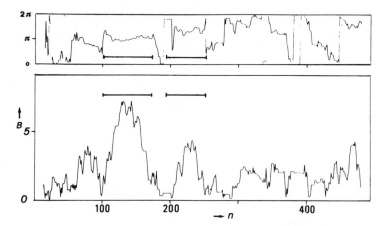

Figure 9. Profiles of bendability and relative phase of kinetoplast DNA along the sequence. The bars represent the fragments with strong anomalies in electrophoretic behavior.

Further, bending of tracts of natural DNAs for which evidence exists for stable curvature has been analyzed. Figure 9 shows the profile of the modulus of bendability B and the relative phase averaged over three turns of kinetoplast DNA circles, isolated from trypanosoma mitochondria (K-DNA) along the sequence. A main tract of about 70 bp and a secondary tract of about 40 bp are easily recognized for both the high bendability values and the practical invariance of the relative phases. The small difference between the phases implicates that K DNA has a curvature confined to a single plane in agreement with experimental data or with a small positive writhe. In principle, in fact, using the difference between successive phase angles as a torsional angle and the modulus of bendability as proportional to the bending angle it is possible to provide a rough representation of a plausible

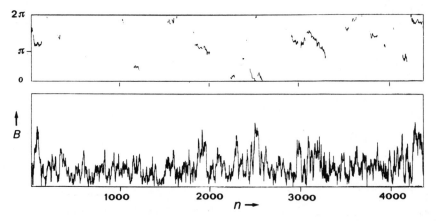

Figure 10. Profiles of bendability and relative phase of pBR322 only the phases corresponding to values of bendability higher than the average value are reported.

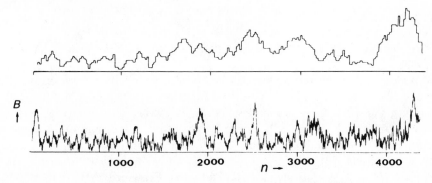

Figure 11. Comparison of the profile of bendability of pBR322 averaged over the length of nucleosomal DNA and the frequency of occurrence of nucleosomes along the sequence.

spatial superstructure of a tract of DNA from its sequence. It is interesting to note that for small angles when the difference of phases is less than the bending angle, flat helicoidal turns are obtained; on the contrary when it is greater, a snaking writhe results. Incidentally, it is interesting to note the existence in K DNA of a superperiodicity of ~80 bp which could be speculated to be in relation with the folding of K DNA in mitochondria. Figure 10 shows the bendability diagram of the circular DNA pBR322, 4360 bp long. For the sake of clarity the bendability was averaged over five turns of B DNA and the relative phase was plotted only in the most significant regions where the bendability is higher than the average bendability. The bendable regions are easily localized along the sequence where the phases are nearly invariant: these regions could indicate the plausible positioning of nucleosomes. It is interesting that the theoretical nucleosome map is in good agreement with the experimental one, recently determined in the laboratory of one of the authors[26] using the method of location in electron micrographs of uncrosslinked bubbles along the sequence resulting after photoreaction with trimethyl-psoralen of histone octamer/DNA complexes and successive denaturation.

Figure 11 shows in fact the direct comparison of the frequency of bubbles along the sequence of pBR322 (corresponding to the frequency of occurrence of nucleosomes) and the bendability diagram averaged over 80 nucleotides, the length of a turn of nucleosome superhelix.

The biological role of the curvature and in general of the superstructures sequence dependent is not yet quite evaluated but it could provide the clue for an understanding of the mechanisms of management by DNA of its genetic information.

Acknowledgments

This research has been supported by the Progetto strategico del CNR "Biotecnologie" and by Grandi Progetti d'Ateneo dell'Università di Roma "La Sapienza" MPI 60%.

References

1. Cold Spring Harbor Symp. on Quant. Biol., **Vol XLVII**, Structures of DNA (1983).

2. S. Arnott and R. Chandrasekaran, "Fibrous polynucleotide duplexes have very polymorphic secondary structures," in Biomolecular stereodynamics (ed. R.H. Sarma), **Vol. I**, Adenine Press, New York, 99 (1981).

3. F.M. Pohl and T. Iovin, Salt-induced cooperative conformational change of a synthetic DNA. Equilibrium and kinetic studies with poly (dG-dC), J. Mol. Biol., **67**, 375 (1972).

4. A.H.J. Wang, G.J. Quigley, F.J. Kolpak, J.L. Crawford, J.H. van Boom, G. van der Marel, and A. Rich, Molecular structure of a left-handed DNA fragment at atomic resolution, Nature, **282**, 680 (1979).

5. E.N. Trifonov, Sequence-dependent variations of B-DNA structure and protein-DNA recognition, in Cold Spring Harb. Symp. on Quant. Biol., **Vol. XLVII**, 271 (1983).

6. J.C. Marini, S.D. Levene, D.M. Crothers and P.T. Englund, A bend helix in kinetoplast DNA, in Cold Spring Harb. Symp. on Quant. Biol., **Vol. XLVII**, 279 (1983).

7. H.S. Koo, H.M. Wu, and D.M. Crothers, DNA bending at adenine, thymine tracts, Nature, **320**, 501 (1986).

8. P.J. Hagerman, Sequence-directed curvature of DNA, Nature, **321**, 449 (1986).

9. P. De Santis, S. Morosetti, A. Palleschi and M. Savino, Conformational and structural constraints in double-helical polynucleotides, Biopolymers, **20**, 1707 (1981).

10. P. De Santis, S. Morosetti, A. Palleschi and M. Savino, Conformational analysis of double-stranded B-type DNA structures, Biopolymers, **20**, 1727 (1981).

11. M. Sundaralingam, Stereochemistry of nucleic acids and their constituents. IV. Allowed and preferred conformations of nucleosides, nucleoside mono-di-tri-tetraphosphates, nucleic acids, and polynucleotides, Biopolymers, **7**, 821 (1969).

12. M. Behe and G. Felsenfeld, Effects of methylation on a synthetic polynucleotide: the B-Z transition in poly (dG-m^5 dC), poly (dG-m^5 dC), Proc. Natl. Acad. Sci. USA, **78**, 1619 (1981).

13. A.H.J. Wang, G.J. Quigley, F.J. Kolpak, G. van der Marel, J.H. Van Boom and A. Rich, Left-handed double helical DNA: variations in the backbone conformation, Science, **211**, 171 (1981).

14. H.R. Drew and R.E. Dickerson, Conformation and dynamics in a Z-DNA tetramer, J. Mol. Biol., **152**, 723 (1981).

15. J.L. Crawford, F.J. Kolpak, A.H.J. Wang, G.J. Quigley, J.H. van Boom, G. van der Marel and A. Rich, The tetramer d(CpGpCpG) crystallizes as a left-handed double helix, Proc. Natl. Acad. Sci. USA, **77**, 4016 (1980).

16. H. Drew, T. Takano, S. Tanaka, K. Itakura and R.E. Dickerson, High-salt d(CpGpCpG), a left-handed Z'DNA double helix, Nature, **286**, 567 (1980).

17. S. Fujii, A.H.J. Wang, G. van der Marel, J.H. Van Boom and A. Rich, Molecular structure of (m^{5}dC-dG)$_3$; the role of the methyl group on 5-methylcytosine in stabilizing Z-DNA (1982).

18. A.H.J. Wang, T. Hakoshima, G. van der Marel, J.H. Van Boom and A. Rich, AT base pairs are less stable than GC base pairs in Z-DNA: the crystal structure of d(m⁵CGTA m⁵CG), Cell, **37**, 321 (1984).

19. M.L. Kopka, A.V. Fratini, H.R. Drew and R.E. Dickerson, Ordered water structure around a B-DNA dodecamer. A quantitative study, J. Mol. Biol., **163**, 129 (1983).

20. A.G.W. Leslie, S. Arnott, R. Chandrasekaran and R.L. Ratliff, Polymorphism of DNA double helices, J. Mol. Biol., **143**, 49 (1980).

21. V.B. Zhurkin, Y.P. Lysov and V.J. Ivanov, Anisotropic flexibility of DNA and the nucleosomal structure, Nucleic Acids Res., **6**, 1081 (1979).

22. R.E. Dickerson, Base Sequence and Helix Structure Variation in B and A DNA, J. Mol. Biol., **166**, 419 (1983).

23. P.J. Hagerman, Evidence for the existence of stable curvature of DNA in solution, Proc. Natl. Acad. Sci. USA, **81**, 4632 (1984).

24. S. Dieckmann and J.C. Wang, On the Sequence Determinants and Flexibility of the Kinetoplast DNA Fragment with Abnormal Gel Electrophoretic Mobilities, J. Mol. Biol., **186**, 1 (1985).

25. H.R. Drew and A.A. Travers, DNA Bending and its Relation to Nucleosome Positioning, J. Mol. Biol., **186**, 773 (1985).

26. E. Caffarelli, L. Leoni, B. Sampaolese and M. Savino, Persistence of cruciform structure and preferential location of nucleosome on some regions of pBR322 and ColE1 DNAs, Eur. J. Biochem., **156**, 335 (1986).

Conformational Peculiarities and Biological Role of Some Nucleotide Sequences

V.I. Poltev
Institute of Biological Physics
USSR Academy of Sciences
142292 Pushchino, Moscow Region

A.V. Teplukhin and V.P. Chuprina
Research Computing Center
USSR Academy of Sciences
142292 Pushchino, Moscow Region

Summary

A conclusion about the dependence of the conformational behavior of nucleic acids on nucleotide sequences has been made on the basis of nonbonded interaction energy calculations using the classical potential functions. The potential functions have been chosen by comparing the results of calculations for model systems (crystals, associates) with the experimental data and with the results of the most rigorous quantum mechanical calculations. Potentials proposed by other authors and the possibilities of further refinement of potential functions are considered.

To study conformational patterns of different nucleotide sequences we have calculated the energy of interaction between base pairs (all combinations have been considered) as a function of the parameters determining their mutual position in the double helix. The calculations have shown, in accordance with the experimental data, that there are sequences for which the energy preferred conformations are the A-like ones (e.g. GG, AT, AC) while for other sequences the preferred conformations are the B-like ones (e.g. AA, TA, CA).

Theoretical conformational analysis of a regular double-helical polynucleotide has shown the existence of two regions of minimal energy values corresponding to the A- and B-families of nucleic acid conformations. Taking into account the possible biological role of sequences containing repeating adenines in one chain and repeating thymines in the other one, we have considered other possible regular conformations of poly(dA) · poly(dT). Two more regions of minimal energy values have been revealed which correspond to the A-like conformation of one chain and the B-like conformation of the other one. The relative energetic advantages of the four regions depends on phosphate group charge neutralization.

Introduction

In the last few years one can observe rapid progress in studies of the atom-molecular mechanisms of functioning of nucleic acids; our image of the structure and dynamics of the DNA double helix is changing qualitatively. New experimental techniques, in particular, progress in chemical synthesis and gene engineering, permit one to obtain polynucleotides and oligonucleotides of a definite chain length and base sequence in amounts necessary for physical studies.

X-ray analysis of nucleic acid fragments gives information on the fine structure of the double helix and on conformational peculiarities of specific sequences in crystals.[1-4] High-resolution NMR spectroscopy yields data on structural details of double helical fragments with a definite nucleotide sequence in solution.[5,6] At the same time, DNA fragments of several nucleotide pairs are too complicated to deduce their three-dimensional structure directly and unambiguously from the experimental data. To obtain coordinates of atoms it is necessary to have a model and then to refine it proceeding not only from the experimental results on this fragment but also from theoretical considerations, in particular, steric and energy criteria. Therefore, experimental studies go in parallel with theoretical ones dealing with calculations of energy of non-bonded interactions of nucleic acid fragments as a function of conformational parameters.

As the systems are complicated, these calculations should be classical, though, of course, they should take into account quantum mechanical considerations and the results of calculations by the most rigorous methods. Such calculations can be used, on the one hand, to refine and interpret the available experimental data and, on the other hand, to predict conformational possibilities and conformational behavior of biomolecules and their complexes which have not yet been studied.

The results of X-ray studies of the crystals of double-helical oligonucleotides suggested some empirical regularities of the dependence of conformational parameters on nucleotide sequence. These regularities, e.g. Calladine-Dickerson rules[7,8] and Calladine-Drew concept on "neutral" and "bistable" sequences,[9] have been established in terms of elastic system mechanics. In our opinion, however, a simplified consideration of base pairs (not taking into account the atomic structure and, consequently, charge distribution and surface topography) cannot yield a quantitative description of the dependence of the double helix conformation on the base sequence. Crystals of oligonucleotides with different base sequences which have been so far subjected to X-ray analysis are not numerous. Therefore, generalization of data for a limited set of individual sequences cannot be considered as a definitive procedure for description of conformational behavior of any sequence. There are no definite answers to the questions of how large are the differences between double-helical conformations of various sequences under different conditions, what conformational parameters mostly reflect these differences, and what is their biological relevance.

In order to find out in what way the nucleotide sequence affects the double helix conformation we have performed systematic calculations of the energy of interaction between nucleic acid components and have searched for the regions of minimum energy values for different sequences. At the last Symposium we reported some results of simulation of the macromolecular structure and biological functioning of nucleic acids by calculating the energy of nonbonded interactions using atom-atom potential functions.[10] Attention was mainly paid to the problem of the mechanisms of fidelity of nucleic acid biosynthesis.

In this paper we shall consider the following problems:

1. Atom-atom potentials for calculating the interactions in nucleic acids, their agreement with the experimental data and the possibilities of their refinement.

2. Possible conformational patterns of different base sequences, conformations preferred by different sequences.

3. Conformational possibilities of fragments of the polynucleotide duplex containing repeating adenines in one chain and thymines in the other one.

Atom-Atom Potentials for Nucleic Acid Calculations

According to the method of atom-atom potential functions, the energy of intermolecular interactions is calculated as a sum of energies of all pairwise atom-atom interactions. We use Eq. 1 to calculate the energy of interaction of hydrogen atoms capable to form hydrogen bonds with the proton-acceptor atoms. To calculate the energy of the other atom-atom interactions, 1-6-12 potentials (Eq. 2) are used:

$$U(r_{ij}) = \frac{e_i \cdot e_j}{\varepsilon \cdot r_{ij}} - A_{ij}^{(10)} \cdot r_{ij}^{-10} + B_{ij}^{(10)} \cdot r_{ij}^{-12} \tag{1}$$

$$U(r_{ij}) = \frac{e_i \cdot e_j}{\varepsilon \cdot r_{ij}} - A_{ij} \cdot r_{ij}^{-6} + B_{ij} \cdot r_{ij}^{-12} \tag{2}$$

In these equations the first term corresponds to electrostatic interactions (E_e) and the sum of the two others — to van-der-Waals interactions (E_v), r_{ij} is the interatomic distance, e_i and e_j are the charges of atoms, ε is the effective dielectric constant which is assumed to be equal to 1 for base-base interactions and to 4 for other interactions (except for special calculations evaluating the influence of the computational procedure for electrostatic interactions on the position and the value of the energy minima), A_{ij} and B_{ij} ($A_{ij}^{(10)}$ and $B_{ij}^{(10)}$) are the parameters depending on the type of atoms, their valency states and (in some cases) neighboring atoms. Torsion potentials for rotation around single bonds and the energy of bond angle distortion are also taken into account in calculations of intramolecular nonbonded interactions. To reduce the number of variables on which the energy function depends, the bond lengths and many bond angles were assumed to be constant.

The atom-atom potential functions which we are using to calculate the energy of the nonbonded interactions of nucleic acids are the result of several refinements performed to obtain a better agreement with the experimental data.[10-13] These potentials permit us to reproduce the data on crystals and associates of aromatic compounds as well as those on conformations of simple fragments of nucleic acids. The choice of parameters for water-nucleic acid and water-water interactions makes it possible to use our set of potential functions for simulation of the behavior of conformationally labile fragments of nucleic acids in aqueous solutions.[14] Nevertheless, potential functions could be ameliorated and a more adequate description of biomolecules could be achieved by taking into account new experimental data and the results of the most rigorous quantum mechanical calculations. We are working in this direction.

The dependence of the energy of atom-atom interactions on the distance (1-6-12 potentials) is used in many other potential functions which have been obtained by choosing the parameters from the experimental data for model systems,[15,16] as well as by approximating quantum mechanical *ab initio* calculations.[17,18] However, the parameters of potential functions proposed by other authors to calculate the energy of nonbonded interactions of nucleic acids differ from ours. We discussed this earlier[12,13] and showed the advantages of our potential functions. Here we shall compare the results obtained for model systems using the most recent modifications of the method of atom-atom potential functions.

Let us first consider the electrostatic component of the energy. Our potentials use the charge values obtained by the simplest, "most empirical" methods of quantum chemistry — the Hückel method for π-electrons and the Del Re method for σ-electrons with Berthod and Pullman parameters.[19] These parameters permit one to obtain the total charges which reproduce the experimentally determined values of dipole moments for a series of aromatic compounds. A number of methods have been used by different authors to calculate charges of base atoms.[16-21] For example, charges obtained in *ab initio* calculations have been reported.[17] Weiner et al. fitted charges of atoms (or groups) to quantum mechanically calculated electrostatic potentials. Rather different charges on base atoms were reported by Rein et al.[21] who used a similar procedure. Lavery et al. reparameterized the Hückel and Del Re methods to reproduce both the potential and the field obtained with precise overlap multipole expansions. It is interesting to note that these new parameters and the calculated charges differ little from the corresponding values obtained in calculations with Berthod and Pullman parameters. The dipole moments calculated with the use of base atom charges obtained in different ways are given in Table I. As seen in the Table, all charge distributions give the dipole moments differing more from the experimental ones than the Berthod and Pullman charges used in our calculations. Therefore, there is no reason to change the procedure which we adopted earlier to determine the charges necessary for calculations of the energy of nonbonded interactions in nucleic acids *via* atom-atom potentials.

Weiner et al. suggested that in calculations of the electrostatic energy the dielectric constant depending on interatomic distance should be taken into account, $\varepsilon = r_{ij}$. It should be noted that thus obtained values of the base interaction energy in

Table I. Dipole moments of nucleic acid bases calculated from atom charges obtained by different methods.

Authors and References	Bases				
	Uracil	Thymine	Cytosine	Guanine	Adenine
Berthod and Pullman[19]	3.9	3.7	7.0	6.8	2.8
Lavery et al.[20]	4.5		7.2	8.7	2.5
Weiner et al.[16]	2.4	2.9	5.6	5.9	2.1
Matsuoka et al.[18]		4.3	6.0	5.8	1.5
Ray et al.[21]		3.5	5.6	6.0	2.5
Experimental[a]	3.9				3.0

[a]The last line presents the experimentally measured moments of 1-alkyl derivatives of pyrimidines and 9-alkyl derivatives of purines.[22]

H-bonded pairs differ from the results of calculations with $\varepsilon = 1$ by only $10-20\%$ while the sequence of energy values in the series of pairs virtually does not change.

To compare the validity of different modifications of the method of atom-atom potential functions for simulation of interactions of nucleic acid components we calculated the energy of intermolecular interactions in crystals of a number of

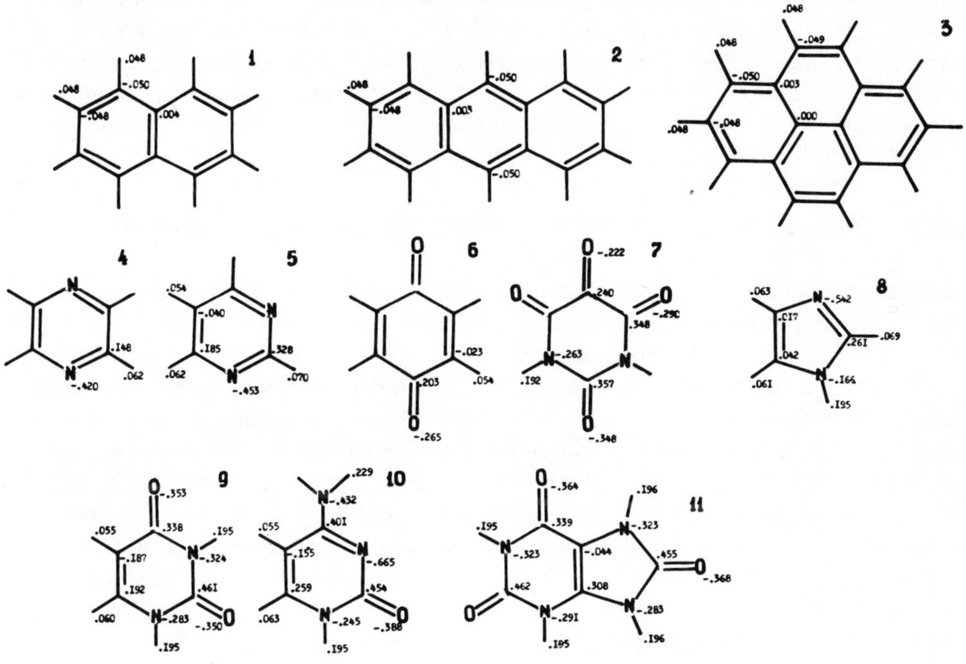

Figure 1. The structure and partial atomic charges (in electronic units) for aromatic compounds whose crystals were considered in this work. 1, naphthalene; 2, antracene; 3, pyrene; 4, pyrazine; 5, pyrimidine; 6, p-benzoquinone; 7, alloxan; 8, imidazole; 9, uracil; 10, cytosine; 11, uric acid.

aromatic compounds. The structure and atomic charges for these compounds are reproduced in Fig. 1. We considered most of these crystals earlier.[11,12]

The energy was calculated as a function of the unit cell parameters so that the energy value can be compared with the heat of sublimation and the calculated values of the unit cell lattice constants (a, b, c and β) in the energy minimum can be compared with those determined in X-ray studies. Table II presents the results of calculations and their comparison with the experimental data. For each crystal it is necessary to calculate the interaction energy of a molecule (0th molecule) with all the others forming a parallelepipded with the sides m·a, n·b, p·c (m, n and p are integers), the 0th molecule being near the center of the parallelepiped. To calculate the total energy with an accuracy of some percent, the numbers m, n and p were chosen to be 2, 3 or 4 depending on the values of a, b and c, so as to take into account at least all interatomic distances smaller than 10 Å. A combination of integers m, n and p, the cell index, is given in Table II.

Calculations have shown that the methods used in our studies reproduce the experimental data for the crystals of aromatic compounds better than Weiner's method.[16] Replacement of the C-H group in aromatic molecules by a united atom (as proposed by Weiner et al.[16]) leads to a worse agreement of the calculated results with the experimental data for all crystals considered. A change of charges on the atoms according to Lavery et al.[20] does not improve the agreement of the calculated results with the experimental data for crystals. As for the energy of association of bases in vacuum, such substitution gives values in worse agreement with the experiment. Agreement with experiment for crystals is not improved by the use of a dielectric constant depending on the distance ($\varepsilon = r_{ij}$). Therefore there seems to be no point in changing the procedure and the parameters of potentials used by us earlier.[10,13]

Dependence of the Double Helix Conformation on the Base Sequence

The dependence of the conformational behavior of the DNA double helix on nucleotide sequence can be considered now as a fact proved experimentally rather than a hypothesis. Some sequences can be transformed into a left-handed Z-form.[4] Besides, conformational parameters for right-handed B- and A-forms can change along the polynucleotide chain according to the base sequence, as shown, for example, by X-ray studies of oligonucleotide crystals.[1-3,8] The conformation of polynucleotides in fibers[40] and the B→A transition in solutions[41] depend on nucleotide sequence. (A:T)$_n$ sequences are "B-philic" while (G:C)$_n$ ones are "A-philic."[42] The probability of nuclease digestion of internucleotide bonds correlates with local variations of the duplex conformation.[43,44] It has been suggested that these changes are recognized by proteins, that there is a "conformational code" of protein-nucleic acid recognition in the transcription processes.

In this relation it is interesting to study the conformational possibilities of the double helix regions with different nucleotide sequences using calculations of non-bonded interaction energy and searching for this energy minima. Such a study of a

Table II. Results of calculations of the energy of intermolecular interactions in crystals of aromatic compounds performed using different modifications of the method of atom-atom potentials as compared with the experimental data.

Crystal	Method	Electro-statics	Hydro-gens	ΔH −E	−E$_e$	a Δa	b Δb	c Δc	β Δβ
naphthalene	E			17.3		8.098	5.953	8.652	124.4
ref. 23,24	P	+	+	17.0	0.4	0.05	−0.17	0.002	2.0
z=2	W	+	+	23.3	0.4	−0.03	0.01	−0.02	2.7
mnp=232	W	−	−	16.8		−0.3	−0.46	−0.54	0.8
antracene	E			24.4		8.443	6.002	11.124	125.6
ref. 25,26	P	+	+	22.4	0.5	−0.01	−0.21	−0.11	−0.6
z=2	W	+	+	32.0	0.5	−0.01	−0.17	−0.10	1.2
mnp=232	W	−	−	22.8		0.03	−0.51	−0.43	2.7
pyrene	E			20.6		13.649	9.256	8.470	100.28
ref. 27,26	P	+	+	21.9	0.2	0.04	−0.31	−0.02	−0.6
z=4	W	+	+	30.7	0.2	−0.06	−0.12	−0.09	0.1
mnp=222	W	−	−	27.1		−0.55	−0.76	−0.71	1.4
pyrazine	E			13.45		9.325	3.733	5.850	
ref. 28,29	P	+	+	11.8	1.3	−0.01	−0.07	−0.24	
z=2	W	−	+	15.6		0.29	−0.27	−0.45	
mnp=243	W	−	−	11.0		−0.11	−0.25	−0.19	
pyrimidine	E					11.555	9.461	3.693	
ref. 30	P	+	+	13.0	2.5	−0.25	0.03	−0.07	
z=4	P	−	+	10.6		−0.11	0.11	−0.10	
mnp=224	W	−	+	14.9		−0.19	0.13	−0.25	
	W	−	−	11.3		−0.60	−0.37	−0.20	
p-benzoquinone	E			15.0		6.763	6.735	5.711	99.55
ref. 31,32	P	+	+	14.9	2.6	0.05	−0.01	−0.05	−0.7
z=2	P	−	+	12.4		0.10	0.01	−0.02	−0.9
mnp=222	W	−	−	15.8		−0.36	0.17	0.98	1.0
alloxan	E					5.886	5.886	14.10	
ref. 33, z=4	P	+	+	23.0	7.0	−0.08	−0.08	0.34	
mnp=332									
imidazole	E			20.4		7.582	5.372	9.79	61.02
ref. 34,35	P	+	+	19.0	8.0	−0.05	−0.14	0.04	0.0
z=4, mnp=232	P	R	+	17.5	6.6	−0.05	−0.12	0.02	−0.2
uracil	E			28.8		11.938	12.376	3.655	120.9
ref.36,37	P	+	+	25.7	7.7	0.08	−0.06	0.08	5.7
z=4	P	R	+	24.9	6.9	0.08	−0.08	0.03	4.7
mnp=224	W	W	−	31.5	11.7	−0.84	0.09	−0.13	2.36
	W	R	−	28.3	10.4	−0.95	0.23	−0.13	1.9
cytosine	E			37.1		13.044	9.496	3.814	
ref. 37,38	P	+	+	35.4	15.6	−0.13	−0.18	−0.14	
z=4	P	R	+	32.2	13.1	−0.16	−0.19	−0.13	
mnp=224	P	L	+	39.8	19.0	−0.18	−0.20	−0.15	
	W	W	−	31.7	13.7	−0.24	−0.13	−0.22	
	W	R	−	31.6	13.9	−0.24	−0.13	−0.22	
uric acid	E					14.464	7.403	6.208	65.1
ref. 39	P	+	+	39.0	11.7	0.18	−0.04	−0.04	−0.4
z=4, mnp=223									

Note: In the column "crystal" references are to the experimental papers where the heat of sublimation and the crystal structure have been determined; z is the number of molecules per unit cell; mnp is the cell index; E, experimental data; P, procedure of Poltev et al.;[10-14] W, potentials of Weiner et al.[16] In the column "electrostatics" + means that the electrostatic energy was calculated by our standard procedure; L, use of the charges according to Lavery et al.;[20] R, calculations taking into account distance-dependent ε; Δa, Δb, Δc, Δβ, deviations of the calculated lattice constants from X-ray data (a, b, c, β). The heat of sublimation (ΔH), the energy of intermolecular interactions (E) and the electrostatic contribution (E$_e$) are in kcal/mol; a, b, c are in Å; β is in degrees.

regular double-helical polynucleotide revealed some regularities important for the understanding of the nucleic acid structure.[10,45-47]

In systematic analysis of possible conformations of the double helix it is convenient to use the parameters describing the mutual position of bases in the helix as independent variables. These parameters were introduced by Arnott to construct double helix models from X-ray data for oriented fibers[48] and then were used in theoretical conformational analysis of regular double-helical polynucleotides.[10,45-47] They are: H, the distance between the pairs along the helix axis; τ, the angle of rotation of a pair relative to its neighbor around the helix axis; TL, the angle of rotation of a pair around the dyad axis; D, the shift of the pair along this axis and TW, the propeller twist in base pairs. Assignment of these parameters, of the sugar ring puckering, of bond lengths and angles, determines completely, though ambiguously, the coordinates of atoms of the regular double helix. Two sugar-phosphate chains have in this case identical conformations.

Having calculated nonbonded interactions of a regular double-helical polynucleotide as a function of 5 Arnott parameters with standard C(2')-endo and C(3')-endo sugar puckering, Zhurkin et al.[40] found several regions of possible conformations of the DNA double helix. Calculations taking into account the possibility of the sugar ring distortion permitted us to describe two extended regions of minimum energy values -- two valleys -- corresponding to A- and B-families of nucleic acid conformations.[10,46,47] Along the bottom of each valley considerable changes of conformational parameters are possible while changes of the energy of nonbonded interactions are rather small (up to 2 kcal/mol). These changes are of the order of degrees (up to 15) for angle variables and $1.5-2$ Å by variables characterizing shifts, the distances between some atomic groups changing by several Å. This possiblity can be important for biological functioning of the DNA double helix.

The presence of extended valleys, on the one hand, allows local variation of conformational parameters according to the base sequence, and, on the other hand, makes possible the existence of a regular (or close to it) helix for any sequence if regularity is imposed from outside, e.g. by a regular environment.

We began the study of conformational patterns of different nucleotide sequences by calculating the energy of base interaction as a function of their mutual position in double helices. Such calculations for a repeating A:U sequence permitted us to considerably decrease the range of variation of conformational parameters in theoretical conformational analysis of a regular double-helical polynucleotide.[46] In some calculations we took into account additional variables RL, BL and SL (see Fig. 2), thus avoiding the constraint for identity of conformations of the two sugar-phosphate backbones. We performed calculations for repeating sequences A:T, A:U, G:C and I:C (hypoxanthine-cytosine) base pairs and for all possible pairwise combinations of A:T and G:C base pairs. These results will be described in detail elsewhere.[49,50] Here we shall summarize the most important results of these calculations.

For each sequence the energy was calculated as a function of two variables — helix parameters H and τ. The variable values were set by a two-dimensional lattice with a 0.05 Å pitch by H and 0.5° by τ in the region of A- and B-families of nucleic acid conformations (2.5 Å $\leq$ H $\leq$ 3.5 Å, 30° $\leq \tau \leq$ 45°). For each pair of H and τ values the minimum energy was searched for with respect to the other variables. In such a way we have found the valley-like regions of minimum values of the energy of base interaction within wide ranges of H and τ. These valleys largely overlap for different sequences but are characteristic of each particular sequence. When H and τ change within these valleys the parameters TL, TW and D (as well as RL, BL and SL) change consistently. For example, a decrease of H is accompanied, as a rule, by an increase of D and TL.

The extended regions of minimum energy values suggest the existence, for each sequence, of specific pathways of changes in conformational parameters along which considerable (by $10-20°$ for angle variables and $2-3$ Å for D) mutual shifts of bases are possible at rather small (up to $1-2$ kcal/mol) changes of energy. Even small shifts ($1-2°$ or $0.2-0.3$ Å) in other directions can lead to a considerable (by several kcal/mol) energy increase. The different parts of the valleys have different width by one or another variable. As a rule, changes of the parameter TL are larger in the region of B-conformations, the D changes being larger in the A-conformation region. On the whole, the valleys in the H and τ region corresponding to the A-family of conformations are narrower than in the B-conformation region.

Table III lists the values of the energy of base interactions and the parameters determining the position of bases in the double helix for some configurations of the repeating base pair sequences. Some of these configurations are presented in Fig. 3. Calculations show that the parameters RL, BL and SL virtually do not affect the value and the position of the energy minimum at chosen H and τ. At the same time, in the regions of conformational space where the energy differs from the minimum at given H and τ by no more than 1 kcal/mol, the parameters RL, BL and SL can considerably change the structure.

The pathway of changes in Arnott parameters during the transition from A- to B-like low-energy conformations and between these conformations within A- and B-families is similar to the pathway of changes in these parameters along the bottom of each valley obtained for a regular double-helical polynucleotide taking into account interactions of all the components.[10,47] This indicates that interaction between bases largely determines DNA helical conformations, though some energy-favorable mutual positions of bases would be eliminated in the DNA because of the constraints imposed by the sugar-phosphate backbone.

It can be assumed that energetical preference of B-like positions of bases in A:T sequences contributes to B-phility of these sequences; this is also one of the reasons why poly(dA) $\cdot$ poly(dT) in fibers is not transformed into the A-form upon a change of the water content and the type of cations.[40] Energetical preference of B-conformations for repeating A:T pairs is ensured by both electrostatic and van-der-Waals interactions (Table III). A:T and A:U sequences are rather similar in the

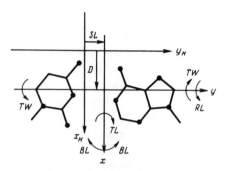

Figure 2. The parameters used to define the position of bases relative to the axis of a regular duplex. x_H, y_H, z_H is the coordinate system associated with the helix axis (z_H is normal to the plane); x, y is the coordinate system associated with the base pair center (the middle of the fragment passing through C(8) of purines and C(6) of pyrimidines). TW is the propeller twist in the pair, the angle of rotation of the pair bases around the y axis in opposite directions; BL is the base pair buckle, the rotation of bases in opposite directions around the x axis; TL and RL are the angles of inclination of the pair to the x and y axes, respectively; D and SL are the shifts of the pair center from the helix axis along the x_H and y_H axes, respectively.

energy behavior as a function of mutual position of bases and differ considerably from the G:C sequence for which A-like positions of bases are preferable.

For repeating G:C sequences all configurations corresponding to the energy minima at a given H and τ have rather large D values (Table III); as a result a pair is shifted relative to its neighbor along the C(6) pyrimidine-C(8) purine axis. This shift (which is also characteristic of a number of other sequences, e.g. AT and AC) can be considered as a feature of the A-like conformation; its presence in the minimum of the B-form, and the H and τ values are indications of the A-philicity of the sequence. As seen in Fig. 3, the mutual position of the bases in conformation 11 (H = 3.4 Å, τ = 36°) has common features with the position of bases in a typical A-conformation 15 (H = 2.8 Å, τ = 32.5°). A decrease of D to the values characteristic of B-conformations is energetically disadvantageous (conformation 12). Energetical preference of A-like conformations for repeating G:C sequences (and other A-philic sequences) is due to electrostatic interactions. When the contribution of these interactions decreases (e.g. assuming ε = 4), B-like conformations become preferable.

The energy of the base interaction for the sequence of repeating I:C pairs in a large range of H and τ values characteristic of B-conformations has two minima. The position of the bases in one of them (Fig. 3f) is close to that in B-conformations of DNA. The position of the bases in the other one (Fig. 3e) is close to that in the energy minimum for the same H and τ for the repeating G:C sequence. The presence of these two minima suggests that depending on conditions, the I:C sequence can display conformational features of either A:T (A:U) or G:C sequences. Two barrier-separated subregions of minimal energy values are characteristic also for a number of sequences containing G:C pairs. We can apply here the term "bistability" proposed by Calladine and Drew.

Table III. Values of the energy and its components at some values of the parameters characterizing the mutual position of bases in repeating sequences of pairs

No.	Pair	$-E$	$-E_e$	$-E^{st}$	$-E_e^{st}$	H	τ	TL	TW	D	RL	BL	SL
1.	A:T	24.6	4.6	13.3	0.4	3.29*	36*	−5.1	5.0	0.7	0*	0*	0*
2.	A:T	24.9	5.0	13.6	0.3	3.29*	36*	−4.3	3.5	0.6	−2.7	7.0	0.1
3.	A:T	23.4	4.0	12.1	1.1	3.0*	30*	6.7	3.7	4.3	0*	0*	0*
4.	A:T	23.1	4.1	11.8	1.0	2.8*	32.5*	10.7	3.9	4.6	0*	0*	0*
5.	A:T	25.1	5.1	14.0	−0.2	3.1	45.2	1.1	6.5	0.2	0*	0*	0*
6.	A:U	23.5	4.7	12.2	0.4	3.29*	36*	−5.5	4.5	0.5	0*	0*	0*
7.	A:U	23.7	4.9	12.4	0.4	3.29*	36*	−5.5	4.0	0.4	−2.0	6.3	0.1
8.	A:U	22.5	4.3	11.1	0.9	3.0*	30*	5.9	1.4	4.3	0*	0*	0*
9.	A:U	22.5	4.4	11.1	0.8	2.8*	32.5*	9.8	1.0	4.5	0*	0*	0*
10.	A:U	24.2	5.4	13.0	−0.3	3.1	46.9	−0.4	5.8	−0.1	0*	0*	0*
11.	G:C	35.4	15.7	10.4	1.3	3.4*	36*	−1.4	−3.9	3.8	0*	0*	0*
12.	G:C	34.4	13.4	9.3	3.9	3.4*	36*	−5.8	1.7	0.5*	0*	0*	0*
13.	G:C	36.6	16.9	11.4	0.1	3.4*	36*	−1.6	−7.6	4.1	−1.6	5.5	0.4
14.	G:C	35.3	15.5	10.2	1.5	3.0*	30*	5.5	−4.0	5.8	0*	0*	0*
15.	G:C	35.1	16.4	10.1	1.7	2.8*	32.5*	8.8	−3.9	5.6	0*	0*	0*
16.	I:C	29.4	13.3	10.8	−0.5	3.4*	36*	−1.4	−5.2	3.9	0*	0*	0*
17.	I:C	29.6	11.5	11.0	1.3	3.4*	36*	−11.1	5.4	0.2	0*	0*	0*
18.	I:C	29.1	12.8	10.5	0.1	3.0*	30*	5.9	−5.0	5.8	0*	0*	0*
19.	I:C	29.0	12.7	10.4	0.2	2.8*	32.5*	9.4	−4.8	5.5	0*	0*	0*

* denotes the parameter values fixed at minimization with respect to other parameters. E, total energy; E_e, electrostatic contribution; E^{st} and E_e^{st}, energy of the stacking interaction and its electrostatic contribution (kcal/mol of base pairs). H, D and SL are in Å; τ, TL, TW, RL and BL are in degrees.

A typical feature of conformations of double-helical DNA fragments in crystals is a distortion of planarity of base pairs resulting in a considerable propeller twist. Calculations show that the propeller twist is promoted by interactions of base pairs, the bases pertaining to the same polynucleotide chain (as noted earlier[1,7-9]) as well as the bases pertaining to different chains. For example, in A:T and A:U sequences an important contribution is made by interactions between bases located at the 5′-ends of pair dimers.

Conformations of DNA Fragments Containing Adenine in One Chain and Thymine in the Other One

Arnott et al. suggested, proceeding from the results of X-ray studies of poly(dA) · poly(dT) oriented fibers, that the structure of this polynucleotide has specific features and is characterized by drastically different conformations of the two chains (the so-called "heteronomous DNA").[51] Double helical fragments containing adenine in one chain and thymine in the other one are thought to play a particular role in DNA functioning.[44] In this respect it was interesting to study extensively the conformational possibilities of the regular double-helical polynucleotide poly(dA) · poly(dT). We used computations and searched for the regions of minimal values of nonbonded interaction energy of this duplex.

In these computations we took the five Arnott parameters as well as RL and SL as the independent variables determining the mutual position of bases (Fig. 2). Thymine CH_3-group was considered as the united atom. In other respects the calculating procedure was the same as in our previous studies on theoretical conforma-

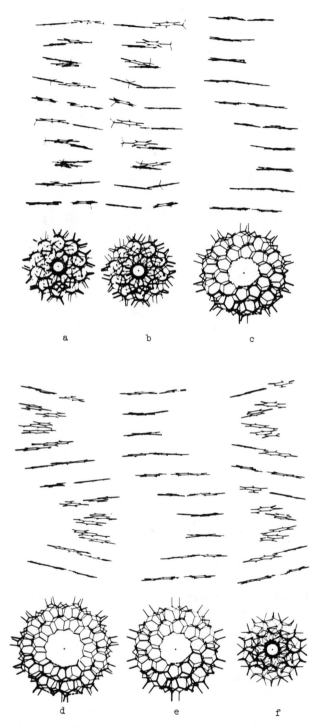

Figure 3. Some mutual positions of bases calculated for repeating nucleotide sequences. Configurations a, b, c, d, e and f correspond to Nos. 1, 2, 11, 15, 16 and 17 from Table III. For each configuration the lower view is along the helix axis, the upper view is in perpendicular direction.

tional analysis of polynucleotides.[10,47] Conformations of the sugar rings of each chain were set by four independent parameters (two dihedral and two bond angles). Another two independent conformational parameters are the glycoside dihedral angles of the two polynucleotide chains. Other dihedral angles were calculated from independent conformational parameters.

Beside the two mentioned valleys corresponding to A- and B-families of nucleic acid conformations the calculations revealed two more regions of minimal energy values of this polynucleotide. In one of the regions the poly(dA) chain has an A-like conformation with the C(3′)-endo sugar while the poly(dT) chain has a B-like conformation. We shall call these structures AB-type ones. They have some features of Arnott's heteronomous DNA, though they differ considerably by the values of conformational parameters. In the other region of minimal energy values the poly(dA) chain has a B-like conformation while the poly(dT) chain has an A-like one. These structures will be called those of the BA-type. According to Rao and Kollman, the most favorable conformation of the $(dA)_6 \cdot (dT)_6$ duplex pertain to such structures.

Table IV lists the characteristics of the four types of low-energy conformations of poly(dA) · poly(dT) corresponding to H = 3.23 Å and $\tau = 36°$, the values obtained in X-ray studies of the oriented fibers of this polynucleotide, and to the energy minima by other conformational parameters. Fig. 4 represents computer stereo drawings of these conformations. The B-family conformation (denoted as BB) has the conformational parameters close to those obtained by us earlier for poly(dA) · poly(dU) at H and τ values corresponding to the B-form of DNA. The A-family conformation (denoted as AA) has a large propeller and rather small (1.7 Å) D values at these H and τ. In an earlier theoretical conformational analysis of poly(dA) · poly(dU) we have already found such "unusual" A-conformations.[10,47] For many other nucleotide sequences, in particular those containing G:C pairs, such conformations are energetically unfavorable. It is interesting that the conformation of the poly(dA) chain in the AB-type structures and that of the poly(dT) chain the the BA-type structures resemble these conformations rather than the "usual" A-form with a moderate propeller and the D values of about 4.5 Å.

It should be noted that the parameters RL and SL are close to zero for conformations of the BB- and AA-types while for conformations of the AB- and BA-types they reach a considerable value (Table IV). When the bases are turned by the RL angle for AB- and BA-conformations, the bases of the B-like chain become almost perpendicular to the helix axis despite a considerable propeller (Fig. 4). RL (and SL) are of opposite signs for the conformations of AB- and BA-types.

Glycoside dihedral angles of the two chains are close for the AA-conformation while for the BB-conformations the angle in the poly(dT) chain is 5° smaller than in the poly(dA) chain. In the double-helical dodecamer crystal differences between the glycoside angles of the purine and pyrimidine nucleotide of the complementary pair have been shown[1,2] to be of the same sign but larger in absolute value. According to our calculations, at H = 3.23 Å, $\tau = 36°$ an increase of the differences between the glycoside angles of the two chains are energetically unfavorable.

Table IV. Conformational parameters of the four types of minimal energy structures of poly(dA) · poly(dT) at H = 3.23 Å and $\tau = 36°$

	Type of Conformation			
	AA	BA	AB	BB
− E	14.4	13.9	12.9	11.8
D	1.7	1.1	1.1	1.3
SL	0.1	0.4	− 0.2	− 0.1
TL	17.5	7.1	1.8	− 0.7
TW	16.7	11.8	10.4	4.5
RL	− 0.7	− 8.4	5.7	− 0.1
χ	96	134	89	129
	97	96	125	124
α	260	289	283	256
	278	282	296	255
β	173	182	172	184
	172	171	180	178
γ	66	58	66	57
	69	68	64	63
δ	86	140	85	135
	85	84	132	131
ε	207	186	202	179
	205	200	185	186
ζ	290	249	294	256
	292	295	255	255
P	9	158	13	147
	18	21	148	144
τ_m	36	40	38	42
	35	37	38	41

Note: for each type of conformation the values of dihedral angles, phases (P) and amplitude (τ_m) of sugar pseudorotation correspond to the two polynucleotide chains: the first value to the poly(dA) and the second one to the poly(dT) chain.

The values of the energy of nonbonded interactions of poly(dA) · poly(dT) listed in Table IV correspond to completely neutralized phosphate groups. When the extent of phosphate group neutralization changes, the relative preference of the four types of conformations changes too. However, the structures of the BA type remain preferable as compared to those of the AB type at total charges of the phosphate groups equal to − 1 as well as to − 0.5.

It is possible to suggest that the capability of the DNA double-helical fragments consisting of repeating A:T pairs to assume specific conformations with different sugar-phosphate backbones could be of biological relevance, in particular in the processes of protein-nucleic acid interactions.

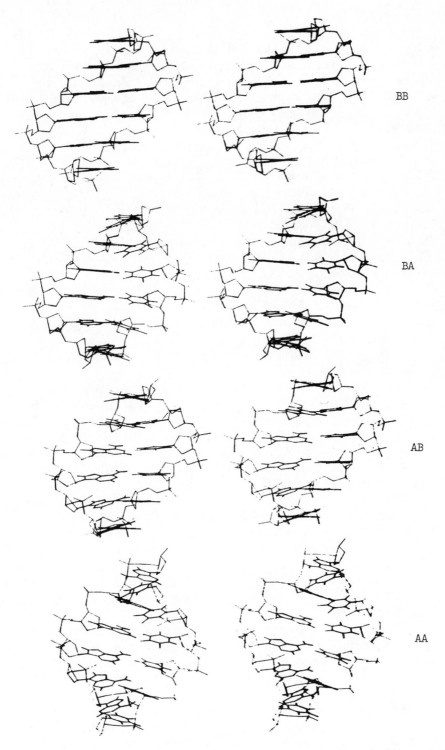

Figure 4. Stereo drawings of different types of conformations of the poly(dA) · poly(dT) double helix.

References

1. R.E. Dickerson, M.L. Kopka and H.R. Drew, Structural correlations in B-DNA, in "Structure and Dynamics: Nucleic Acids and Proteins," E. Clementi and R.H. Sarma, eds., Adenine Press, N.Y., p. 149 (1983).
2. R.E. Dickerson, M.L. Kopka and P. Pjura, Base sequence, helix geometry, hydration and helix stability in B-DNA, in "Biological Macromolecules and Assemblies," F.A. Jurnak and A. McPherson, eds., Wiley, N.Y., 237 (1985).
3. Z. Shakked and O. Kennard, The A form of DNA, ibid., 1.
4. A.H.-J. Wang and A. Rich, The structure of the Z form of DNA, ibid., 127.
5. D.J. Patel, S.A. Kozlowski and R. Bhatt, Sequence dependence of base-pair stacking in right-handed DNA in solution: Proton nuclear Overhauser effect NMR measurements, Proc. Natl. Acad. Sci. USA, **80**, 3908 (1983).
6. G.M. Clore and A.M. Gronenborn, Interproton distance measurements in solution for a double-stranded DNA undecamer comprising a portion of the specific target site for cyclic AMP receptor protein in the gal operon. A nuclear Overhauser enhancement study, FEBS Lett., **175**, 117 (1984).
7. C.R. Calladine, Mechanics of sequence-dependent stacking of bases in B-DNA, J. Mol. Biol., **161**, 343 (1982).
8. R.E. Dickerson, Base sequence and helix structure variation in B and A DNA, J. Mol. Biol., **166**, 414 (1983).
9. C.R. Calladine and H.R. Drew, A base-centered explanation of the B-to-A transition in DNA, J. Mol. Biol., **178**, 773 (1984).
10. V.I. Poltev and V.P. Chuprina, Relation of macromolecular structure and dynamics of DNA to the mechanisms of fidelity and errors of nucleic acid biosynthesis, in "Structure and Motion: Membranes, Nucleic Acids and Proteins," E. Clementi et al., eds., Adenine Press, N.Y., 433 (1985).
11. V.I. Poltev, Simulation of intermolecular and intramolecular interactions of nucleic acid subunits by means of atom-atom potential functions, Int. J. Quantum Chem., **16**, 863 (1979).
12. V.B. Zhurkin, V.I. Poltev and V.L. Florentiev, Atom-atom potential functions for conformational analysis of nucleic acids, Mol. Biol. USSR, **14**, 1116 (1980).
13. V.I. Poltev and N.V. Shulyupina, Stimulation of interactions between nucleic acid bases by refined atom-atom potential functions, J. Biomol. Struct. Dyn., **3**, 739 (1986).
14. V.I. Poltev, T.I. Grokhlina and G.G. Malenkov, Hydration of nucleic acid bases studied using novel atom-atom potential functions, J. Biomol. Struct. Dyn., **2**, 413 (1984).
15. F.A. Momany, L.M. Carruthers, R.F. McGuire and H.A. Scheraga, Intermolecular potentials from crystal data. 3. Determination of empirical potentials and application to the packing conformation and lattice energies in crystals of hydrocarbons, carboxylic acids, amines and amindes, J. Phys. Chem., **78**, 1595 (1974).
16. S.J. Weiner, P.A. Kollman, D.A. Case, U.C. Singh, C. Chio, G. Alagona, S. Profeta and P. Weiner, A new force field for molecular mechanical simulation of nucleic acids and proteins, J. Amer. Chem. Soc., **106**, 765 (1984).

17. R. Scordamaglia, F. Cavallone and E. Clementi, Analytical potentials from *ab initio* computations for the interaction between biomolecules. 2. Water with the four bases of DNA, J. Amer. Chem. Soc., **99**, 5545 (1977).

18. O. Matsuoka, C. Tosi and E. Clementi, Conformational studies on polynucleotide chains. 1. Hartree-Fock energies and description of nonbonded interactions with Lennard-Jones potentials, Biopolymers, **17**, 33 (1978).

19. H. Berthod and A. Pullman, Sur le calcul des caracteristiques de la squelette des molecules conjuguees, J. Chim. Phys., **62**, 942 (1965).

20. R. Lavery, K. Zakrzewska and A. Pullman, Optimized monopole expansions for the representation of the electrostatic properties of the nucleic acids, J. Comput. Chem., **5**, 363 (1984).

21. N.K. Ray, M. Shibata, G. Bolis and R. Rein, Potential-derived point-charge model study of electrostatic interactions in DNA base components, Chem. Phys. Lett., **109**, 352 (1984).

22. H. DeVoe and I. Tinoco, The stability of helical polynucleotides: Base contributions, J. Mol. Biol., **4**, 500 (1962).

23. V.I. Ponomaryov, O.S. Filipenko and L.O. Atovmyan, Crystal and molecular structure of naphthalene at $-150°C$, Krystallografia USSR, **21**, 393 (1976).

24. R.S. Bradley, T.G. Cleasby, The vapour pressure and lattice energy of some aromatic ring compounds, J. Chem. Soc., **6**, 1690 (1953).

25. R. Mason, The crystallography of antracene at 95 K and 290 K, Acta Cryst., **17**, 547 (1964).

26. J.D. Kelley and F.O. Rice, The vapour pressures of some polynuclear aromatic hydrocarbons, J. Phys. Chem., **68**, 3794 (1964).

27. A. Camerman and J. Trotter, The crystal and molecular structure of pyrene, Acta Cryst., **18**, 636 (1965).

28. P.J. Wheatley, The crystal and molecular structure of pyrazine, Acta Cryst., **10**, 182 (1957).

29. J. Tjebbes, The heats of combustion and formation of the three diazines and their resonance energies, Acta Chem. Scand., **16**, 916 (1962).

30. P.J. Wheatley, The crystal and molecular structure of pyrimidine, Acta Cryst., **13**, 80 (1960).

31. J. Trotter, A three-dimensional analysis of the crystal structure of p-benzoquinone, Acta Cryst., **13**, 86 (1960).

32. A.S. Coolidge and M.S. Coolidge, The sublimation pressures of substituted quinones and hydroquinones, J. Amer. Chem. Soc., **49**, 100 (1927).

33. W. Bolton, The crystal structure of alloxan, Acta Cryst., **17**, 147 (1964).

34. S. Martinez-Carrera, The crystal structure of imidazole at $-150°C$, Acta Cryst., **20**, 783 (1966).

35. H. Zimmermann and H. Geisenfelder, Uber die Mesomerieenergie von Azolen, Z. Electroch., **65**, 368 (1961).

36. R.F. Stewart and L.H. Jensen, Redetermination of the crystal structure of uracil, Acta Cryst., **23**, 1102 (1967).

37. I.K. Yanson, B.I. Verkin, O.I. Shklyarevsky and A.B. Teplitsky, Sublimation heats of nitrogen bases of nucleic acids, Studia Biophys., **46**, 29 (1974).

38. D.L. Barker and R.E. Marsh, The crystal structure of cytosine, Acta Cryst., **17**, 1581 (1964).

39. H. Ringertz, The molecular and crystal structure of uric acid, Acta Cryst., **20**, 397 (1966).
40. A.G.W. Leslie, S. Arnott, R. Chandrasekaran and R.L. Ratliff, Polymorphism of DNA double helices, J. Mol. Biol., **143**, 49 (1980).
41. V.I. Ivanov, V.B. Zhurkin, S.K. Zavriev, Yu. P. Lysov, L.E. Minchenkova, E.E. Minyat, M.D. Frank-Kamenetskii and A.K. Schyolkina, Conformational possibilities of double-helical nucleic acids: Theory and experiment, Int. J. Quant. Chem., **16**, 189 (1979).
42. V.I. Ivanov, Possible relevance of the B-A transition in DNA to transcription, Comments Mol. Cell. Bioph., **2**, 333 (1985).
43. G.P. Lomonossoff, P.J.G. Butler and A. Klug, Sequence-dependent variation in the conformation of DNA, J. Mol. Biol., **149**, 745 (1981).
44. H.R. Drew and A.A. Travers, DNA structural variations in the E. coli tyrT promoter, Cell, **37**, 491 (1984).
45. V.B. Zhurkin, Yu.P. Lysov and V.I. Ivanov, Different families of double-stranded conformations of DNA as revealed by computer calculations, Bio-polymers, **17**, 377 (1978).
46. V.E. Khutorsky and V.I. Poltev, Conformations of double-helical nucleic acids, Nature, **264**, 483 (1976).
47. V.P. Chuprina, V.E. Khutorsky and V.I. Poltev, Theoretical refinement of A- and B-conformation models of regular polynucleotides, Studia Biophys., **85**, 81 (1981).
48. S. Arnott, The geometry of nucleic acids, Progr. Biophys. Mol. Biol., **21**, 265 (1970).
49. V.I. Poltev and A.V. Teplukhin, Interaction of nucleic acid bases and con-formational behaviour of repeating nucleotide sequences, Mol. Biol. USSR, **20**, No. 6, in press (1986).
50. V.I. Poltev and A.V. Teplukhin, in preparation.
51. S. Arnott, R. Chandrasekaran, I.H. Hall and L.C. Puigjaner, Heteronomous DNA, Nucleic Acids Res., **11**, 4141 (1983).
52. S.N. Rao and P.A. Kollman, On the role of uniform and mixed sugar puckers in DNA double-helical structures, J. Amer. Chem. Soc., **107**, 1611 (1985).

Multifrequency ESR of an Isotopically Enriched Copper System in the Immobilized Phase: A Monte Carlo Approach

R. Basosi
Institute of Physical Chemistry
University of Siena, Italy

M. Pasenkiewicz-Gierula and W. Froncisz
Dept. of Biophysics, Institute of Molecular Biology
Jagiellonian University, 31-120, Krakow, Poland

W.E. Antholine, A. Jesmanowicz and J.S. Hyde
National Biomedical ESR Center, The Medical College of Wisconsin
Milwaukee, Wisconsin 53226, U.S.A.

Abstract

Often the only ESR parameters reported from the spectrum of an immobilized copper complex are $g_\parallel$ and $A_\parallel^{Cu}$ and the presence or absence of nitrogen hyperfine lines. Here, experimental ESR spectra are used to show the improvement in resolution if a single isotope of copper, ^{63}Cu or ^{65}Cu, and ^{15}N donor atoms are incorporated into the complex. The second derivative spectrum enhances the resolution of the nitrogen hyperfine structure. The Fourier transform of the ESR spectrum separates peaks for hyperfine lines with large couplings from lines with small couplings. Multifrequency spectra and their simulations lead to a full set of ESR parameters which include g-values, A-values, line widths, line positions, and strain parameters.

Introduction

X-band ESR spectra of cupric complexes provide indirect information about the number of nitrogen, oxygen, and sulfur donor atoms through comparison of the $g_\parallel$ and $A_\parallel$ parameters with the values for model complexes.[1] Recently Froncisz and Hyde determined that the minimum line width for the low field lines in the $g_\parallel$ region for a non-blue Type II cupric complex is obtained at about 2 GHz.[2,3] At this frequency the $M_I = -1/2$ line in the $g_\parallel$ region is often narrow enough to resolve nitrogen hyperfine structure. If this line is well resolved, one can determine the number of nitrogens coordinated to cupric ion in the square planar configuration without referring to model studies.[4] By contrast, although the $g_\perp$ region is more

Supported by NIH Grants GM35472 and RR01008.

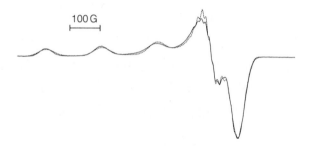

Figure 1. ESR spectrum (solid line) and simulated spectrum (dashed line) for cupric ion (69% ^{63}Cu and 31% ^{65}Cu) in the presence of excess histidine. Spectrometer conditions: microwave freq. 9.098 GHz, mod. amp. 5G, mod. freq. 100 KHz, temp. 77 K, microwave power 5 mW. Parameters for simulation: $g_x = g_y = 2.035$, $g_z = 2.220$, $A_x^{Cu} = A_y^{Cu} = 11$ G, $A_z^{Cu} = 175$ G, $A_x^N = A_y^N = A_z^N = 13$ G, line width $W_x = W_y = 5.7$ G, $W_z = 3.8$ G, strain parameters $C_z(1) = -0.004$, $C_z(2) = 3.25$ MHz.

intense and has well resolved hyperfine structure, the hyperfine pattern usually consists of more lines than are resolved. The lines in the $g_\perp$ region overlap with the $g_\parallel$ region and "overshoot lines," which arise from the particular angular dependence of the copper hyperfine line position and often have appreciable intensity. Furthermore, the hyperfine coupling constant for copper, $A_\perp^{Cu}$, is similar in magnitude to the hyperfine constant for nitrogen, $A_\perp^N$, and the patterns in the $g_\perp$ region are difficult to analyze. Thus computer simulations are necessary to determine a full set of

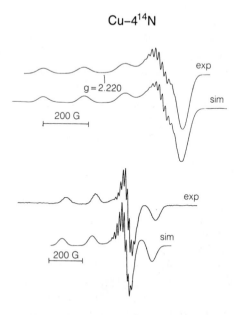

Figure 2. Experimental (exp) and simulated (sim) X-band (top two spectra) and S-band (bottom two spectra) for ^{65}Cu(II)(His)$_4$ in D$_2$O at a pD of 7.3 at 77 K. [L-His] = 0.1 M and [^{63}Cu^{+2}] = 0.008 M. Parameters for simulations: $g_x = g_y = 2.035$, $g_z = 2.220$, $A_x^{Cu} = A_y^{Cu} = 18$ G, $A_z^{Cu} = 182$ G, $A_x^N = A_y^N = 15$ G, $A_z^N = 13$ G, line width $w_x = w_y = 6$ G, $w_z = 4$ G, strain parameters $C_x(1) = C_y(1) = 0.001$, $C_z(1) = 0.004$, $C_x(2) = C_y(2) = 9$ MHz, $C_z(2) = 10$ MHz.

ESR parameters. This paper focuses on the methods we have used to obtain better resolved spectra and better fits between experimental and simulated spectra.

Incorporation of Copper and Nitrogen Isotopes

Naturally abundant copper consists of ^{63}Cu (69%) and ^{65}Cu (31%). Either of the pure isotopes is available from Oak Ridge National Laboratory, Oak Ridge, Tennessee. Substitution of a single isotope often improves the resolution for X-band spectra in the $g_\perp$ region and in both the $g_\perp$ and the $M_I = -1/2$ line in the $g_\parallel$ region for S-band spectra (Figs. 1 and 2). If ^{15}N can be substituted for the ^{14}N in the ligand, less lines will be observed ($I = 1/2$ for ^{15}N and $I = 1$ for ^{14}N). These lines may be better resolved because the hyperfine coupling constant for ^{15}N is 1.4 times the coupling constant for ^{14}N (Fig. 3).

Second Derivative Display

The second derivative spectrum emphasizes sharp features like the nitrogen hyperfine structure and does not emphasize features such as a poor baseline (Fig. 4). The separation between lines in the perpendicular region becomes easier to measure because shoulders in the first derivative display become peaks in the second derivative display. This display is most useful when little resolution is obtained in the first derivative spectrum. It should be noted that the signal to noise ratio is poorer for the second derivative. Often it is possible to turn up the modulation amplitude to 20 or 40 G for the low frequency modulation (270 Hz) but not the high frequency (4G, 100 kHz) and retain resolution of the narrow lines. High frequency noise can be cut off using Fourier transforms to improve the signal to noise.

Fourier Transforms and Reverse Fourier Transforms

Fourier transforms and reverse Fourier transforms were calculated with an IBM-9000 computer using a fast Fourier transform program. A significance plot[5] for the spectra in Fig. 4 gives at least two well separated peaks (Fig. 5). The reverse Fourier transform of the outermost peak at X-band gives a pattern expected for four equivalent ^{15}N donor atoms (Fig. 5). Presumably the intensities of the lines from the reverse Fourier transform of the X-band spectrum are not identical to the intensities of the lines from the pattern for four equivalent ^{15}N donor atoms because of some interference with the copper lines. The pattern from the S-band data is more complicated due to overlap of ^{15}N and ^{65}Cu lines in the perpendicular region (Fig. 5). This difference in the patterns at X and S-band was found to hold for ^{14}N donor atoms (data not shown). Thus the reverse transforms are often helpful for analyzing the hyperfine structure. A major problem in analyzing hyperfine structure of the copper nitrogen spectrum appears when the nitrogen stick pattern is deconvoluted from the pattern in the $g_\perp$ region of the spectrum. The problem arises from division by zeros which appear in the Fourier transform

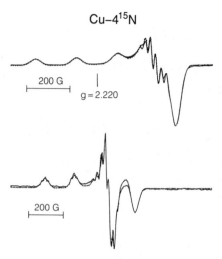

Figure 3. Experimental (solid line) X-band (top) and S-band (bottom) spectra of $^{65}Cu(II)(L-His)_4$ for which ^{15}N is substituted for ^{14}N in the imidazole ring (histidine-1,3-$^{15}N_2$) in D_2O at pD 7.3 at 77 K compared with simulations (dotted lines). ESR conditions and parameters given in figure legend 2 except $A_x^N = A_y^N = 21$ G, $A_z^N = 18$ G, $w_x = w_y = w_z = 8$ G.

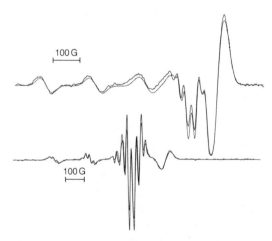

Figure 4. Experimental second derivative (solid line) X-band (top) and S-band (bottom) spectra of $^{65}Cu(II)(L-His)_4$ for which ^{15}N is substituted for ^{14}N in the imidazole ring (histidine-1,3-$^{15}N_2$) in D_2O at pD 7.3 at 77 K compared with simulations (dotted lines). ESR conditions and parameters given in figure legends 1 and 2 except the second modulation frequency is 270 Hz, mod amplitude 10 G.

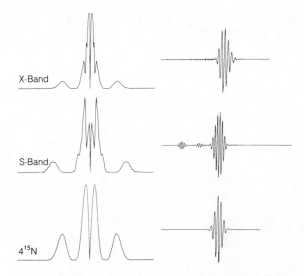

X-Band

S-Band

4¹⁵N

Figure 5. Fourier (left) and reverse Fourier transforms (right) for simulated spectra of ^{65}Cu(II)(L-His)$_4$ for which ^{15}N is substituted for ^{14}N in the imidazole ring for the X-band spectrum (upper), the S-band spectrum (middle), and a simulated spectrum for four equivalent ^{15}N donor atoms (bottom). The Fourier transforms are plotted as intensity versus position number. In this figure there are 512 positions for the magnetic field which correspond one-to-one to Fourier transform space. Note, the range for X-band and S-band are not the same. The hyperfine couplings, after correction for the range are almost identical for the X and S band data.

of the stick diagram for four nitrogen atoms. Large spikes in the reverse transform obliterate the deconvoluted patterns.

Simulations

In general, X and Q-band spectra are more sensitive to g-values and S-band spectra are more sensitive to A-values.[6] The simulation program, described in Ref. 7, takes into account g- and A-strain contributions to the line width and generates the spectrum in the frequency-swept domain. The equation for the Gaussian half-width is

$$\sigma_v = \left\{ \sigma_R^2 \left[\frac{\Delta g}{g} v_0(H) + \Delta A M_I \right] 2 \right\} 1/2$$

where σ_R is the residual line width and $\Delta g/g$ and ΔA represent g- and A-strain contributions to the line width. This program adds six strain parameters to the set of ESR parameters. Without these strain parameters the line widths cannot be fit very well. Because of the large number of variables, spectra (Figs. 1-4) were simulated using a Monte Carlo method.[8] Initial g and A values were determined from the experimental spectra, and adjusted after a few simulations. Then these parameters were fixed and the strain parameters were randomly varied within defined limits to minimize chi square. Usually 100 simulations were performed on each spectrum to obtain the strain parameters.

References

1. J. Peisach and W.E. Blumberg, Arch. Biochem. Biophys., **165**, 691 (1974).
2. W. Froncisz and J.S. Hyde, J. Chem. Phys., **73**, 1 (1980).
3. J.S. Hyde and W. Froncisz, Ann. Rev. Biophys. Bioeng., **11**, 391 (1981).
4. J.S. Hyde, W.E. Antholine, W. Froncisz and R. Basosi, Proc. Intl. Symp. Adv. Mag. Res. Techniques in Systems of Molecular Complexity, Siena, Italy, May 15-18, 1985.
5. S.J. Brumby, J. Magn. Reson., **35**, 357 (1979).
6. J.S. Hyde, W.E. Antholine and R. Basosi, Sensitivity Analysis of Multifrequency EPR Spectroscopy, in "Biological and Inorganic Copper Chemistry," K.D. Karlin and J. Zubieta, eds., Adlenine Press, New York, p. 239 (1985).
7. G. Rakhit, W.E. Antholine, W. Froncisz, J.S. Hyde, J.R. Pilbrow, G.R. Sinclair and B. Sarkar, J. Inorg. Biochem., **25**, 217 (1985).
8. G. Giugliarelli and S. Cannistrao, Il Nuovo Cimento, **4D, N.2**, 194 (1984).

Neutron Scattering from Agarose Gels

F. Cavatorta and A. Deriu
Department of Physics
University of Parma
Parma, Italy

H.D. Middendorf
Clarendon Laboratory
University of Oxford
Oxford, U.K.

Abstract

The potential of cold neutron scattering for the study of aqueous gels is discussed. Advanced elastic and inelastic neutron techniques can provide much new information on the structure and dynamics of polysaccharide gels in parameter domains not accessible by other methods. Elastic coherent scattering ($0.02 < Q_0 < 3.6$ Å^{-1}) from agarose gels at different concentration and H_2O/D_2O contrast shows that it is possible to get information on the dimensions of the interstitial water volumes contained by the three-dimensional network structure. Inelastic and quasielastic measurements from a hydrogenous gel using a time-focusing spectrometer indicate that one can quantify the extent of dynamic water structuring at wavenumbers below 100 cm^{-1}.

1. Introduction

The aqueous gels formed by polysaccharides are of great interest because of their highly variable water content, unusual phase diagrams, extraordinary viscoelastic behavior, and selectivity as molecular sieves.[1-4] polysaccharide gels Polysaccharide gels are excellent model systems for the study of biomolecular interactions involving helical structures, and an understanding of their molecular properties is important for numerous practical applications. Although we know a good deal about chain conformation and association in slightly hydrated polysaccharides, relatively little is known about the size distribution of chain aggregates, the connectivity of networks, and the occurrence of higher-order structure in the gel state proper. Optical and NMR techniques have given us considerable insight into the dynamical properties of gels containing bulk like water, but space and time resolved data on interactions in the crucial 5 to 500 Å region are scarce.

The scattering of cold neutrons, both elastic and inelastic, can contribute much new information to the study of polysaccharide gels in parameter domains not accessible

75

by other techniques. Using diffractometers and time-of-flight spectrometers at Rome and Grenoble, we have begun to explore the potential of neutron scattering for work of this kind and present here some results of our first experiments on agarose gels.

2. Basic Considerations

With regard to biological origin and large-scale structural organization, we may distinguish between three major classes of gel-forming polysaccharides as follows: (i) the glycosaminoglycans (usually linked to specific proteins); (ii) the agaroids, carrageenans, and similar substances extracted from red seaweeds; and (iii) the cation-chelating polysaccharides (such as alginates, pectic substances, and certain bacterial polysaccharides). To discuss how neutron scattering experiments can contribute to our understanding of the properties of polysaccharide gels, we focus on thermo-reversible, agarose-type gels as the most widely studied class. From X-ray diffraction and a variety of spectroscopic techniques, we have a good picture of the molecular organization and mobility in such gels over parts of the parameter domain shown in Fig. 1. The classical X-ray work on hydrated fibers and films of polysaccharides of the agar-carrageenan family has established atomic details of their double-helical structure and has, in conjunction with optical data, provided evidence for chain association and network formation in gels.[5] The primary mode of interchain association appears to be the formation of hydrogen-bonded bundles of helices. The current picture is that of a three-dimensional network of such bundles connected through junction zones (Fig. 2). Because of the importance of the random coil ⇔ double-helix transition in molecular biology, the temperature course and nature of gel formation have been investigated extensively by circular dichroism, quasi-elastic light scattering, photon correlation and NMR techniques mainly in the 0.5 to 10% concentration range. Much valuable work on the role of H-bonding, the effect of D_2O substitution, and the extent of supramolecular ordering in agarose gels has in recent years been done at Palermo.[6,7]

While many facets of the structure and dynamics of polysaccharide gels can be studied in this way, it will be essential for a more comprehensive understanding of the relation between macroscopic and microscopic properties to measure dynamic structure factors S_{coh} (Q, ω) and S_{inc} (Q, ω) for momentum transfers $\hbar Q$ in the $0.001 < Q < 10$ Å^{-1} region and for energy transfers $\hbar\omega < k_B T$ (Q_0 is the scattering vector defined as $(4\pi/\lambda_0) \sin \theta$, λ_0 being the incident wavelength). The first of these energy resolved structure factors describes two particle correlations and therefore relates to interference effects; the second represents self-correlations and is related to single-particle dynamics. They are space-time Fourier transforms of the van Hove correlation functions $G(\underline{r}, t)$ and $G_S(\underline{r}, t)$, respectively. The van Hove functions are central to a substantial body of theoretical work on the time-averaged and time-dependent statistical mechanics of interacting particles, and they provide a firm basis for the interpretation of experimentally accessible $S(\underline{Q}, \omega)$ functions together with their associated time correlation functions.[8,9]

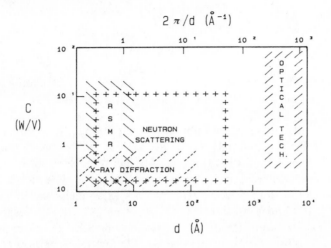

Figure 1. Concentration vs. scale length regimes of radiation scattering techniques applied to the study of polysaccharide gels and hydrated fibers. Concentration C in g of dry polysaccharide per cm³ of water; RSMR = Rayleigh Scattering of Mössbauer Radiation.

In a real experiment one actually measures an integral of $S(\underline{Q}, \omega)$ over the momentum and energy resolution function of the technique used. In terms of $S(\underline{Q}, \omega)$, the structure factor familiar from conventional diffraction work is given by

$$S(\underline{Q}) = \int_{-\infty}^{+\infty} S(\underline{Q}, \omega) \, d\omega \tag{1}$$

Experimental techniques capable of energy resolution allow at least a separation of the elastic from the total inelastic scattering to be performed. The elastic structure factor can then be expressed as

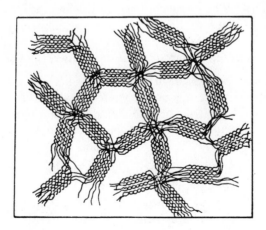

Figure 2. A schematic representation of the agarose gel network (from Ref. 5, with permission).

$$S_{el}(\underline{Q}) = \int_{-\Delta E/2}^{+\Delta E/2} S(\underline{Q}, \omega)\, d\omega \qquad (2)$$

where ΔE denotes the experimental energy resolution. The remaining part is identified with the total inelastic scattering, i.e. $S_{in}(\underline{Q}) = S(\underline{Q}) - S_{el}(\underline{Q})$.

The elastic structure factor can be obtained, for example, from Rayleigh scattering of Mössbauer radiation (RSMR) where the extreme sharpness of the Fe-57 γ-line ($\Delta E \sim 10^{-5} cm^{-1}$) is exploited. Recently RSMR measurements have been performed on agarose gels at different levels of water content.[10,11] For polysaccharides the $S_{el}(\underline{Q})$ and $S_{in}(\underline{Q})$ functions deduced from RSMR experiments are mainly due to the heavy nuclei (C,O) and they show broad peaks at Q values corresponding to the main repeating distances of the double helix. It is possible therefore to derive from $S_{el}(\underline{Q})$ the mean square displacements $<u^2>$ of the scattering centers contributing to the observed peaks in the diffraction pattern. It turns out that in agarose gels the $<u^2>$ measured at Q values appropriate to the characteristic distances of the long range structure (e.g. helix diameter) are larger than those of the chain bonds. This "anisotropy," which is small in dry agarose, increases markedly with decreasing agarose concentration below 4 g/cm^3 and may be an indication of the onset of a new vibrational regime in the polymer backbone as a result of solvent interactions.

As regards neutron scattering, a decisive advantage is that it is possible first to examine the time averaged properties of a sample in diffraction experiments, and then to use the same radiation to probe the time-dependent properties by means of measuring $S(\underline{Q}, \omega)$ for coherent and/or incoherent scattering, depending on the isotopic composition of the sample. The utility of neutron scattering for the study of strongly hydrated macromolecular systems is due to a unique combination of three factors:

(a) The spatiotemporal information accessible with modern instruments extends over 8 decades in time $\{10^{-6} > (2\pi/\omega) > 10^{-14}\ s\ \}$, and spatially over three decades $\{0.5 < (2\pi/Q) < 500\ \text{Å}\}$. The frequency region thus reaches from energy transfers of the order of $10k_BT$ down to $10^{-7}k_BT$ at ordinary temperatures, and covers the transition from quantized vibrational modes to low-energy diffusive processes which is of prime interest for systems intermediate between solids and liquids.

(b) The large contrast between hydrogen and deuterium with respect to both coherent and incoherent scattering can be exploited very effectively in structural[12] and spectroscopic[13] work. It is possible to create a wide range of H/D contrast by partial or full covalent deuteration of the polymeric component, and by changing the H_2O/D_2O composition of the water of hydration.

(c) The simplicity of point-like nuclear scattering facilitates the quantitative interpretation of Q,ω-dependent data in terms of fundamental correlation functions, and makes neutron scattering by far the most informative counterpart to numerical simulation studies of macromolecular dynamics and hydration.[14]

Despite these strong assets, biomolecular applications of neutron spectroscopy[15,16] have developed only very slowly compared with structural studies using neutrons.[17] While the Q,ω-range and the resolving power of instruments available for macromolecular spectroscopy are excellent, the flux levels of most neutron sources are still orders of magnitude below the corresponding level of photon sources. This limitation is compounded by the fact that globally there are only a few centralized research establishments equipped with spectrometers of advanced design. Substantial efforts have been made in recent years at all major research centers to ease the pressure on neutron scattering facilities and to develop new ways of satisfying the demand for more intense sources. Apart from numerous projects to upgrade existing fission sources, proton accelerators are now being used to produce pulsed beams of neutrons by means of spallation in heavy-metal targets.[18] The first high-intensity spallation neutron source (ISIS) has recently become operational at the Rutherford Appleton Laboratory.[19,20]

3. Diffraction Experiments

We have performed diffraction experiments at the Casaccia Research Center (Rome) using 2.05 Å neutrons in the range $0.4 < Q < 3.6$ Å^{-1}. At the Institut Laue-Langevin (I.L.L.), we used the D16 diffractometer ($\lambda_0 = 4.52$ Å) for measurements at both low angles ($0.02 < Q < 0.2$ Å^{-1}, slab geometry) and higher angles ($0.3 < Q < 2.1$ Å^{-1}, annular geometry).

Pure (low electroendosmosis) agarose powder was purchased from Pharmacia Fine Chemicals. Agarose gels were prepared by cooling a solution of 1% weight of agarose with H_2O/D_2O ratios 0, 0.11, 0.25, 0.54, 1.0, 1.5, 4.0 from 95°C down to room temperature. Samples with different agarose concentration C (g of polysaccharide/cm^3 of water) were obtained by a method of applying moderate pressure and absorbing water successively.

The differential neutron cross-sections with respect to angle were measured for an annular sample of diameter 9 mm (thickness 0.4 mm) at higher angles, and with slabs of the same thickness for the small-angle measurements. The results for samples with concentration C in the range $0.5 - 1.0$ g/cm^3 are summarized in Fig. 3a$-$e together with the differential cross-section of the starting 1% fully deuterated gel (Fig. 3g). The diffraction pattern of Fig. 3g closely resembles that of pure D_2O showing only a broad maximum at $Q = 1.95$ Å^{-1}. At higher agarose concentrations some broad peaks appear at $Q < 2.0$ Å^{-1} superimposed on the solvent profile; the Q-values of these are 1.35, 0.95, 0.67 Å^{-1}. The first one corresponds to a distance $d = 2\pi/Q$ of 4.8 Å and can be assigned to the inner diameter of the agarose double helix. The second one corresponds to a repeat distance of about 6.6 Å and can be attributed to the axial translation per disaccharide residue.[5] The one at the lowest Q-value ($d = 9.4$ Å) can be assigned to the translation period of the double helix. These peak positions compare very well with those obtained from agarose fibers at very low hydration[5] and with recent X-ray and RSMR results on agarose gels at concentrations similar to those employed in the present work.[10,11]

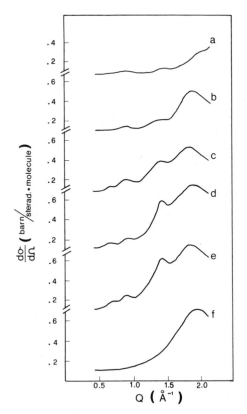

Figure 3. Differential cross sections for samples of agarose gel with concentration C in the range 0.5 − 1.0 g/cm³ and contrast (H₂O/D₂O) values equal to: a) 4.00; b) 1.00; c) 0.54; d) 0.11; e) 0.00. Curve f is relative to a gel with concentration C = 0.01 and contrast 0.00.

The small-angle measurements were performed on samples prepared with pure D₂O with concentration C ranging from 0.16 to 10 g/cm³. In this case the scattering from agarose is dominated by the incoherent proton signal, thus giving a constant background. The coherent contribution is mainly due to the deuterated solvent; it is therefore possible to deduce information on the dimensions of the interstitial water volumes of the three-dimensional gel network.

The small-angle data have been analyzed in the Guinier approximation[17] which, in the low-Q region, predicts that the scattering intensity from a particle or interstitial volume is given by

$$I(Q) = I(0) \exp(-Q^2 < R^2 >)$$ (3)

where $I(0)$ is the scattering intensity at zero angle and $<R^2>$ is an average second moment of the scattering density distribution in the particle or volume element. For spherical geometry, for example, $<R^2> = R^2/3$.

Table I. Dependence of the radius of gyration R_g and of the average radius R_v on agarose concentration C.

C (g/cm³)	R_g(Å)	R_v(Å)
0.16	47.0	49.3
0.28	44.1	42.6
0.39	43.3	39.9
0.52	41.3	35.5
0.75	33.3	25.9
1.00	32.5	23.8
1.22	24.6	22.8
1.33	16.4	19.4
2.13	15.2	16.9
10.11	12.9	12.4

For gels at low to intermediate polysaccharide concentration it is probably a reasonable approximation to assume a polyhedral shape for the average interstitial water volume. It would be possible then, on the basis of neutron Guinier data in conjunction with the known X-ray parameters for polysaccharide helices, to calculate the number of helices in a bundle making up the edges of such a polyhedron. For the intermediate to high concentration range of the present study, we have adopted a spherical shape as the simplest approximation. The Guinier plots

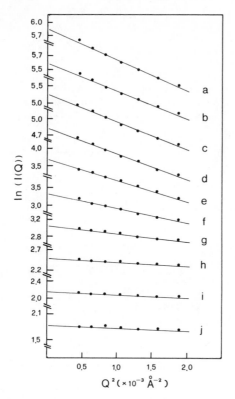

Figure 4. Guinier plots obtained from low Q scattering intensity for samples of agarose gel with contrast = 0.00 and concentration C equal to: a) 0.16; b) 0.28; c) 0.39; d) 0.52; e) 0.75; f) 1.00; g) 1.22; h) 1.33; i) 2.13; j) 10.11.

(ln(I(Q)) vs. Q^2) are shown in Fig. 4; from the slope of the straight lines the values for the radii of gyration R_g have been determined and are reported in Table I. The dependence of R_g on C is given in Fig. 5a; it shows a marked increase with decreasing concentration below 2 g/cm³.

Information on the dimension of the enclosed water volume can also be obtained from the concentration dependence of the coherent intensity extrapolated towards Q=0. Indeed in our case one must have, for N identical particles of volume V,

$$I(0) = A N (\rho_D - \rho_A)^2 V \tag{4}$$

where I(0) is the coherent intensity scattered at zero angle, and ρ_D and ρ_A are the scattering densities of D₂O and of agarose respectively. From the I(0) data, by adopting a spherical equivalent shape for the interstitial volumes, an average radius R_v can be obtained. The R_v values are reported in Table I and agree well with the radii of gyration deduced from the Guinier plots (Fig. 5b).

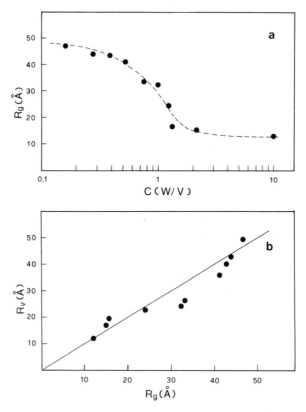

Figure 5. a) Dependence of the radius of gyration (Rg) on agarose concentration C (g/cm³). The dashed curve is only a guide to the eye.
b) Correlation between the radii of gyration derived from the Guinier plots (Rg) and the average radii (Rv) derived from the values of I(0).

Although the qualitative picture of bundles of helices connecting junction zones is consistent with the present knowledge of the structure of gels, it is difficult to see how in detail the network structure can accommodate varying amounts of water without modifications of the connectivity (i.e. density of junction zones, number of helices per bundle). Further neutron diffraction measurements, for an extended concentration range and at Q values lower than those reported here, are likely to shed light on this question.

4. Inelastic Scattering

The inelastic scattering data that we were able to collect so far consist of sets of time-of-flight spectra for an agarose-H_2O gel with concentration C = 1 g/cm^3 at four temperatures between 296 and 337 K. These are from short runs on the time-focusing spectrometer IN6 at I.L.L. In its standard configuration, this instrument produces 19 t.o.f. spectra from a total of 227 ^{3}He-detectors distributed over scattering angle 2θ between 10° and 115°, corresponding to elastic momentum transfers $\hbar Q_0$ = 0.2 to 2.1 Å^{-1}. Each spectrum represents the accumulated output of a 512-channel analyzer recording the number and time of arrival of scattered neutrons within successive 4.9 ms-long time intervals between pulses incident on the sample. The range of energy transfers extends from $\hbar\omega$ = −15 cm^{-1} through the broadened elastic peak ($\hbar\omega$=0) up to 1000 cm^{-1} on the energy-gain side, the incident wavelength being 5.1 Å. The Q,ω-domain covered is constrained by the conservation equations for momentum and energy such that along each 2θ=const spectrum $\hbar\omega$ and $\hbar Q$ are related by

$$Q^2 = k_0^2 \left[2 + \overline{\omega} - 2 \cdot (1 + \overline{\omega})^{1/2} \cos 2\theta \right] \qquad (5)$$

where $k_0 = 2\pi/\lambda_0$ and $\overline{\omega} = \hbar\omega/E_0$ is the nondimensional energy transfer.

Neutron scattering from any natural (i.e. not covalently deuterated) biomolecular sample is predominantly incoherent because of the large proton cross-section of 80 barn vs. an average of 5−6 barn for C,O,N,S (1 barn = 10^{-24} cm^2). This is a fortiori true for H_2O-hydrated samples, and the double differential cross-section measured for inelastic scattering ($\hbar\omega \gtrsim 10$ cm^{-1}) may, in the single-phonon approximation, be written as

$$\frac{d^2\sigma}{d\theta d\omega} = N_p \frac{k}{k_0} \sigma_{inc} \sum_p e^{-2W_p} \sum_{j,\underline{q}} (\overline{n}_j + \frac{1}{2} \mp \frac{1}{2})(Q \cdot \underline{C}_{pj})^2 (2m\omega_j)^{-1} \cdot \delta(\omega \mp \omega_j) \qquad (6)$$

Here, p = 1,2,...,N_p labels the protons in the sample, the double sum extends over all modes $(\underline{q}, j)$, σ_{inc} represents the proton incoherent cross section, k_0 and k are the incident and scattered wavevectors, respectively, $2W_p$ is the Debye-Waller factor, $\underline{C}_{pj}$ is the normalized amplitude vector of nucleus p vibrating in mode j, $\overline{n}_j = [\exp(\hbar\omega_j/k_BT) - 1]^{-1}$ is the thermal population factor, and the delta-function ensures conservation of energy. The information obtained relates to the proton dynamics and is therefore complementary to that derived from RSMR which provides information on heavier atoms (O,C,S).

There are, broadly, three characteristic regions in the t.o.f. spectra from the agarose gel studies (Fig. 6a). The prominent peak centered on $\hbar\omega \simeq 500$ cm^{-1} is a feature of all such spectra from molecular systems at ordinary temperatures. It results from the competing effect of intensity increases $\sim Q^2$ for $2\theta = $const spectra and the thermal population factor $\bar{n}_j$ which becomes proportional to $e^{-\hbar\omega/k_BT}$ when $\hbar\omega_j > 2$ to 3. Secondly, a broad, and at low angles, weak band between 30 and 100 cm^{-1} picks up intensity with increasing 2θ, and the minimum between this feature and the rising wings of the quasi-elastic peak increases by a factor of about 2. Thirdly, in the 100 to 300 cm^{-1} region the pattern of angle-dependent changes is less clear and there are marked crossover effects. These appear to arise from two ledges or ridge-like intensity increases in all $\omega = $const sections between $\hbar\omega \simeq 10$ and 300 cm^{-1}, one at low Q (0.4Å^{-1}) and the other around $Q = 1$Å^{-1}. These are not observed in pure water spectra and they must be due to the polysaccharide network and/or the closely associated water fraction. Although the spectra shown are corrected counts and not interpolated constant-Q representations, it is obvious from Fig. 6a that these features will be equally noticeable in fully reduced $S(Q,\omega)$ plots.

The frequency distributions that can be derived from incoherent scattering data of the kind shown in Fig. 6b are weighted with respect to $(\underline{u}_p \cdot Q)^2/Q^2$ and averaged over all protons in the system. The simplest approach to the interpretation of neutron spectra from biopolymers in the presence of bulk-like water is to assume a composite scattering law according to

$$S_{inc}(Q, \omega) = fS_{inc}^{(b)}(Q, \omega) + (1 - f)S_{inc}^{(m)}(Q, \omega) \qquad (7)$$

where f is the fraction of water protons that is regarded as immobile or "bound" on the time scale in question, and superscripts (b) and (m) refer to "bound" and "mobile" protons, respectively. The motions of the former may be described by a Debye-Waller factor, whereas various more complicated scattering laws for rotational and translational modes may be written for $S_{inc}^{(m)}(Q, \omega)$ and tested against the spectra observed. In an agarose-H_2O gel with $C = 1$ g/cm^3, there are 18 polysaccharide protons and 68 water protons per chemical repeat unit (i.e. $C_{12}H_{18}O_9$). At frequencies $\hbar\omega > k_BT$ we expect a 30 to 40% contribution of the polysaccharide protons to $d^2\sigma/d\theta d\omega$; towards lower frequencies the amplitude weighting will increasingly favor the water protons so that Eq. 7 becomes a very good approximation for $\hbar\omega \lesssim 100$cm^{-1}. The proton-amplitude weighted frequency distributions (Fig. 6b) are defined as

$$P(Q_0, \omega) = \frac{\omega}{Q^2} \sinh(\frac{\hbar\omega}{2k_BT})\tilde{S}(Q, \omega) \qquad (8)$$

where $\tilde{S}(Q, \omega)$ is the symmetrized scattering law which can be expressed in terms of the differential cross section as

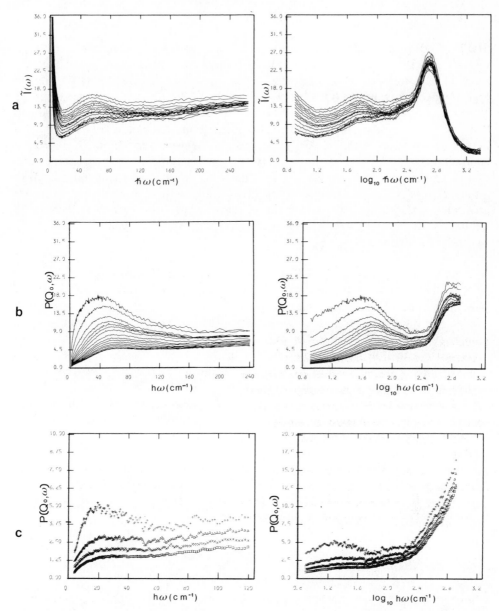

Figure 6. Time-of-flight spectra and frequency distributions for an agarose H_2O gel with $C = 1$ g/cm^3. The abscissae in the upper graphs are linear in cm^{-1} while in the lower graphs the scale is $\log_{10}$ of $\hbar\omega$ in cm^{-1}.

a) Time-of-flight spectra from $Q_0 = 0.25$ Å^{-1} (lowest curve) to $Q_0 = 1.74$ Å^{-1} (highest curve).

b) Partial frequency distributions ($P(Q_0,\omega)$) from $Q_0 = 0.25$ Å^{-1} (highest curve) to 1.74 Å^{-1} (lowest curve).

c) Difference frequency distributions for a gel at 337 K relative to that at 296 K.

$$\tilde{S}(Q, \omega) = \frac{4}{N_p\sigma_{inc}} \frac{k_0}{k} \exp (\hbar\omega/2k_BT) \frac{d^2\sigma}{d\theta d\omega} \qquad (9)$$

The $P(Q_0,\omega)$ spectra show significant intensity increases with decreasing Q, i.e. with increasing scale lengths $2\pi/Q$ (Fig. 6b). The maximum is seen to shift from 65 cm^{-1} to 40 cm^{-1} as the interaction distances probed increase from about 5 to 25 Å. The high-frequency region in $P(Q_0,\omega)$ is rather similar to that of t.o.f. spectra except for an enhanced plateau-like region between $\hbar\omega \simeq 500$ and 900 cm^{-1}. Beyond 800 to 900 cm^{-1} the resolution of the t.o.f. technique deteriorates rapidly and the $P(Q_0,\omega)$ curves become very sensitive to changes in the Debye-Waller factor, background subtraction, and other corrections. The angle and energy dependent properties discussed so far are very similar for the analyzed gel at four temperatures from ambient up to 337 K except that total intensity increases appear to grow at a rate somewhat higher than that given by an Arrhenius-type law. Partial frequency distributions thus provide a means of quantifying the extent of dynamic water structuring below 100 cm^{-1}.

The quasi-elastic peaks lie immediately to the left of the $d^2\sigma/d\theta d\omega$ curves shown in Fig. 6a; they are 10- to 100-fold more intense than the inelastic scattering and only part of their outer wings are seen in the upper frame of Fig. 6a. For the sample at 296 K, the difference widths $\Delta E(Q)$ relative to the increase from 0.1 cm^{-1} at Q = 0.25 Å^{-1} to 0.23 cm^{-1} at Q = 1.5 Å^{-1} (Fig. 7). At 337 K the broadenings are systematically higher by 25 to 100%. Although the overall increase of ΔE vs. Q^2 is roughly linear (corresponding to effective diffusion coefficients between 1 and $2 \times 10^{-6} cm^2/s$), there is some structure especially at low Q which deserves to be investigated in future contrast experiments. A detailed lineshape analysis on the basis of explicit models for translational and rotational diffusive motions is beyond the scope of the present paper.

The above results demonstrate that the inelastic and quasi-elastic scattering of neutrons from aqueous gels are a valuable tool to characterize their spatiotemporal properties, but a larger data base covering a wider concentration and H/D contrast range is needed to make more detailed and quantitative comparisons.

Concluding Remarks

We have discussed some theoretical and experimental aspects of the application of neutron scattering to the study of polysaccharide gels, and have presented results, from initial experiments on agarose gels. The diffraction results, although limited to relatively concentrated gels in the present work, demonstrate clearly that one can obtain unique geometrical information on the polysaccharide network over distances up to about 200 Å. We have not yet used one of the more powerful small-angle neutron cameras equipped with large area detectors (such as D11 at I.L.L.) for which Q $\simeq$ 0.002 Å^{-1}. The Q-range over which data can be recorded for fixed detector geometry is very wide on pulsed-source diffractometers employing the Laue time-of-flight technique (such as LOQ at R.A.L.). There is thus considerable

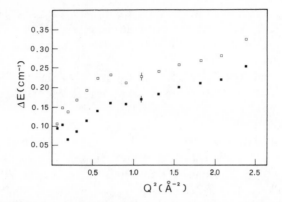

Figure 7. Difference half-widths $\Delta E(Q^2)$ {FWHM} of the quasi-elastic peaks relative to a vanadium scattering standard. Sample as in Fig. 5 at temperatures 296 K (■) and 337 K (□). Typical error bars are shown.

scope for more comprehensive structural studies on polysaccharide gels, both by extending the scale lengths covered to 2000 Å and by using faster instruments to obtain the counting statistics needed for studying gels at lower concentrations. Similar considerations apply to the time-dependent measurements. Here we have only shown data for a simple hydrogenous gel in the Q,ω-range covered by the time-of-flight technique. Using backscattering (0.01 to 10 cm^{-1}) and spin-echo (10^{-5} to 10^{-2} cm^{-1}) spectrometers, the energy range can be expanded significantly in the direction of both lower energies and smaller momentum transfers ($Q \simeq 0.02$ Å^{-1}). Using advanced quasi-elastic spectrometers which are gradually becoming faster and more accessible, it will be possible to extend the pioneering work of Trantham et al.[21] very substantially. Altogether, we can look forward to neutron scattering results that will allow us to develop a much more detailed and quantitative picture of the spatiotemporal properties of gels in parameter regimes that were previously inaccessible.

Acknowledgments

We wish to thank the E.N.E.A. (Casaccia) and the Institut Laue-Langevin (Grenoble) for the use of neutron facilities, Drs. P. Bisanti, L. Braganza and G. Zaccai for discussions and advice on experimental aspects, and Mr. J.M. Reynal for technical assistance. This work was supported by the Centro Interuniversitario di Struttura della Materia and the U.K. Science and Engineering Research Council.

References

1. G.O. Aspinall, ed., "The Polysaccharides," Vols. 1 and 2 (1982), Vol. 3, Academic Press, N.Y. (1985).

2. G.A. Ackers, Molecular Sieve Methods of Analysis, in "The Proteins," H. Neurath and R.L. Hill, eds., Academic Press, N.Y. (1982).

3. F. Family and D.P. Landau, "Kinetics of Aggregation and Gelation," North-Holland, Amsterdam (1984).

4. D.A.D. Parry and E.N. Baker, Rep. Prog. Phys., **47**, 1133 (1984).

5. S. Arnott, A. Fulmer, W.E. Scott, I.C.M. Dea, R. Moorhouse, and D.A. Rees, J. Mol. Biol., **90**, 269 (1974).

6. M.B. Palma-Vittorelli and M.U. Palma, in "Structure and Motion: Membranes, Nucleic Acids and Proteins," E. Clementi, G. Corongiu, M.H. Sarma and R.H. Sarma, eds., Adenine Press, N.Y. (1985).

7. P.L. San Biagio, F. Madonia, F. Sciortino, M.B. Palma-Vittorelli and M.U. Palma, J. de Phys., **45**, C7-225 (1984).

8. S.W. Lovesey, "Theory of thermal neutron scattering," Vol. 1, Oxford University Press (1985).

9. G. Albanese and A. Deriu, Rivista del Nuovo Cimento, **2**, n.9 (1979); G. Albanese in "Applications of the Mössbauer Effect," Vol. 1 (Proc. ICAME Conf., 1983, Alma-Ata, USSR), Gordon & Breach, N.Y., p. 63 (1985).

10. G. Albanese, A. Deriu and F. Ugozzoli in "Applications of the Mössbauer Effect," Vol. 2 (Proc. ICAME Conf., 1983, Alma-Ata, USSR), Gordon & Breach, N.Y., p. 1561 (1985).

11. G. Albanese, A. Deriu, F. Ugozzoli, and C. Vignali, J. Mol. Biol., to be published.

12. H.B. Stuhrmann and A.J. Miller, Appl. Cryst., **11**, 325 (1978).

13. J.W. White, Proc. R. Soc. Lond., **A345**, 119 (1975).

14. G. Corongiu, S.L. Fornili and E. Clementi, Intl. J. Quant. Chem., Quant. Biol. Symp., **10**, 277 (1983).

15. H.D. Middendorf and J.T. Randall, in "Structure and Motion: Membranes, Nucleic Acids and Proteins," E. Clementi, G. Corongiu, M.H. Sarma and R.H. Sarma, eds., Adenine Press, N.Y. (1985).

16. H.D. Middendorf, Ann. Rev. Biophys. Bioeng., **13**, 425 (1984).

17. G. Zaccai and B. Jacrot, Ann. Rev. Biophys. Bioeng., **12**, 139 (1983).

18. C.G. Windsor, "Pulsed Neutron Scattering," Taylor & Francis, London (1981).

19. B.F. Fender, L.C.W. Hobbis and G. Manning, Phil. Trans. R. Soc. Lond. B., **290**, 657 (1980).

20. A.J. Leadbetter, in "Neutron Scattering in the Nineties," I.A.E.A. Wien, p. 219 (1985).

21. E.C. Trantham, H.E. Rorschach, J.S. Clegg, C.F. Hazlewood, R.M. Nicklow and N. Wakabayashi, Biophys. J., **45**, 927 (1984).

Non-empirical Pair Potentials for the Interaction Between Amino Acids

J.A. Sordo,* M. Probst, S. Chin,**
G. Corongiu and E. Clementi
IBM Corporation
Data Systems Division, Dept. 48B/MS 428
Neighborhood Road
Kingston, New York 12401

Abstract

A pair potential describing the potential hypersurface between interacting aliphatic amino acids (without sulphur) is presented. This pair potential has been derived entirely from *ab initio* calculations at the Hartree-Fock level. Almost two thousand SCF calculations have been performed and used as input for a nonlinear least-squares fitting in order to obtain the parameters for the atom-atom analytical pair potential.

Introduction

The rigorous theoretical study of biological systems is still far removed from the present day computational facilities.[1] In fact, the theoretical background is available but its practical implementation for large molecular systems demands both very powerful machines and very long computing times. In order to obtain information on these kinds of systems some approximate treatment must be used.[2] Some of the many semi-empirical methods available in the literature can be used; however, many problems still remain when using such methods. In fact, in spite of the remarkable reduction in computing time when passing from *ab initio* to semi-empirical methods, this factor continues to be a very serious drawback when dealing with large systems. A very interesting alternative is provided by analytical pair potentials.[3-10] (See the Appendix for a list of some *ab initio* analytical pair potentials available at the present time.) Once an analytical expression representing the potential hypersurface is available, the interaction energy between very large

---

* Permanent address: Departamento de Quimica Fisica, Universidad de Oviedo, Oviedo, Principado de Asturias, Spain.

** Present address: Institute of Inorganic and Analytical Chemistry, University of Innsbruck, A-6020, Innsbruck, Austria.

systems can be evaluated relatively fast. In addition, an analytic expression brings about the possibility of using statistical mechanics; thus providing the ability to predict thermodynamic properties of the system under study.[11,12]

The parameters included in the analytical pair potential may be obtained from experimental data (empirical pair potentials),[5,6] from semi-empirical calculations (semi-empirical pair potentials),[3,4] or from calculations based on first principles (*ab initio* pair potentials).[7,8]*

One major advantage of the latter pair potentials compared to the others is that information concerning any point on the potential hypersurface is accessible from calculations, but not always from experiments.[13] In this paper we present an *ab initio* pair potential for the interaction between aliphatic amino acids (without sulphur). This research is a continuation of the work on *ab initio* analytical pair potentials for the interaction between water molecules,[14-16] and between amino acids and water[7,8] that has been systematically developed at this laboratory. In addition, work is currently in progress to derive the corresponding analytical pair potentials for the amino acids containing sulphur, as well as for the non aliphatic amino acids.

At the completion of this work, a rather comprehensive set of analytical pair potentials for amino acids, and for amino acids and water will have been derived. This set of pair potentials will provide a very powerful tool for theoretical studies of systems of biological interest.

At the present time, the application of pair potentials, like those presented in this paper, to perform Monte Carlo (MC) and Molecular Dynamics (MD) simulations on systems containing molecules of biological interest has proved to be a very useful tool in providing structural and energetic information of such systems.[17-34] The development of a pair potential for the interaction between amino acids will facilitate additional studies of these type of systems, like for example the determination of the 3-D structure of proteins, a subject of fundamental importance in biochemistry.

The procedure for obtaining the analytical pair potentials from *ab initio* calculations is well-defined in previous works.[7,8] However, an important point arises as a result of the greater complexity (degrees of freedom) of the molecules considered in this paper: the configurational space for these types of systems becomes very prolix. Therefore, since the reliability of an *ab initio* pair potential depends strongly on the completeness of the configurations sampled and selected for the fitting procedure, special consideration has been exercised in this work on this regard (see next section).

---

* There are also pair potentials that cannot be assigned just to one of these three classes (see, for example, Refs. 9 and 10).

Method

The procedure for obtaining an *ab initio* pair potential may be divided into four steps: (a) selection of a functional form, (b) selection of conformations, (c) *ab initio* calculations and (d) fitting procedure.

Analytical pair potential

As in our previous work[7,8] we used an analytical pair potential of the form

$$V_{MN} = \sum_i \sum_{i>j} - (A_{ij}^{ab})^2/r_{ij}^6 + (B_{ij}^{ab})^2/r_{ij}^{12} + C_{ij}^{ab} q_i q_j/r_{ij} \tag{1}$$

where V_{MN} is the total interaction energy (Kcal/mol) between molecules M and N, i is the i-th atom in molecule M belonging to the class a, j is the j-th atom in molecule N belonging to the class b and r_{ij} is the distance in Å between the two atoms i and j; and A_{ij}^{ab}, B_{ij}^{ab}, and C_{ij}^{ab} are the fitting parameters describing the interaction between atoms in class a and in class b. Actually, based on our previous experience, the parameter C_{ij}^{ab} has been kept constant and equal to one in all the cases.

q_i, and q_j represent the charges associated to atoms i and j, respectively. These are not fitting parameters but they have a clear physical meaning. Mulliken population analysis (MPA) provides appropriate values for such charges. However, it is well-known that the values provided by MPA depend, for a given basis set, on the conformation considered. In order to observe the variation in charges in different conformations, the corresponding MPA's have been computed for all the conformations considered in this work. By comparing the MPA for many different conformations, it was observed that the estimation of the charges q_i, q_j from the MPA on the monomers is a reasonable approximation to those computed in the pairs of amino acids.

In fact, in Table I we collect, as an example, the values of the charges q_i for all the atoms in alanine as calculated from the MPA on the monomer as well as from the MPA on the dimer, this last in a conformation in which the interaction energy was very strong (-16 Kcal/mol) and therefore the differences between the two sets of MPA's are expected to be the greatest (of course, the sets become identical when there is no interaction between amino acids). From Table I it can be concluded that taking the charges from MPA on the monomer is a reasonable choice.

The different classes of atoms for the amino acids have been adopted from our previous work,[7,8] with the exception that the old classes 2 and 3 (see the first of the papers in Ref. 7) corresponding to different aliphatic hydrogens have been unified in the present work. In fact, hydrogens belonging to classes 2 and 3 exhibit quite similar chemical behavior (as measured by means of the values of both partial charges and molecular orbital valency values).[7] This point was fully confirmed by carrying out two different fits. In the first one the classes were exactly as defined in

Table I Charges (au.) from Mulliken population analysis for atoms in alanine as calculated from monomer (ALA) and from a very stable conformation in the dimer (ALA − ALA) (see Ref. 7 for notation).

Atom	q_i(ALA)	Molecule 1 q_i(ALA − ALA)	Molecule 2 q_i(ALA − ALA)
O(1)	8.4333	8.4870	8.4866
O(2)	8.5632	8.5778	8.5783
N	7.6104	7.6098	7.6098
C′	5.4843	5.4600	5.4610
C(A)	6.1401	6.1413	6.1413
C(B)	6.6379	6.6391	6.6390
H(A)	0.7737	0.7682	0.7683
H(1)	0.7153	0.7170	0.7171
H(2)	0.7275	0.7273	0.7274
H(B1)	0.7886	0.7895	0.7895
H(B2)	0.7814	0.7776	0.7777
H(B3)	0.7660	0.7634	0.7635
H(O2)	0.5783	0.5412	0.5412

our previous work,[7,8] and in the second one the same class was used to represent all the aliphatic hydrogens appearing in the amino acids. Both fits led practically to the same results. Table II collects the classes of atoms considered in this work.

Selection of conformations

As stated above, the selection of the conformations used in the fitting procedure is a very crucial point in this work. This has been pursued with two goals in mind. First, there are relevant chemical groups in the aliphatic amino acids not containing sulphur; these are the carboxylic and aminic groups (present in all these amino acids), and the alcohol group (present only in some of them). Some of the interactions involving these groups give rise to very stable conformations. For example, those in which the two carboxylic groups, one on each interacting molecule, form a double hydrogen-bond, or those in which the aminic group in one molecule interacts with the − OH in the carboxylic group of the other molecule, and so forth. Therefore, in order to ensure an adequate description of the local minima regions on the hypersurface, it is necessary to include such conformations into the fit. The generation of these special conformations has been performed by using graphics facilities. These facilities allow one to recover the final coordinates of two or more molecules once an appropriate set of movements (rotations and/or translations) have been performed (in order to achieve a given relative orientation of the molecules).

On the other hand, large portions of the configurational space still remain uncovered. In particular, there are many conformations in which the interaction between the above mentioned chemical groups are not directly involved. Several procedures have been designed to systematically generate these other conformations. In such procedures one of the molecules is kept fixed with the center of mass at the origin; then a sphere of radius r (centered at the origin) is constructed and a spherical grid formed by a set of regularly spaced points is defined on its surface. The initial radius, r, is increased step by step (increments of 0.5 Å, up to 4.5 Å are taken) and equivalent spherical grids are defined on each one of the generated spherical surfaces. Then the center of mass of the second molecule is positioned on each one of the grid points and a rotation defined by three Eulerian angles is performed on this molecule.* The starting radius, r, is chosen in such a way that the distance between the two nearest atoms (one belonging to the first molecule and the other belonging to the second molecule) is on the order of the length of a hydrogen-bond ($\simeq$ 2.5 au.). The three Eulerian angles are systematically chosen in order to account for the largest number of different relative orientations between both molecules.

Finally, further additional conformations are generated from feedback provided by the fitting procedure itself (see the subsection below on Fitting Procedure).

Following the above schemes a total of almost two thousand conformations were generated.

Ab initio Calculations

The magnitude of the problem forced us to use standard 7/3(SZ) basis sets.[36] Indeed, we had to perform almost two thousand SCF calculations at the Hartree-Fock level for systems with an average of more than 100 electrons, the typical average time being about 90 minutes of CPU time for each SCF calculation (integral evaluation, self-consistent field calculation and — see below — basis set superposition correction) on a FPS-164. It is well-known that a minimal basis set gives errors due to the basis set superposition error (BSSE). In order to minimize these errors the counterpoise (CP) method was applied to every SCF point. A discussion on the limitations of the 7/3(SZ) basis sets in constructing pair potentials can be found in the section *Results and Discussion* at the end of this paper.

Since the two amino acids alanine and serine contain all the classes of atoms collected in Table II, the pairs alanine-alanine, alanine-serine and serine-serine were selected as the pairs on which the SCF calculations were performed. However, due to the chemical similarities among all amino acids these parameters should be transferable to other amino acids. As expected,[8] transferability of the parameters of the analytical pair potential (based on the concept of *class of atoms*[7]) was quite good. This is described below.

---

*   This procedure is quite similar to that described in Ref. 10.

Table II. Notation for the classes of atoms considered in this work.

OXYGEN
| OCBL | carbonylic oxygen: $R_2 - C = O$ |
| OCBX | hydroxylic oxygen: $R - COOH$, $R - OH$ |

NITROGEN
| NNH2 | aminic nitrogen: $R - NH_2$ |

CARBON
CALP	alpha carbon: $R_2 - CH - NH_2$
CCBX	carboxylic carbon: $R - COOH$
CCH3	aliphatic carbon: $R - CH_3$
CCH2	aliphatic carbon: $R_2 - CH_2$

HYDROGEN

HNH2	aminic hydrogen: $R - NH_2$
HCBX	hydroxylic hydrogen: $R - COOH$, $R - OH$
HALI	aliphatic hydrogen: $R_3 - CH$, $R_2 - CH_2$, $R - CH_3$

Fitting Procedure

The fitting procedure was carried out in several steps:

1. The interaction energies of the configurations generated according to the procedure described in a previous subsection (on Selection of Configurations) were fitted to the analytical pair potential in Eq. 1 and a set of parameters $\{A_{ij}^{ab}, B_{ij}^{ab}\}$ were obtained (recall C_{ij}^{ab} is kept fixed and equal to 1.0).

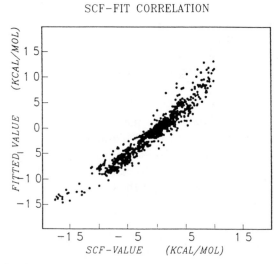

SCF−FIT CORRELATION

Figure 1. Correlation between interaction energies (corresponding to points included into the fitting) calculated *via* Hartree-Fock (with BSSE) and those derived from the parameters in Table III.

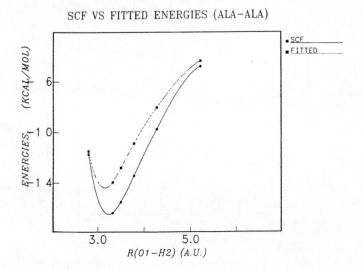

Figure 2. Potential curves from Hartree-Fock (Kcal/mol) (with BSSE) (full line) and from parameters in Table III (dashed line) for alanine-alanine interacting through the carboxylic groups forming double hydrogen-bonded associations. R(O1 − H2) represents the distance between the carbonyl oxygen in one of the molecules and the hydrogen belonging to the − OH group in the second molecule.

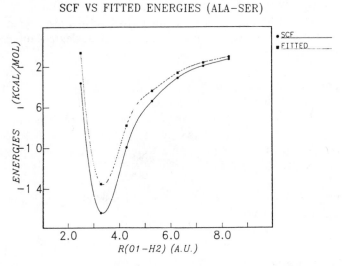

Figure 3. Potential curves from Hartree-Fock (Kcal/mol) (with BSSE) (full line) and from parameters in Table III (dashed line) for alanine-serine interacting through the carboxylic groups forming double hydrogen-bonded associations. R(O1 − H2) represents the distance between the carbonyl oxygen in one of the molecules and the hydrogen belonging to the − OH group in the second molecule.

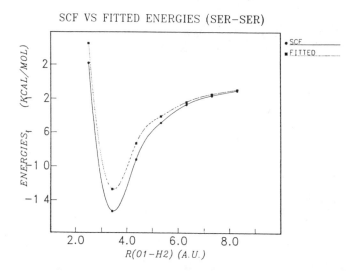

Figure 4. Potential curves from Hartree-Fock (Kcal/mol) (with BSSE) (full line) and from parameters in Table III (dashed line) for serine-serine interacting through the carboxylic groups forming double hydrogen-bonded associations. R(O1 − H2) represents the distance between the carbonyl oxygen in one of the molecules and the hydrogen belonging to the − OH group in the second molecule.

2. With the above set of parameters the following procedure was used to calculate a number of iso-energy surfaces. One amino acid was kept fixed at the origin and the C^{α} of the second amino acid was moved to the intersection points of a two-dimensional grid. The size of such a grid was chosen to be 8x8 (au.). For each grid point (a total of 1024 points were considered), we minimized the interaction energy with respect to the three rotational degrees of freedom of the second amino acid.

3. We searched for local minima of the energy in the iso-energy map. Some of such local minima conformations corresponded to those already used in the fitting procedure (those giving strong stabilization energies) but some others were new. SCF calculations for new local energy minima that had not previously been used in the fitting procedure were performed, and these points were included in the fitting procedure.

4. The fitting in (1) was carried out to convergence, giving the final parameters for the analytical pair potential.

Results and Discussion

Table III shows the values of the parameters of the analytical pair potential (see Table II for notation).

Table III. Parameters for the analytical pair potential.

CLASSES		A_{ab}	B_{ab}
OCBL	OCBL	−0.0017994601	449.2353812660
OCBL	OCBX	−15.3452747440	515.0417781913
OCBX	OCBX	10.2615562006	501.2204561285
OCBL	NNH2	−0.1777267156	652.3369320899
OCBX	NNH2	0.0054988482	585.8908448713
NNH2	NNH2	0.3954541294	1950.1260186359
OCBL	CCBX	−27.7291797104	38.8901461999
OCBX	CCBX	−0.0554031759	272.8995452033
NNH2	CCBX	0.3143993909	−46.3888030020
CCBX	CCBX	−0.0105631100	−22.9443897783
OCBL	CALP	−0.0291505401	−661.2449788459
OCBX	CALP	0.1321841970	1055.5766730801
NNH2	CALP	0.3409143670	1184.3104939666
CCBX	CALP	0.1189005438	−3837.1956543764
CALP	CALP	0.0796205069	−1835.4782780307
OCBL	CCH3	−0.0054418292	−850.6637778957
OCBX	CCH3	12.2235541553	1070.6348760156
NNH2	CCH3	−0.1415573497	1693.9367402666
CCBX	CCH3	0.0709358113	−2346.9554249521
CALP	CCH3	−0.0744790912	91.4693918423
CCH3	CCH3	3.9035487943	117.7007948924
OCBL	HALI	0.0100104936	58.1347999171
OCBX	HALI	−0.0057355691	37.1133783696
NNH2	HALI	−7.6147662630	−20.0010033956
CCBX	HALI	−0.0012647782	20.0009367646
CALP	HALI	0.0371962248	−140.8167632125
CCH3	HALI	−0.0435116003	254.7811251236
HALI	HALI	−0.0303826003	21.1202810881
OCBL	HNH2	−0.0333453216	85.5294974513
OCBX	HNH2	0.0003559460	20.0010001890
NNH2	HNH2	19.6970116907	20.0009999804
CCBX	HNH2	−0.0104359360	202.0605094874
CALP	HNH2	0.0350531116	289.8723957001
CCH3	HNH2	−0.0778772419	20.2080140941
HALI	HNH2	0.0036263653	22.3910902142
HNH2	HNH2	−0.0049268319	34.9710677754
OCBL	HCBX	−7.6392620992	20.0010000584
OCBX	HCBX	−8.6064261253	29.4571763687
NNH2	HCBX	13.6188255460	34.6657375591
CCBX	HCBX	0.0107943699	−285.5197897867
CALP	HCBX	−0.0932920547	197.8740089961
CCH3	HCBX	3.0621747073	−181.4816889928
HALI	HCBX	−0.0102485460	20.0860012656
HNH2	HCBX	−0.0081038376	−60.0811141771
HCBX	HCBX	0.0070152362	−59.5188516025
OCBL	CCH2	0.2806730512	−1364.6974723225
OCBX	CCH2	−0.0658998774	1253.4982577744
NNH2	CCH2	0.0351181737	−1603.2337106975
CCBX	CCH2	−0.5532250871	−39.6422378380
CALP	CCH2	−0.4234282007	−2927.3728178322
CCH3	CCH2	−3.5935800461	794.4216861577
HALI	CCH2	0.2509957681	21.6249354028
HNH2	CCH2	−0.0156636988	28.1540870410
HCBX	CCH2	14.5961501700	−23.1168006778
CCH2	CCH2	0.0110613490	−1910.1337193013

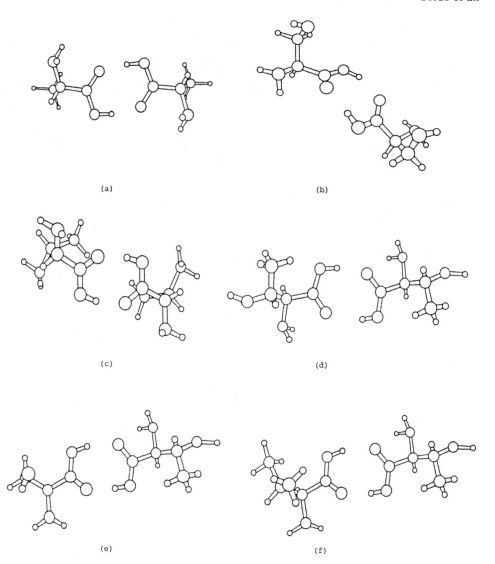

(a)

(b)

(c)

(d)

(e)

(f)

Figure 5. Conformations corresponding to local minima for several pairs of amino acids as obtained from parameters in Table III. Alanine-alanine (insets a,g), serine-serine (insets b,i), valine-valine (insets c,h), threonine-threonine (insets d,j), serine-threonine (insets e,l), valine-threonine (insets f,k).

The standard deviation obtained from the fitting procedure was 1.08 Kcal/mol; this is good considering that almost two thousand points have been fitted to a relatively simple functional form (Eq. 1). Figure 1 shows the correlation between the SCF calculations and the fitted values as calculated with Eq. 1 and the parameters in Table III. The regression coefficient was 0.93 and therefore the correlation may be classified as good.

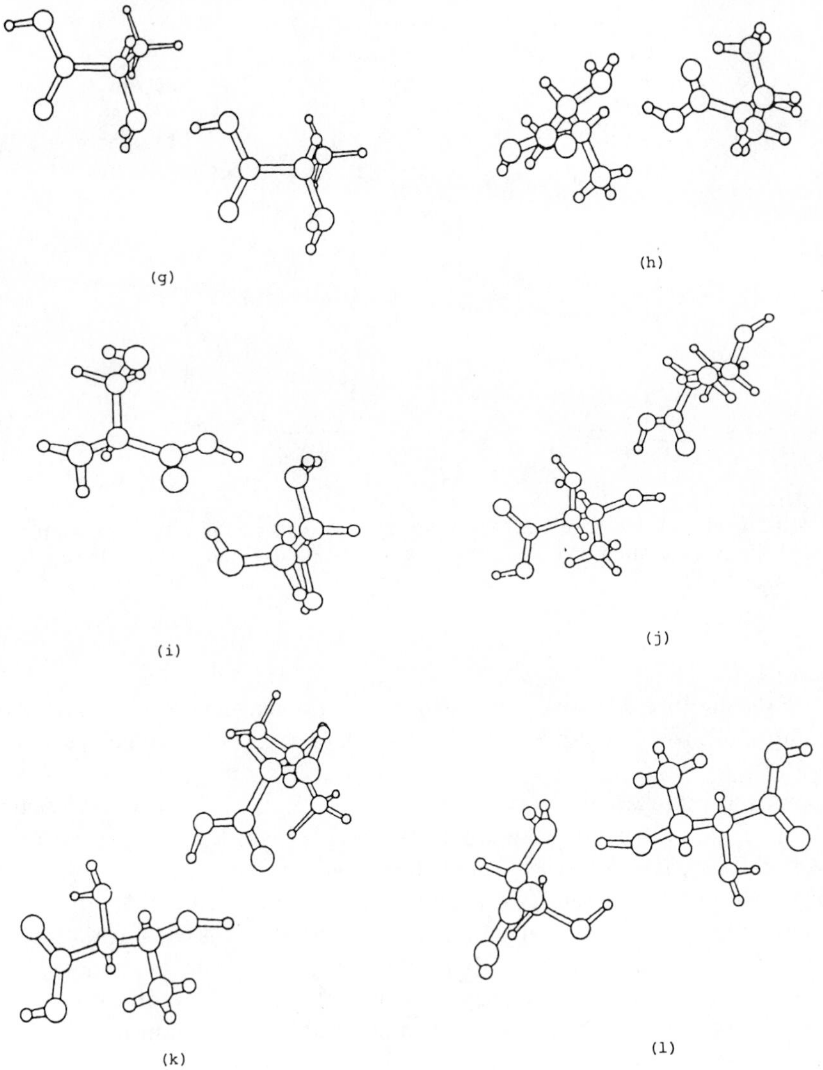

(g)

(h)

(i)

(j)

(k)

(l)

Figure 5. Continued.

Figures 2,3 and 4 exhibit further evidence of the general reliability of our analytical pair potential. As described previously, some of the conformations were generated in order to describe those kinds of interactions between functional groups of special relevance in amino acids ($-COOH$, $-NH_2$ and $-OH$). For some of these conformations a series of SCF points at varying intermolecular distances were computed (keeping the relative orientations of the two molecules fixed). The resulting potential energy curves may be compared with those provided by the analytical pair potential. Figures 2,3 and 4 show the corresponding results for alanine and serine when both molecules interact through the carboxylic groups (see Fig. 5, insets (a)

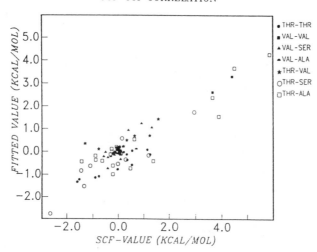

Figure 6. Correlation between interaction energies (corresponding to points not included into the fitting) calculated *via* Hartree-Fock (with BSSE) and those derived from the parameters in Table III.

to (b)). From both qualitative and quantitative points of view the analytical pair potential provides a reasonably good description of such potential curves.

As a check on the reliability of the analytical pair potential a systematic search for local minima between pairs of amino acids was performed. From a chemical point of view is easy to make an *a priori* prediction on the existence of a set of conformations in which a large stabilization should be expected because of the formation of hydrogen-bonding. Samples of such conformations involving the interaction between $-COOH$, $-NH_2$ and $-OH$ groups for the pairs alanine-alanine, alanine-serine and serine-serine were included into the fitting. The question which arises is whether or not the analytical pair potential is able to reproduce these types of interactions for other pairs of amino acids.

Figure 5 collects a representative sample of the local minima conformations for some of these other pairs of amino acids as obtained from the analytical pair potential proposed in this work. Indeed many of the conformations computed to be the most stable ones are those one might expect based on chemical intuition. For all the pairs, the structures involving a double hydrogen-bonding ($-COOH$ ... $HOOC-$) between carboxylic groups in amino acids are the most stable ones (Figure 5, insets (a) to (f)). This agrees with the well-known fact of the dimeric structure observed in carboxylic acids.

However, the above mentioned associations are not the only ones providing stable dimers. In fact, the high basicity of the aminic nitrogen allows for the possibility of very stable conformations involving hydrogen-bonding between this nitrogen in one of the amino acid and the carboxylic hydrogen in the other ($-H_2N$ ... $HOOC-$).

Table IV. Total energies, monomer energies (with CP correction to BSSE) (au), and interaction energies (Kcal/mol) for three selected geometries of the system Alanine-Alanine as calculated with different basis sets.

	STO-3G	ΔE	7/3(SZ)	ΔE	9/5(DZ)	ΔE
Geometry #1						
	− 635.40304	0.7	− 641.38408	− 11.7	− 643.36098	− 9.4
	− 317.70201		− 320.68273		− 321.67306	
	− 317.70218		− 320.68268		− 321.67300	
Geometry #2						
	− 635.40228	− 6.1	− 641.38897	− 16.4	− 643.36719	− 15.5
	− 317.69623		− 320.68142		− 321.67127	
	− 317.69637		− 320.68140		− 321.67121	
Geometry #3						
	− 635.39946	− 6.5	− 641.38677	− 15.5	− 643.36642	− 15.4
	− 317.69447		− 320.68102		− 321.67092	
	− 317.69462		− 320.68103		− 321.67092	

Conformations (g) and (h) (Figure 5) are examples of this kind of association, and are predicted with the proposed pair potentials.

The existence of the alcohol group in some of the amino acids considerably increases the number of stable dimeric associations as shown in Figure 5 (insets (i) to (l)). Such associations are of two classes. On one hand are those in which the carboxylic group in one of the amino acids interacts with both aminic and alcohol groups in the other amino acid. These associations may also be classified as double hydrogen-bonding associations, but they are not as strong as in the case of the associations involving two carboxylic groups (Figure 5, insets (a) to (f)) because of

Table V. Total energies, monomer energies (with CP correction to BSSE), (au), and interaction energies (Kcal/mol) for three selected geometries of the system Alanine-Serine as calculated with different basis sets.

	STO-3G	ΔE	7/3(SZ)	ΔE	9/5(DZ)	ΔE
Geometry #4						
	− 709.21227	7.2	− 715.99157	− 3.6	− 718.15745	− 3.0
	− 391.52149		− 395.30254		− 396.47963	
	− 317.70217		− 320.68331		− 321.67306	
Geometry #5						
	− 709.22083	− 6.9	− 716.00732	− 16.4	− 718.17283	− 16.2
	− 391.51330		− 395.29978		− 396.47606	
	− 317.69655		− 320.68138		− 321.67092	
Geometry #6						
	− 709.21122	− 5.4	− 715.99199	− 9.9	− 718.16291	− 11.6
	− 391.50819		− 395.29702		− 321.47477	
	− 317.69444		− 320.67917		− 321.66966	

Table VI. Total energies, monomer energies (with CP correction to BSSE), (au), and interaction energies (Kcal/mol) for three selected geometries of the system Serine-Serine as calculated with different basis sets.

	STO-3G	ΔE	7/3(SZ)	ΔE	9/5(DZ)	ΔE
Geometry #7						
	−783.02974	11.8	−790.60246	2.2	−792.95458	2.4
	−391.52293		−395.30335		−396.47981	
	−391.52560		−395.30266		−396.47863	
Geometry #8						
	−783.03920	−6.4	−790.62413	−15.3	−792.97617	−15.1
	−391.51395		−395.29987		−396.47607	
	−391.51502		−395.29985		−396.47607	
Geometry #9						
	−783.02509	−5.0	−790.60916	−9.2	−792.96708	−10.8
	−391.50839		−395.29718		−396.47492	
	−391.50879		−395.29737		−396.47491	

steric reasons. Thus, for example, in the case of the interaction between threonine and valine (see Figure 5, inset (k)) the spatial conformation of the hydrogen in the alcohol group of threonine prevents the formation of a double hydrogen-bond involving the carboxylic group in valine and both aminic and alcohol groups in threonine. The interaction energy for this last type of association is on the order of −10 Kcal/mol and should be compared with a value of about −14 Kcal/mol corresponding to the interaction energy between the two carboxylic groups in a double hydrogen-bonded association (Figure 5, insets (a) to (j)).

On the other hand, the interaction between the aminic and alcohol groups in one of the amino acids and the same groups in the other amino acid also give rise to very stable conformations (Figure 5, inset (l)).

Therefore, Figure 5 shows that the analytical pair potential developed in this paper provides a good general description of the kinds of interactions which give rise to stable associations of pairs of amino acids, even for those pairs of amino acids not explicitly included into the fitting procedure.

As an additional check on the transferability of our potential, SCF energies of one hundred conformations in which at least one of the interacting amino acids was not

Table VII. Atomic energies (au) for the basis sets discussed in this work.

Element	STO-3G	7/3(SZ)	9/5(DZ)	HF
H	−0.46658	−0.49928	−0.49994	−0.5
C	−37.19839	−37.61599	−37.68519	−37.688619
N	−53.71901	−54.28441	−54.39535	−54.400934
O	−73.80415	−74.62784	−74.80040	−74.809398

one of those used in determining the parameters in Table III were computed. The correlation between the SCF values and the fitted values is shown in Figure 6. The correlation is quite good (the standard deviation for these additional calculations is 0.58 Kcal/mol) and is indeed a strong support to the concept of *class of atoms* around which all of this research is being developed. In other words, our parameters reproduce the hypersurface of other chemically equivalent molecules not originally included in the derivation of these parameters. Therefore, the parameters appear to be quite transferable, and reliable.

Basis Sets Limitations

As mentioned in a previous section, the calculations have been carried out at the SCF level using the 7/3(SZ) *minimal* basis sets. In this section we shall address questions concerning the accuracy (degree of confidence) of the results reported in this work.

In previous work aimed at developing analytical pair potentials, special care was exercised in examining the dependence of the results on the quality of the basis sets used.[17-19] The analysis of the results supported the use of 7/3(SZ)-type minimal basis sets in the derivation of pair potentials. In addition, subsequent applications of these pair potentials in studies of structural and energetic properties of several chemical and biochemical systems *via* MC and MD simulations have been reported in the last few years. The agreement found with available experimental data supports the use of 7/3(SZ) basis sets for constructing such potentials[20-34], particularly in dealing with systems of such complexity as to be computationally unfeasible to use larger basis sets.

In order to confirm the validity of the 7/3(SZ) basis sets in this study of the interactions between amino acids, SCF calculations for several hydrogen-bonded geometries for the systems alanine-alanine, alanine-serine and serine-serine were performed using two other basis sets: the *subminimal* STO-3G basis sets developed by Pople et al.[35] and 9/5 basis sets of *double-zeta* quality.[36]

Table IV, V and VI collect the total energies for these associations, the monomer energies, and the interaction energies. In Figures 7, 8 and 9 we plot the corresponding potential curves as calculated using STO-3G, 7/3(SZ) and 9/5(DZ) basis sets.

It is clear from Figs. 7-9 that 7/3(SZ) and 9/5(DZ) basis sets, provide comparable interaction energies from both qualitative and quantitative points of view. The STO-3G basis sets show substantial deviations, even if able to provide some qualitative information (i.e. except for Fig. 7, the relative order of the interaction energies remains the same as the order calculated with 9/5(DZ) basis sets). However, quantitative accuracy was not achieved with the STO-3G bases. We note that this is what might be expected by examining the atomic energies (see Table VII) for the different basis sets (HF energies have been included for comparison). The improvement of the 7/3(SZ) *minimal* basis sets over the STO-3G *subminimal* basis sets is

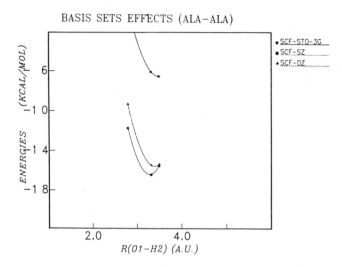

Figure 7. Potential curves from Hartree-Fock (Kcal/mol) as calculated with different basis sets for alanine-alanine interacting through the carboxylic groups forming double hydrogen-bonded associations. R(O1 − H2) represents the distance between the carbonyl oxygen in one of the molecules and the hydrogen belonging to the − OH group in the second molecule.

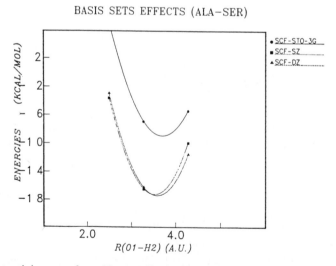

Figure 8. Potential curves from Hartree-Fock (Kcal/mol) as calculated with different basis sets for alanine-serine interacting through the carboxylic groups forming double hydrogen-bonded associations. R(O1 − H2) represents the distance between the carbonyl oxygen in one of the molecules and the hydrogen belonging to the − OH group in the second molecule.

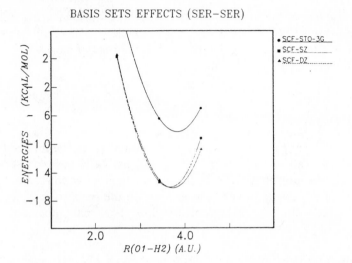

Figure 9. Potential curves from Hartree-Fock (Kcal/mol) as calculated with different basis sets for serine-serine interacting through the carboxylic groups forming double hydrogen-bonded associations. $R(O1 - H2)$ represents the distance between the carbonyl oxygen in one of the molecules and the hydrogen belonging to the $-OH$ group in the second molecule.

very clear. As it is known, this is a direct consequence of the care exercised in balancing and optimizing (the contraction coefficients were variationally optimized) the 7/3(SZ) basis sets[36] (in general, the 7/3(SZ) basis sets are somewhat better than the Slater single-zeta and not too different from the Slater double-zeta basis sets[17,36]). Molecular calculations reported in Tables IV-VI and Figs. 7-9 fully corroborate this point.

All the above clearly confirms previous findings concerning the usefulness of the 7/3(SZ) basis sets in creating a library of analytical pair potentials for associations involving molecules of biological interest.

Before finishing this section, let us make some comments on the CP method used in this work to correct for the error arising from the superposition of the basis sets.

Doubts have been reported in the literature concerning the use of the CP method to correct for the BSSE.[37,38] However, other recent literature also gives some examples in which the use of the CP method results in much improved interaction energies when calculated with small basis sets.[39-41] Therefore, at this moment, a definitive conclusion about the validity of the use of the CP method cannot be determined.[42] Furthermore, most of the above studies[37-41] have been carried out on small systems and we feel that extrapolations to larger systems are not straightforward. In fact, based on the origin of the BSSE it seems reasonable to infer that as long as the number of atoms *directly involved* in the molecular association grows, an increase in the BSSE should be expected.

The above discussion, combined with previous experience in using 7/3(SZ) basis sets (see for example refs. 17 and 43 for a discussion concerning this point) have been the reasons to include the CP correction in the present work.

Conclusions

By performing an extensive number of SCF calculations at the Hartree-Fock level, an *ab initio* analytical pair potential for the computation of the interaction energies between aliphatic amino acids (without sulphur) has been derived. Because of the complexity of the potential hypersurfaces for the interaction between pairs of amino acids, special attention has been focused on the generation of the conformations from which the parameters of the analytical pair potential have been obtained.

Since parameters for the interactions between amino acids and water, as well as for the interaction between water molecules are already available in the literature, the present-day pair potential library provides a powerful tool for dealing with a wide variety of problems involving molecules of biological interest.[44] In particular amino acids are the units from which proteins are built-up. In this regard, the determination of the 3-D structure of proteins is a most important topic that could be tackled by using the analytical pair potential developed in this paper. Some preliminary research is being conducted in order to test the appropriateness of our intermolecular pair potential for dealing with structural problems in proteins involving intramolecular interactions.

Acknowledgment

One of us (J.A.S) thanks the Ministerio de Educacion y Ciencia (Spain) for a fellowship as well as IBM-Kingston for its very generous hospitality and support. Portions of this contribution will be published in the *Journal of the American Chemical Society*.

Appendix

In this appendix we compile the analytical pair potentials obtained from *ab initio* calculations that have been published to date.

1. Clementi, E. and Popkie, H., J. Chem. Phys. **57**, 1078 (1972).

    Li(+) : Water

2. Popkie, H., Kistenmacher, H. and Clementi, E., J. Chem. Phys. **59**, 1325 (1973).

    Water : Water

3. Matsuoka, O., Clementi, E. and Yoshimine, M., J. Chem. Phys., **64**, 1351, (1976).

   Water : Water (CI level)

4. Clementi, E., Cavallone, F. and Scordamaglia R., J. Am. Chem. Soc., **99**, 5531, (1977).

   Amino acids : Water

5. Scordamaglia, R., Cavallone, F. and Clementi, E. J. Am. Chem. Soc., **99**, 5545, (1977).

   DNA Bases : Water

6. Matsuoka, O., Tosi, C., and Clementi, E., Biopolymers **17**, 33 (1978).

   Sugar : Phosphate : Sugar

7. Corongiu, G., and Clementi, E., Gazz. Chim. Ital. **108**, 687 (1978).

   Diethylphosphate ion : Water

8. Corongiu, G., Clementi, E., Gazz. Chim. Ital. **108**, 273 (1978).

   Sugar(Ribose derivative) : Water

9. Tosi, C., Clementi, E., and Matsuoka, O., Biopolymers **17**, 51 (1978).

   Sugar : Phosphate : Sugar (Intramolecular interactions)

10. Carozzo, L., Corongiu, G., Petrongolo, C., and Clementi, E., J. Chem. Phys. **68**, 787 (1978).

    Serine(zwitterion) : Water
    Glycine(zwitterion) : Water

11. Corongiu, G., Clementi, E., Pretsch, E., and Simon, W., J. Chem. Phys. **70**, 1266 (1979).

    Na(+) : Ether
    Na(+) : Thioether
    Na(+) : Amide

12. Ragazzi, M., Ferro, D.R., and Clementi, E., J. Chem. Phys. **70**, 1040, (1979).

    Formyl-triglycyl : Water

13. Clementi, E., Corongiu, G., and Lelj, F., J. Chem. Phys. **70**, 3726 (1979).

    Phosphate ion : Water

14. Corongiu, G., Clementi, E., Dagnino, M., and Paoloni, L., Chem. Phys. **40**, 439 (1979).

    4-OH-pyridine : Water
    4-pyridone : Water

15. Corongiu, G., Clementi, E., Pretsch, E., and Simon, W., J. Chem. Phys. **72**, 3096 (1980).

    Li(+) : Ether
    Li(+) : Thioether
    Li(+) : Amide

16. Clementi, E., Corongiu, G., Jonsson, B., and Romano, S., J. Chem. Phys. **72**, 260, (1980).

    Zn(2 +) : Water
    Zn(2 +) : CO2

17. Kolos, W., Ranghino, G., Clementi, E., and Novaro, O., Int. J. Quantum. Chem. **17**, 429 (1980).

    Methane : Methane (Dispersion included)

18. Clementi, E., Corongiu, G., and Ranghino, G., J. Chem. Phys. **74**, 578 (1981).

    Na(+) : Glutamate
    Na(+) : Aspartate
    Na(+) : Alanine

19. Pretsch, E., Neszmelyi, A., Simon, W., Corongiu, G., and Clementi, E., IBM Tech. Rep. POK-7 (1981).

    Ionophore : Li(+)
    Ionophore : Na(+)

20. Welti, M., Pretsch, E., Clementi, E., and Simon, W., Helv. Chim. Acta **65**, 1996 (1982).

    Ionophore : Ca(2 +)
    Ionophore : Mg(2 +)

21. Ranghino, G., Clementi, E., and Romano, S., Biopolymers **22**, 1449, (1983).

    Lysine : Water
    Arginine : Water
    Glutamate : Water
    Aspartate : Water

22. Corongiu, G., Fornili, S.L., and Clementi, E., Int. J. Quantum Chem.**10**, 277, (1983).

    Agarose : Water

23. Clementi, E. and Corongiu, G., Int. J. Quantum Chem.: Quantum Biology Symposium,**10**, 31, (1983).

    Water : Water (Three-body Interactions included)

24. Bolis, G., Clementi, E., Scheraga, H.A., Wertz, H., and Tosi, C., J. Am. Chem. Soc.**105**, 355, (1983).

    Methane : Water (Dispersion included)
    Methanol : Water (Dispersion included)

25. Fornili, S.L., Vercauteren, D.P., and Clementi, E., J. Biomolecular Struct. and Dynamics**1**, 1281, (1984).

    Gramicidin A : Water

26. Detrich, J., Corongiu, G. and Clementi, E., Chem. Phys. Lett., **112**, 426, (1984).

    Water : Water (Four-body Interactions included)

27. Kim, K.S., Vercauteren, D.P., Welti, M., Fornili, S.L., and Clementi, E., IBM Tech. Rep. POK-42 (1984).

    Gramicidin A : Na(+)

28. Kim, K.S., Vercauteren, D.P., Welti, M., Chin, S., and Clementi, E., Biophysical J.,**47**, 327, (1985).

    Gramicidin A : K(+)

References

1. E. Clementi, "Lecture Notes in Chemistry" (and references therein), **19**, Springer Verlag (1980).

2. A. Pullman, ed., "From Diatomics to Biopolymers," John Wiley & Sons (1978).
3. A. Pullman, Fortschr. Chem. Forsch., **31**, 45, (1972).
4. B. Melly and A. Pullman, C.R. Acad. Sci., **274**, 1371, (1972).
5. A. Momamy, R.F. McGuire, A.W. Burguess and H.A. Scheraga, J. Chem. Phys., **78**, 1595 (1974); *ibid.,* **79**, 2361, (1975); L.L. Shipman, A.W. Burgess and H.A. Scheraga, Proc. Natl. Acad. Sci. U.S.A., **72**, 543, 854 (1975).
6. W.L. Jorgensen and J. Gao, J. Am. Chem. Soc. (and references therein), **90**, 2174, (1986).
7. E. Clementi, F. Cavallone and J. Scordamaglia, J. Am. Chem. Soc., **99**, 5531 (1977); *ibid.* **99**, 5545 (1977).
8. G. Bolis and E. Clementi, J. Am. Chem. Soc., **99**, 5550 (1977).
9. S. Fraga, J. Comput. Chem., **3**, 329 (1982); Comput. Phys. Commun., **29**, 351 (1983).
10. J.A. Sordo, M. Klobukowski and S. Fraga, J. Am. Chem. Soc., **107**, 7569 (1985).
11. Romano, S. and Clementi, E., Gazz. Chim. Ital.,**108**, 319, 1978.
12. S. Romano and E. Clementi, Int. J. Quantum Chem., **14**, 839 (1978); *ibid.* **17**, 1007 (1980).
13. J. Šnir, R.A. Nemenoff and H.A. Scheraga, J. Chem. Phys., **82**, 2498 (1978).
14. O. Matsuoka, E. Clementi and M. Yoshimine, J. Chem. Phys., **64**, 1351 (1976).
15. E. Clementi and G. Corongiu, Int. J. Quantum Chem.: Quantum Biology Symposium, **10**, 31 (1983).
16. J. Detrich, G. Corongiu and E. Clementi, Chem. Phys. Lett., **112**, 426 (1984).
17. G. Bolis, E. Clementi, H.A. Scheraga, H. Wertz and C. Tosi, J. Am. Chem. Soc., **105**, 355 (1983).
18. C.A. Venanzi, H. Weinstein, G. Corongiu and E. Clementi, Int. J. Quantum Chem., **9**, 355 (1982).
19. E. Clementi, G. Corongiu, B. Jonsson and S. Romano, J. Chem. Phys., **72**, 260 (1980).
20. G. Corongiu, S.L. Fornili and E. Clementi, Int. J. Quantum Chem., **10**, 277 (1983).
21. G. Ranghino, E. Clementi and S. Romano, Biopolymers, **22**, 1449 (1983).
22. M. Ragazzi, D.R. Ferro and E. Clementi, J. Chem. Phys., **70**, 1040 (1979).
23. G. Bolis and E. Clementi, Chem. Phys. Letters, **82**, 147 (1981).
24. G. Bolis, G. Corongiu and E. Clementi, Chem. Phys. Letters, **86**, 299 (1982).
25. S.L. Fornili, D.P. Vercauteren and E. Clementi, J. Biomolecular Struct. and Dynamics, **1**, 1281 (1984).
26. K.S. Kim, D.P. Vercauteren, M. Welti, S. Chin and E. Clementi, Biophysical J., **47**, 327 (1985).
27. L. Carozzo, G. Corongiu, C. Petrongolo, J. Chem. Phys., **68**, 787 (1978).
28. E. Clementi and G. Corongiu, J. Chem. Phys., **72**, 3979 (1980).
29. E. Clementi and G. Corongiu, Biopolymers, **18**, 2431 (1979).
30. E. Clementi and G. Corongiu, Int. J. Quantum Chem., **16**, 897 (1979).
31. E. Clementi and G. Corongiu, Chem. Phys. Letters, **60**, 175 (1979).
32. E. Clementi and G. Corongiu, Biopolymers, **21**, 175 (1982).
33. E. Clementi and G. Corongiu, Proceedings of the second SUNYA Conversation in the Discipline Biomolecular Stereodynamics, Vol. I, R.H. Sharma, ed., Adenine Press, NY (1982).

34. K.S. Kim and E. Clementi, J. Am. Chem. Soc., **102**, 227 (1985).
35. W.J. Hehre, R.F. Stewart and J.A. Pople, J. Chem. Phys., **51**, 2657 (1969).
36. L. Gianolio, R. Pavani and E. Clementi, Gazz. Chim. Ital., **108**, 181 (1978).
37. M.J. Frisch, J.E. Del Bene, J.S. Binkley and HF Shaefer III, J. Chem. Phys., **84**, 2279 (1986).
38. D.W. Schwenke and D.G. Truhlar, J. Chem. Phys., **82**, 2418 (1985).
39. H.H. Greenwood and J.S. Plant, Int. J. Quantum Chem., **30**, 127 (1986).
40. M.M. Szczesniak and S. Scheiner, J. Chem. Phys., **84**, 6328 (1986).
41. R. Bonaccorsi, R. Cammi and J. Tomasi, Int. J. Quantum Chem., **29**, 373 (1986).
42. For an extended review on basis sets we refer to the very recent work by E.R. Davidson and D. Feller, Chem. Rev., **86,**, 681 (1986).
43. W. Kolos, Theoret. Chim. Acta (Berl.), **51**, 219 (1979).
44. E. Clementi, J. Phys. Chem., **89**, 4426 (1985).

Molecular Dynamics of Proteins

Martin Karplus
Department of Chemistry
Harvard University
Cambridge, MA 02138, U.S.A.

Dynamics of macromolecules of biological interest began in 1977 with the publication of a paper on the simulation of a small protein, the bovine pancreatic trypsin inhibitor.[1] Although the trypsin inhibitor is rather uninteresting from a dynamical viewpoint—its function is to bind to trypsin—experimental and theoretical studies of this model system—the "hydrogen atom" of protein dynamics—served to initiate explorations in this field.

Before focusing on dynamical studies of biomolecules, it is useful to set the field in perspective relative to the more general development of molecular dynamics. Molecular dynamics has followed two pathways which come together in the study of biomolecular dynamics. One of these, usually referred to as trajectory calculations, has been concerned primarily with the study of gas phase scattering of atoms and molecules. Much has been done in applying the classical trajectory method to a wide range of chemical reactions.[2,3] The classical studies have also been supplemented by semiclassical and quantum mechanical calculations since quantum effects sometimes play an important role.[3,4] Now the focus of trajectory studies is on more complex molecules, their redistribution of internal energy, and the role of this on their reactivity.

The other pathway in molecular dynamics has been concerned with physical rather than chemical interactions and with the thermodynamic and average dynamic properties of a large number of particles at equilibrium, rather than the detailed trajectories of a few particles. Although the basic ideas go back to van der Waals and Boltzmann, the modern area began with the work of Alder and Wainright on hard sphere liquids in the late 1950s.[5] The paper by Rahman[6] in 1964, on a molecular dynamics simulation of liquid argon with a soft sphere (Lennard-Jones) potential represented an important next step. Simulations of complex liquids followed; the now classic study of liquid water by Stillinger and Rahman was published in 1974.[7] Since then, there have been many studies on the equilibrium and nonequilibrium behavior of a wide range of systems.[8,9]

This background set the stage for the development of molecular dynamics of biomolecules. The size of an individual molecule, composed of five hundred or more atoms for even a small protein, is such that its simulation in isolation can determine approximate equilibrium properties, as in the molecular dynamics of fluids, though detailed aspects of the atomic motions are of considerable interest, as in trajectory calculations. A basic assumption in initiating these studies was that potential functions could be constructed which were sufficiently accurate to give meaningful results for systems as complex as proteins or nucleic acids. In addition, it was necessary to assume that for such inhomogeneous systems, in contrast to the homogeneous character of even "complex" liquids like water, simulations of an attainable time scale (10 ps in the initial studies) could provide a useful sample of the phase space in the neighborhood of the native structure. For neither of these assumptions was there strong supporting evidence. Nevertheless, it seemed worthwhile in 1975 to apply the techniques of molecular dynamics with the available potential functions to the internal motions of proteins with known crystal structures.[1,10]

The most important consequence of the first simulations of biomolecules was that they introduced a conceptual change. Although to chemists and physicists it is self-evident that polymers like proteins and nucleic acids undergo significant fluctuations at room temperature, the classic view of such molecules in their native state had been static in character. This followed from the dominant role of high-resolution x-ray crystallography in providing structural information for these complex systems. The remarkable detail evident in crystal structures led to an image of biomolecules with every atom fixed in place. D. C. Phillips, who determined the first enzyme crystal structure, wrote recently "the period 1965 − 75 may be described as the decade of the rigid macromolecule. Brass models of DNA and a variety of proteins dominated the scene and much of the thinking."[11] Molecular dynamics simulations have been instrumental in changing the static view of the structure of biomolecules to a dynamic picture. It is now recognized that the atoms of which biopolymers are composed are in a state of constant motion at ordinary temperatures. The x-ray structure of a protein provides the average atomic positions, but the atoms exhibit fluid-like motions of sizable amplitudes about these averages. Crystallographers have acceded to this viewpoint and have come so far as to sometimes emphasize the parts of a molecule they do not see in a crystal structure as evidence of motion or disorder.[12] The new understanding of protein dynamics subsumes the static picture in that use of the average positions still allows discussion of many aspects of biomolecule function in the language of structural chemistry. However, the recognition of the importance of fluctuations opens the way for more sophisticated and accurate interpretations.

Simulation studies in this area, as in others, have the possibility of providing the ultimate detail concerning motional phenomena. The primary limitation of simulation methods is that they are approximate. It is here that experiment plays an essential role in validating the simulation methods; that is, comparisons with experimental data can serve to test the accuracy of the calculated results and to provide criteria for improving the methodology. When experimental comparisons indicate that the simulations are meaningful, their capacity for providing detailed results often makes it possible to examine specific aspects of the atomic motions far more easily than by making measurements.

At the present stage of the molecular dynamics of biomolecules, there is a general understanding of the motion that occurs on a subnanosecond time scale; that is, the types of motion have been demonstrated, their characteristics evaluated and the important factors determining their properties delineated. Simulation methods have shown that the structural fluctuations are sizable; particularly large fluctuations are found where steric constraints due to molecular packing are small (e.g., in the exposed sidechains and external loops), but substantial mobility is also found in the interior of a macromolecule. Local atomic displacements in the interior are correlated in a manner that tends to minimize disturbances of the global structure. This leads to fluctuations larger than would be permitted in a rigid protein matrix.

For motions on a longer time scale, our understanding is more limited. When the motion of interest can be described in terms of a reaction path (e.g., hinge-bending, local activated events), methods exist for examining the nature and the rate of the process. However, for the motions that are slow due to their complexity and involve large-scale structural changes, extensions of the available approaches are required. Harmonic and simplified model dynamics, as well as reaction-path calculations, can provide information on some of the slower motions, such as opening fluctuations and helix-coil transitions.

In applying molecular dynamics to physical studies of biomolecules (such as x-ray, nuclear magnetic resonance, infra-red, Raman, inelastic neutron scattering, fluorescence depolarization, and so on), a number of aspects are important. There is of course the direct comparison between the results of calculations and the experimental data. More interesting is the possibility of extending the interpretation of experiments. Also, experimental data can be generated by simulations and analyzed as would real data to test the method used. Finally, new effects can be predicted from the simulation as a stimulus for additional experimental investigations.

In what follows, applications of molecular dynamics that illustrate each of these points are outlined.

A Test: X-Ray Diffraction

Since atomic fluctuations are the basis of protein dynamics, it is important to have experimental tests of the accuracy of the simulation results concerning them. For the magnitudes of the motions, the most detailed data are provided, in principle, by an analysis of the Debye-Waller or temperature factors obtained in crystallographic refinements of x-ray structures.

It is well known from small molecule crystallography that the effects of thermal motion must be included in the interpretation of the x-ray data to obtain accurate structural results. Detailed models have been introduced to take account of anisotropic and anharmonic motions of the atoms and these models have been applied to high resolution data for small molecules.[13] In protein crystallography, the limited data available relative to the large number of parameters that have to be determined, have made it necessary to assume that the atomic motions are isotropic and harmonic. In that case the structure factor, $F(Q)$, which is related to the measured intensity by $I(Q) = |F(Q)|^2$, is given by

$$F(Q) = \sum_{j=1}^{N} f_j(Q) \, e^{iQ \cdot <r_j>} \, e^{W_j(Q)} \tag{1}$$

where Q is the scattering vector, $<r_j>$ is the average position of atom j with atomic scattering factor $f_j(Q)$ and the sum is over the N atoms in the asymmetric unit of the crystal. The Debye-Waller factor, $W_j(Q)$, is defined by

$$W_j(Q) = -\frac{8}{3}\pi^2 <\Delta r_j^2> s^2 = -B_j s^2 \tag{2}$$

where $s = |Q|/4\pi$. The quantity B_j is usually referred to as the temperature factor, which is directly related to the mean-square atomic fluctuations in the isotropic harmonic model. More generally, if the motion is harmonic but anisotropic, a set of six parameters

$$B_j^{xx} = <\Delta x_j^2>, \ B_j^{xy} = <\Delta x_j \Delta y_j>, \dots B_j^{z} = <\Delta z_j^2>$$

is required to fully characterize the atomic motion. Although in the earlier x-ray studies of proteins, the significance of the temperature factors was ignored (presumably because the data were not at a sufficient level of resolution and accuracy), more recently attempts have been made to relate the observed temperature factors to the atomic motions. In principle, the temperature factors provide a very detailed measure of the motions because information is available for the mean square fluctuation of each heavy atom. In practice, there are two types of difficulties in relating the B factors obtained from protein refinements to the atomic motions. The first is that, in addition to thermal fluctuations, any static (lattice) disorder in the crystal contributes to the B factors; i.e., since a crystal is made up of many unit cells, different molecular geometries in the various cells have the same effect on the average electron density, and therefore the B factor, as atomic motions. In only one case, the iron atom of myoglobin, has there been an experimental attempt to determine the disorder contribution.[14] Since the Mossbauer effect is not altered by static disorder (i.e., each nucleus absorbs independently) but does depend on atomic motions, comparisons of Mossbauer and x-ray data have been used to estimate a disorder contribution for the iron atom; the value obtained is

$$<\Delta r_{Fe}^2> = 0.08 \ \text{Å}^2$$

Although the value is only approximate, it nevertheless indicates that the observed B factors (e.g., on the order of 0.44 Å² for backbone atoms and 0.50 Å² for sidechain atoms) are dominated by the motional contribution. Most experimental B factor values are compared directly with the molecular dynamics results (i.e., neglecting the disorder contribution) or are rescaled by a constant amount (e.g., by setting the smallest observed B factor to zero) on the assumption that the disorder contribution is the same for all atoms.[15]

Second, since simulations have shown that the atomic fluctuations are highly anisotropic and, in some cases, anharmonic, it is important to determine the errors introduced into the refinement process by the assumption of isotropic and harmonic motion. A direct experimental estimate of the errors is difficult because sufficient data are not yet available for protein crystals. Moreover, any data set includes other errors which would obscure the analysis. As an alternative to an experimental analysis of the errors in the refinement of proteins, a purely theoretical approach can be used.[16] The basic idea is to generate x-ray data from a molecular dynamics simulation of a protein and to use these data in a standard refinement procedure. The error in the analysis can then be determined by comparing the refined x-ray structure and temperature factors with the average structure and the mean square fluctuations from the simulation. Such a comparison, in which no real experimental results are used, avoids problems due to inaccuracies in the measured data (exact calculated intensities are used), to crystal disorder (there is none in the model), and to approximations in the simulation (the simulation is exact for this case). The only question about such a comparison is whether the atomic motions found in the simulation are a meaningful representation of those occurring in proteins. As has been shown,[1,15,17] molecular dynamics simulations provide a reasonable picture of the motions in spite of errors in the potentials, the neglect of the crystal environment and the finite time classical trajectories used to obtain the results. However, these inaccuracies do not affect the exactitude of the computer "experiment" for testing the refinement procedure that is described below.

In this study,[16] a 25 ps molecular dynamics trajectory for myoglobin was used.[18] The average structure and the mean square fluctuations from that structure were calculated directly from the trajectory. To obtain the average electron density, appropriate atomic electron distributions were assigned to the individual atoms and the results for each coordinate set were averaged over the trajectory. Given the symmetry, unit cell dimensions and position of the myoglobin molecule in the unit cell, average structure factors, $< F(Q) >$, and intensities, $I(Q) = | < F(Q) > |^2$, were calculated from the Fourier transform of the average electron density, $< \rho(r) >$, as a function of position r in the unit cell. Data were generated at 1.5 Å resolution, as this is comparable to the resolution of the best x-ray data currently available for proteins the size of myoglobin.[19,20] The resulting intensities at Bragg reciprocal lattice points were used as input data for the widely applied crystallographic program, PROLSQ.[21] The time-averaged atomic positions obtained from the simulation and a uniform temperature factor provide the initial model for refinement. The positions and an isotropic, harmonic temperature factor for each atom were then refined iteratively against the computer generated intensities in the standard way. Differences between the refined results for the average atomic positions and their mean square fluctuations and those obtained from the molecular dynamics trajectory are due to errors introduced by the refinement procedure.

The overall rms error in atomic positions ranged from 0.24 Å to 0.29 Å for slightly different restrained and unrestrained refinement procedures.[16] The errors in backbone positions (0.10 − 0.20 Å) are generally less than those for sidechain atoms (0.28 − 0.33 Å); the largest positional errors are on the order of 0.6 Å. The backbone errors, though small, are comparable to the rms deviation of 0.21 Å between the positions of the backbone atoms in the refined experimental structures of oxymyoglobin and carboxy myoglobin.[19,20] Further, the positional errors are not uniform over the whole structure. There is a strong correlation between the positional error and the magnitude of the mean square fluctuation for an atom, with certain regions of the protein, such as loops and external sidechains, having the largest errors.

The refined mean square fluctuations are systematically smaller than the fluctuations calculated directly from the simulation. The magnitudes and variation of temperature factors along the backbone are relatively well reproduced, but the refined sidechain fluctuations are almost always significantly smaller than the actual values. The average backbone B factors from different refinements are in the range 11.3 to 11.7 Å^2, as compared with the exact value of 12.4 Å^2; for the sidechains, the refinements yield 16.5 to 17.6 Å^2, relative to the exact value of 26.8 Å^2. Regions of the protein that have high mobility have large errors in temperature factors, as well as in the positions. Examination of all atoms shows that fluctuations greater than about 0.75 Å^2 (B = 20 Å^2) are almost always underestimated by the refinement. Moreover, while actual mean square atomic fluctuations have values as large as 5 Å^2, the x-ray refinement leads to an effective upper limit of about 2 Å^2. This arises from the fact that most of the atoms with large fluctuations have multiple conformations and that the refinement procedure picks out one of them.

To do refinements that take some account of anisotropic motions for all but the smallest proteins, it has been necessary to introduce assumptions concerning the nature of the anisotropy. One possibility is to assume anisotropic rigid body motions for sidechains such as tryptophan and phenylalanine.[22,23] An alternative is to introduce a "dictionary" in which the orientation of the anisotropy tensor is related to the stereochemistry around each atom;[21] this reduces the six independent parameters of the anisotropic temperature factor tensor $\underline{B}_j$ to three parameters per atom. Analysis of a simulation for BPTI[24] has shown that the actual anisotropies in the atomic motions are generally not simply related to the local stereochemistry; an exception is the mainchain carbonyl oxygen which has its largest motion perpendicular the C = 0 bond. Thus, use of stereochemical assumptions in the refinement can yield incorrectly oriented anisotropy tensors and significantly reduced values for the anisotropies. The large scale motions of atoms are collective and sidechains tend to move as a unit so that the directions of largest motion are not related to the local bond direction, and have similar orientations in the different atoms forming a group that is undergoing correlated motions. This means that it is necessary to use the full anisotropy tensor to obtain meaningful results. To do so requires proteins that are particularly well ordered so that the diffraction data extend to better than 1 Å resolution.

An Extension: Nuclear Magnetic Resonance

Nuclear magnetic resonance (NMR) is an experimental technique that has played an essential role in the analysis of the internal motions of proteins.[10,25] Like x-ray diffraction, it can provide information about individual atoms; unlike x-ray diffraction, NMR is sensitive not only to the magnitude but also to the time scale of the motions. Nuclear relaxation processes are dependent on atomic motions on the nanosecond to picosecond time scale. Although molecular tumbling is generally the dominant relaxation mechanism for proteins in solution internal motions contribute as well; for solids, the internal motions are of primary importance. In addition, NMR parameters, such as nuclear spin-spin coupling constants and chemical shifts, depend on the protein environment. In many cases different local conformations exist but the interconversion is rapid on the NMR time scale, here on the order of milliseconds, so that average values are observed. When the interconversion time is on the order of the NMR time scale or slower, the transition rates can be studied; an example is provided by the reorientation of aromatic rings.[26,27]

In addition to supplying data on the dynamics of proteins, NMR can also be used to obtain structural information. With recent advances in techniques it is now possible to obtain a large number of approximate interproton distances for proteins by the use of nuclear Overhauser effect measurements.[28] If the protein is relatively small and has a well resolved spectrum, a large portion of the protons can be assigned and several hundred distances for these protons can be determined by the use of two-dimensional NMR techniques.[29] Clearly these distances can serve to provide structural information for proteins, analogous to their earlier use for organic molecules.[28,30] Of great interest is the possibility that enough distance information can be measured to actually determine the high resolution structure of a protein in solution to supplement results from crystallography, particularly for proteins that are difficult to crystallize. In what follows we consider two questions related to this possibility. The first concerns the effect of motional averaging on the accuracy of the apparent distances obtained from the NOE studies and the second, whether the number of distances that can be obtained experimentally are sufficient for a structure determination.

For spin-lattice relaxation, such as observed in nuclear Overhauser effect (NOE) measurements, it is possible to express the behavior of the magnetization of the nuclei being studied by the equation[31,32]

$$\frac{d(I_z(t) - I_o)_i}{dt} = -\rho_i(I_z(t) - I_o)_i - \sum_{i \neq j} \sigma_{ij}(I_z(t) - I_o)_j \tag{3}$$

where $I_z(t)_i$ and I_{oi} are the z components of the magnetization of nucleus i at time t and at equilibrium, ρ_i is the direct relaxation rate of nucleus i, and σ_{ij} is the cross relaxation rate between nuclei i and j. The quantities ρ_i and σ_{ij} can be expressed in terms of spectral densities

$$\rho_i = \frac{6\pi}{5} \gamma_i^2 \gamma_j^2 \hbar^2 \sum_{i \neq j} \left[1/3 J_{ij}(\omega_i - \omega_j) + J_{ij}(\omega_i) + 2J_{ij}(\omega_i + \omega_j) \right] \tag{4}$$

$$\sigma_{ij} = \frac{6\pi}{5} \gamma_i^2 \gamma_j^2 \hbar^2 \left[2J_{ij}(\omega_i + \omega) - 1/3 J_{ij}(\omega_i - \omega_j) \right] \tag{5}$$

The spectral density functions can be obtained from the correlation functions for the relative motions of the nuclei with spins i and j,[31,33]

$$J_{ij}^n(\omega) = \int_0^\infty \frac{< Y_n^2(\Theta_{lab}(t)\phi_{lab}(t)) Y_n^{2*}(\Theta_{lab}(0)\phi_{lab}(0)) >}{r_{ij}^3(0) r_{ij}^3(t)} \cos{(\omega t)} \, dt \tag{6}$$

where $Y_n^2 (\Theta(t)\phi(t))$ are second-order spherical harmonics and the angular brackets represent an ensemble average which is approximated by an integral over the molecular dynamics trajectory. The quantities $\Theta_{lab}(t)$ and $\phi_{lab}(t)$ are the polar angles at time t of the internuclear vector between protons i and j with respect to the external magnetic field and r_{ij} is the interproton distance. In the simplest case of a rigid molecule undergoing isotropic tumbling with a correlation time τ_o this reduces to the familiar expression

$$J_{ij}(\omega) = \frac{1}{4\pi r_{ij}^6} \left[\frac{\tau_o}{1 + (\omega\tau_o)^2} \right] \tag{7}$$

The nuclear Overhauser effect (NOE) corresponds to the selective enhancement of a given resonance by the irradiation of another resonance in a dipolar coupled spin system. Of particular interest for obtaining motional and distance information are measurements that provide time dependent NOE's from which the cross relaxation rates σ_{ij} (see Eq. 5) can be determined directly or indirectly by solving a set of coupled equations (Eqs. 3−5). Motions on the picosecond time scale are expected to introduce averaging effects that decrease the cross relaxation rates by a scale factor relative to the rigid model. A lysozyme molecular dynamics simulation[34] has been used to calculate dipole vector correlation functions[31] for proton pairs that have been studied experimentally.[35,36] Four proton pairs on three sidechains (Trp 28, Ile 98, and Met 105) with very different motional properties were examined. Trp 28 is quite rigid, Ile 98 has significant fluctuations, and Met 105 is particularly mobile in that it jumps among different sidechain conformations during the simulation. The rank order of the scale factors (order parameters) is the same in the theoretical and experimental results. However, although the results for the Trp 28 protons agree with the measurements to within the experimental error, for both Ile 98 and Met 105 the motional averaging found from the NOE's is significantly greater than the calculated value. This suggests that these residues are undergoing rare fluctuations involving transitions that are not adequately sampled by the simulation.

If nuclear Overhauser effects are measured between pairs of protons whose distance is not fixed by the structure of a residue, the strong distance dependence of the cross relaxation rates (i.e., $1/r^6$) can be used to obtain estimates of the interproton distances.[29,35-37] The simplest application of this approach is to assume that proteins are rigid and tumble isotropically. The lysozyme molecular dynamics simulation was used to determine whether picosecond fluctuations are likely to introduce important errors into such an analysis.[31] The results show that the presence of the motions will cause a general decrease in most NOE effects observed in a protein. However, because the distance depends on the sixth root of the observed NOE, motional errors of a factor of two in the latter lead to only a 12% uncertainty in the distance. Thus, the decrease is usually too small to produce a significant change in the distance estimated from the measured NOE value. This is consistent with the excellent correlation found between experimental NOE values and those calculated using distances from a crystal structure.[36] Specific NOE's can, however, be altered by the internal motions to such a degree that the effective distances obtained are considerably different from those predicted for a static structure. Such possibilities must, therefore, be considered in any structure determination based on NOE data. This is true particularly for distances involving averaging over large scale fluctuations.

Because of the inverse sixth power of the NOE distance dependence, experimental data so far are limited to protons that are separated by less than 5 Å. Thus, the long-range information required for a direct protein structure determination is not available. To overcome this limitation it is possible to introduce additional information provided by empirical energy functions.[38] One way of proceeding is to do molecular dynamics simulations with the approximate interproton distances introduced as restraints in the form of skewed biharmonic potentials[37,39] with the force constants chosen to correspond to the experimental uncertainty in the distance.

A model study of the small protein crambin[39] was made with realistic NOE restraints. Two hundred forty approximate interproton distances less than 4 Å were used, including 184 short-range distances (i.e., connecting protons in two residues that were less than 5 residues apart in the sequence) and 56 long-range distances. The molecular dynamics simulations converged to the known crambin structure from different initial extended structures. The average structure obtained from the simulations with a series of different protocols had rms deviations of 1.3Å for the backbone atoms, and 1.9Å for the sidechain atoms. Individual converged simulations had rms deviations in the range 1.5 to 2.1Å and 2.1 to 2.8Å for the backbone and sidechain atoms, respectively. Further, it was shown that a dynamics structure with significantly larger deviations (5.7Å) could be characterized as incorrect, independent of a knowledge of the crystal structure because of its higher energy and the fact that the NOE restraints were not satisfied within the limits of error. The incorrect structure resulted when all NOE restraints were introduced simultaneously, rather than allowing the dynamics to proceed first in the presence of only the short-range restraints followed by introduction of the long-range restraints. Also of interest is the fact that although crambin has three disulfide bridges it was not necessary to introduce information concerning them to obtain an accurate structure.

The folding process as simulated by the restrained dynamics is very rapid. At the end of the first 2 ps the secondary structure is essentially established while the molecule is still in an extended conformation. Some tertiary folding occurs even in the absence of long-range restraints. When they are introduced, it takes about 5 ps to obtain a tertiary structure that is approximately correct and another 6 ps to introduce the small adjustments required to converge to the final structure.

It is of interest to consider whether the results obtained in the restrained dynamics simulation have any relation to actual protein folding. That correctly folded structures are achieved only when the secondary structural elements are at least partly formed before the tertiary restraints are introduced is suggestive of the diffusion-collision model of protein folding.[40] Clearly, the specific pathway has no physical meaning since it is dominated by the NOE restraints. Also, the time scale of the simulated folding process is approximately 12 orders of magnitude faster than experimental estimates. About 6 to 8 orders of magnitude of the rate increase are due to the fact that the secondary structure is stable once it is formed, in contrast to a real protein where the secondary structural elements spend only a small fraction of time in the native conformation until coalescence has occurred. The remainder of the artificial rate increase presumably arises from the fact that the protein follows a single direct path to the folded state in the presence of the NOE restraints, instead of having to go through a complex search process.

A Prediction: Structural Role of Active Site Waters in Ribonuclease A

To achieve a realistic treatment of the solvent-accessible active sites, a molecular dynamics simulation method, called the stochastic boundary method, has been implemented.[41-43] It makes possible the simulation of a localized region, approximately spherical in shape, that is composed of the active site with or without ligands, the essential portions of the protein in the neighborhood of the active site, and the surrounding solvent. The approach provides a simple and convenient method for reducing the total number of atoms included in the simulation, while avoiding spurious edge effects.

The stochastic boundary method for solvated proteins starts with a known x-ray structure; for the present problem the refined high-resolution (1.5 to 2 Å) x-ray structures provided by G. Petsko and co-workers were used.[44,45] The region of interest (here the active site of ribonuclease A) was defined by choosing a reference point (which was taken at the position of the phosphorus atom in the CpA inhibitor complex) and constructing a sphere of 12 Å radius around this point. Space within the sphere not occupied by crystallographically determined atoms was filled by water molecules, introduced from an equilibrated sample of liquid water. The 12 Å sphere was further subdivided into a reaction region (10 Å radius) treated by full molecular dynamics and a buffer region (the volume between 10 and 12 Å) treated by Langevin dynamics, in which Newton's equations of motion for the non-hydrogen atoms are augmented by a frictional term and a random-force term; these additional terms approximate the effects of the neglected parts of the system and permit energy transfer in and out of the reaction region. Water molecules diffuse freely between the reaction and buffer regions but are prevented from escaping by an average boundary force.[42] The protein atoms in the buffer region are constrained by harmonic forces derived from crystallographic temperature factors.[43] The forces on the atoms and their dynamics were calculated with the CHARMM program;[38] the water molecules were represented by the ST2 model.[46]

One of the striking aspects of the active site of ribonuclease is the presence of a large number of positively charged groups, some of which may be involved in guiding and/or binding the substrate.[47] The simulation demonstrated that these residues are stabilized in the absence of ligands by well-defined water networks. A particular example includes Lys-7, Lys-41, Lys-66, Arg-39 and the doubly protonated His-119. Bridging waters, some of which are organized into trigonal bipyramidal structures, were found to stabilize the otherwise very unfavorable configuration of near-neighbor positive groups because the interaction energy between water and the charged $C\text{-}NH_n^+$ (n = 1, 2, or 3) moieties is very large; e.g., at a donor-acceptor distance of 2.8 Å, the $C\text{-}NH_3^+ - H_2O$ energy is -19 kcal/mol with the empirical potential used for the simulation,[38] in approximate agreement with accurate quantum mechanical calculations[48] and gas-phase ion-molecule data.[49] The average stabilization energy of the charged groups (Lys-7, Lys-41, Lys-66, Arg-39, and His-119) and the 106 water molecules included in the simulation is -376.6 kcal/mol. This energy is calculated as the difference between the simulated system and a system composed of separate protein and bulk water. Unfavorable protein-protein charged-group interactions are balanced by favorable water-protein and water-water interactions. The average energy per molecule of pure water from an equivalent stochastic boundary simulation[42] was -9.0 kcal/mol, whereas that of the waters included in the active site simulation was -10.2 kcal/mol; in the latter a large contribution to the energy came from the interactions between the water molecules and the protein atoms. It is such energy differences that are essential to a correct evaluation of binding equilibria and the changes introduced by site-specific mutagenesis.[50]

During the simulation, the water molecules involved in the charged-group interactions oscillated around their average positions, generally without performing exchange. On a longer time scale, it is expected that the waters would exchange and that the sidechains would undergo larger-scale displacements. This is in accord with the disorder found in the x-ray results for lysine and arginine residues (e.g., Lys-41 and Arg-39),[44,51] a fact that makes difficult a crystallographic determination of the water structure in this case. It is also of interest that Lys-7 and Lys-41 have an average separation of only 4 Å in the simulation, less than that found in the x-ray structure. That this like charged pair can exist in such a configuration is corroborated by experiments that have shown that the two lysines can be cross-linked;[52] the structure of this compound has been reported recently[53] and is similar to that found in the native protein.

In addition to the role of water in stabilizing the charged groups that span the active site and participate in catalysis, water molecules make hydrogen bonds to protein polar groups that become involved in ligand binding. A particularly clear example is provided by the adenine-binding site in the CpA simulation. The NH_2

group of adenine acted as a donor, making hydrogen bonds to the carbonyl of Asn-67, and the ring N^{1A} acted as an acceptor for a hydrogen bond from the amide group of Glu-69. Corresponding hydrogen bonds were present in the free ribonuclease simulation, with appropriately bound water molecules replacing the substrate. These waters, and those that interact with the pyrimidine-site residues Thr-45 and Ser-123, help to preserve the protein structure in the optimal arrangement for binding. Similar substrate "mimicry" has been observed in x-ray structures of lysozyme[54] and of penicillopepsin,[55] but has not yet been seen in ribonuclease.

Acknowledgments

I gratefully acknowledge the work done by many collaborators in the research described in this review. Their essential contributions are made clear by the citations in the reference list. This contribution is essentially the same as one submitted to a special issue of the *Israel Journal of Chemistry*.

References

1. J.A. McCammon, B.R. Gelin, and M. Karplus, Nature, **267**, 585 (1977).
2. R.N. Porter, Ann. Rev. Phys. Chem., **25**, 371 (1974).
3. R.B. Walker and J.C. Light, Ann. Rev. Phys. Chem., **31**, 401 (1980).
4. G.C. Schatz and A. Kuppermann, J. Chem. Phys., **62**, 2502 (1980).
5. B.J. Alder and T.E. Wainright, J. Chem. Phys., **31**, 459 (1959).
6. A. Rahman, Phys. Rev., **A136**, 405 (1964).
7. F.H. Stillinger and A. Rahman, J. Chem. Phys., **60**, 1545 (1974).
8. W.W. Wood and J.J. Erpenbeck, Ann. Rev. Phys. Chem., **27**, 319 (1976).
9. W.G. Hoover, Ann. Rev. Phys. Chem., **34**, 103 (1983).
10. For early reviews of experimental and theoretical developments, see F.R.N. Gurd and J.M. Rothgeb, Adv. Prot. Chem., **33**, 73 (1979); and M. Karplus and J.A. McCammon, CRC Crit. Rev. Biochem., **9**, 293 (1981).
11. D.C. Phillips in *Biomolecular Stereodynamics*, ed. R.H. Sarma, Adenine, New York, 1981, p. 497.
12. M. Marquart, J. Deisendorfer, R. Huber, and W. Palm, J. Mol. Biol., **141**, 369 (1980).
13. U.H. Zucker and H. Schulz, Acta Cryst., **A38**, 563 (1982).
14. H. Hartmann, F. Parak, W. Steigemann, G.A. Petsko, D.R. Ponzi, and H. Frauenfelder, Proc. Natl. Acad. Sci. USA, **79**, 4967 (1982).
15. G.A. Petsko and D. Ringe, Ann. Rev. Biophys. & Bioeng., **13**, 331 (1984).
16. J. Kuriyan, G.A. Petsko, R.M. Levy, and M. Karplus, J. Mol. Biol. (in press).
17. M. Karplus and J.A. McCammon, Ann. Rev. Biochem., **52**, 263 (1983).
18. R.M. Levy, R.P. Sheridan, J.W. Keepers, G.S. Dubey, S. Swaminathan, and M. Karplus, Biophys. J., **48**, 509 (1985).
19. J. Kuriyan, G.A. Petsko, and M. Karplus, J. Mol. Biol. (in press).
20. S.E.V. Phillips, J. Mol. Biol., **142**, 531 (1980).
21. J.H. Konnert and W.A. Hendrickson, Acta Cryst., **A36**, 344 (1980).

22. I. Glover, I. Haneef, J. Pitts, S. Wood, D. Moss, I. Tickle, and T. Blundell, Biopolymers, **22**, 293 (1983).

23. P.J. Artymiuk, C.C.F. Blake, D.E.P. Grace, S.J. Oatley, D.C. Phillips, and N.J.E. Sternberg, Nature, **280**, 563 (1979).

24. H. Yu, M. Karplus, and W.A. Hendrickson, Acta Cryst., **B41**, 191 (1985).

25. I.D. Campbell, C.M. Dobson, and R.J.P. Williams, Adv. Chem. Phys., **39**, 55 (1978).

26. I.D. Campbell, C.M. Dobson, G.R. Moore, S.J. Perkins, and R.J.P. Williams, FEBS Lett., **70**, 96 (1976).

27. G. Wagner, A. DeMarco, and K. Wuthrich, Biophys. Struct. Mech., **2**, 139 (1976).

28. J.H. Noggle and R.E. Schirmer, *The Nuclear Overhauser Effect*, Academic Press, New York, 1971.

29. G. Wagner and K. Wuthrich, J. Mol. Biol., **160**, 343 (1982).

30. B. Honig, B. Hudson, B.D. Sykes, and M. Karplus, Proc. Natl. Acad. Sci. USA, **68**, 1289 (1971).

31. E.T. Olejniczak, C.M. Dobson, M. Karplus, and R.M. Levy, J. Am. Chem. Soc., **106**, 1923 (1984).

32. I. Solomon, Phys. Rev., **99**, 559 (1955).

33. R.M. Levy, M. Karplus, and P.G. Wolynes, J. Am. Chem. Soc., **103**, 5998 (1981).

34. T. Ichiye, B. Olafson, S. Swaminathan, and M. Karplus, Biopolymers (in press).

35. E.T. Olejniczak, F.M. Poulsen, and D.M. Dobson, J. Am. Chem. Soc., **103**, 6574 (1981).

36. F.M. Poulsen, J.C. Hoch, and C.M. Dobson, Biochemistry, **19**, 2597 (1980).

37. G.M. Clore, A.M. Gronenborn, A.T. Brunger, and M. Karplus, J. Mol. Biol., **186**, 435 (1985).

38. B.R. Brooks, R.E. Bruccoleri, B.D. Olafson, D.J. States, S. Swaminathan, and M. Karplus, J. Comp. Chem., **4**, 187 (1983).

39. A.T. Brunger, G.M. Clore, A.M. Gronenborn, and M. Karplus, Proc. Natl. Acad. Sci. USA, **83**, 380 (1986).

40. D. Bashford, D.L. Weaver, and M. Karplus, J. Biomol. Struct. Dyn., **1**, 1243 (1984).

41. C.L. Brooks and M. Karplus, J. Chem. Phys., **79**, 6312 (1983).

42. A. Brunger, C.L. Brooks, and M. Karplus, Chem. Phys. Lett., **105**, 495 (1984).

43. C.L. Brooks, A. Brunger, and M. Karplus, Biopolymers, **24**, 843 (1985).

44. W.A. Gilbert, A.L. Fink, and G.A. Petsko, Biochemistry (in press).

45. R.L. Campbell and G.A. Petsko, Biochemistry (in press).

46. F.H. Stillinger and A. Rahman, J. Chem. Phys., **60**, 1545 (1974).

47. J.B. Matthew and F.M. Richards, Biochemistry, **21**, 4989 (1982).

48. D.J. Desmeules and L.C. Allen, J. Chem. Phys., **72**, 4731 (1980).

49. P. Kebarle, Ann. Rev. Phys. Chem., **28**, 445 (1977).

50. A.R. Fersht, J.-P. Shi, J. Knill-Jones, D.M. Lowe, A.J. Wilkinson, D.M. Blowq, P. Brick, P. Carter, M.M.Y. Waye, and G. Winter, Nature (London), **314**, 235 (1985).

51. A. Wlodawer in *Biological Macromolecules and Assemblies: Volume 2, Nucleic Acids and Interactive Proteins*, eds., F.A. Jurnak and A. McPherson, Wiley, New York, 1985, p. 394.

52. P.S. Marfey, M. Uziel, and J. Little, J. Biol. Chem., **240**, 3270 (1965).

53. P.C. Weber, F.R. Salemme, S.H. Lin, Y. Konishi, and H.A. Scheraga, J. Mol. Biol., **181**, 453 (1985).

54. C.C.F. Blake, W.C.A. Pulford, and P.J. Artymiuk, J. Mol. Biol., **167**, 693 (1983).

55. M.N.G. James and A.R. Sielecki, J. Mol. Biol., **163**, 299 (1983).

Proton NMR Studies of Protein Dynamics and Folding: Applications of Magnetization Transfer NMR

Christopher M. Dobson and Philip A. Evans
Inorganic Chemistry Laboratory
University of Oxford
South Parks Road
Oxford OX1 3QR

Abstract

When a protein exists at equilibrium in more than one conformational state it may be possible to observe separately in the NMR spectrum resonances corresponding to the different states. Provided that interconversion between these states occurs at suitable rates, magnetization transfer techniques may be used to detect it, and in favorable cases to obtain rate constants for specific conformational transitions. Results of one- and two-dimensional ^{1}H NMR experiments with lysozyme and staphylococcal nuclease are used to illustrate the potentials of such an approach to studying the dynamics and folding of proteins.

NMR and Protein Structure

NMR provides an experimental basis for investigating in considerable detail the structures and dynamics of biological molecules in solution.[1] Through theoretical studies and analysis of the spectra of molecules of known structure parameters obtained in an NMR experiment can often be related to specific internuclear distances and angles. Of particular importance in this regard are nuclear Overhauser enhancement effects (NOEs) which depend on dipolar coupling between nuclear spins and are strongly dependent on the distances between spins.[2] NOEs between specific pairs of protons can be observed either by intensity changes in selective irradiation experiments or by observation of cross-peaks in a suitable two-dimensional experiment.[3] Also of major importance are spin-spin coupling constants which can be detected either from the splitting of resonances into multiplets or in the appearance of cross-peaks in two-dimensional correlated NMR experiments.[3] The coupling is transmitted via the electrons of chemical bonds, and the magnitude falls off rapidly with the number of bonds involved. Three-bond couplings in ^{1}H NMR spectroscopy are of primary importance because these can be related directly to bond torsion angles.[4]

It is now possible to make extensive assignments of ^{1}H and ^{13}C resonances of proteins with molecular weight values of up to about 20,000. Once this has been achieved, NMR experiments can be used in various ways to provide detailed structural information.[5] One approach is to make use of the NMR data directly to provide independent structural information. Both NOE effects and spin-spin coupling constants are specific interactions between defined pairs of nuclei and, provided sufficient such data are available, computer aided search procedures can generate detailed structural models for small proteins.[6-8] Another approach, which can be used even for quite large proteins, is to compare experimental NMR data with those calculated from the average structure of the protein determined in the crystalline state through X-ray diffraction studies.[9-11] This enables a comparison of the structures of the protein in the two states to be made; any differences can then be examined in detail.

It is, however, well established that protein molecules experience a wide variety of fluctuations.[12,13] Frequently, the magnitude of such fluctuations is such that an average structure can be defined which provides a good description of the overall molecular conformation. The assumption of such a structure is implicit in almost all structural studies of proteins whether by diffraction in crystals or by NMR in solution. The high degree of correlation found between experimental NMR parameters and those predicted from X-ray diffraction data for globular proteins such as lysozyme, basic pancreatic trypsin inhibitor and cytochrome c suggests that for internal residues at least the conformational fluctuations are limited. Indeed, experimental and theoretical studies of globular proteins suggest that typical rms fluctuations in atomic positions are of the order of 0.1 nm, and in torsion angles of the order of 20°.[13] Fluctuations of this magnitude are sufficiently fast that averaged NMR signals are observed, weighted according to the populations of the rapidly interconverting conformers. Nevertheless, analysis of such parameters from protein NMR spectra has provided important information about the nature of the rapid fluctuations that do take place.[14,15] Additional information about protein dynamics has come from analysis of the rates of solvent exchange of individual labile ^1H atoms for ^2H, monitored by measurement of the intensities of proton signals as a function of time following dissolution in ^2H_2O.[15-17] These have provided important data about local and cooperative motions within proteins. The ability to follow exchange at virtually every amide or side-chain NH in the molecule is an exciting consequence of the success of assignment studies.

A limitation of many of these studies arises from the difficulty of obtaining in many cases an unambiguous definition of both the magnitude and rate of the conformational fluctuations. An alternative approach to the study of dynamic processes in proteins arises when significantly different conformations or conformational states of a protein exist in equilibrium with each other and interconvert slowly on the NMR timescale (usually about 10^{-3}s). Under these conditions instead of averaged signals (see above) separate resonances may be observed for the different conformers in the NMR spectrum.[5] This provides an opportunity to distinguish and identify specific states of the protein and to determine their relative populations. Further, by means of a variety of 1-D and 2-D NMR techniques, it is possible to follow the transfer of nuclear magnetization between the different states.[18] This offers the possibility of investigating directly the pathways and kinetics of inter-conversion of defined conformations of proteins.

Magnetization Transfer NMR Studies of Proteins

In principle magnetization transfer techniques can be applied to a wide range of cases where interconversion between different states of a protein occurs, provided that distinct resonances from the individual states can be identified and that the kinetics of interconversion are not slow compared with the nuclear spin lattice relaxation rates.[19] Although rather few studies have so far been reported, those that have been serve to demonstrate the wide applicability of the approach. In studies of cytochrome c, for example, magnetization transfer experiments have been used to study both the kinetics of the localized rotation of a tyrosine side-chain about its C_β-C_γ bond,[20] and those of the cooperative interconversion between the reduced and oxidized protein.[21,22] In this paper we discuss recent studies of two proteins, lysozyme and staphylococcal nuclease, which have been primarily concerned with conformational changes related to protein folding and unfolding.

Lysozyme from hen egg white is a protein of 129 amino acid residues. It has been studied in detail by NMR and ^1H resonances of nearly every residue in the protein have now been assigned.[23,24] Figure 1 shows part of the ^1H NMR spectrum of lysozyme at different temperatures; the resonances between 6.0 and 9.0 ppm are all from aromatic protons. At high temperature the spectrum differs considerably from that at lower temperature, as many of the local interactions giving rise to characteristic chemical shifts have been lost on unfolding. At temperatures where both folded and unfolded species are present in significant concentrations resonances from both can be detected in the spectrum. This is shown most clearly for the $H^{\varepsilon 1}$ resonance of the single histidine residue (His 15) close to 8.8 ppm, and for resonances of the three tyrosine residues at about 6.8 ppm. Analysis of the temperature dependence of the intensities of various resonances corresponding to the two states has permitted the cooperativity of the unfolding process to be demonstrated, and values for the denaturation temperature and folding enthalpy to be measured.[25,26]

Magnetization transfer between the folded and unfolded state has been studied in lysozyme by means of both 1-D and 2-D techniques.[26,27] The simplest form of 1-D experiment involves selective irradiation at the resonant frequency of a given nucleus for a fixed length of time prior to the accumulation of the NMR spectrum. Provided that interconversion takes place between the different states of the molecule, and that the nucleus in question has a different resonant frequency in each state, the saturation caused by the irradiation can be transferred to the resonance of the nucleus in the different state. An example of such an experiment with lysozyme is shown in Fig. 2 in which the resonance of His 15 $H^{\varepsilon 1}$ in the folded state was saturated for different lengths of time. By following the dependence of the extent of magnetization transfer as a function of the time of saturation the rate of folding can be determined. Similarly, by saturation of the resonances of His 15 $H^{\varepsilon 1}$ in the unfolded state thee rat of unfolding can be determined. Full details of these and related experiments have been described elsewhere;[26] measurements of the temperature dependencies of the exchange rates, see Fig. 3, have been of particular interest in establishing the validity of the experimental procedures. These results established for a protein the feasibility of quantitative analysis of such experimental data at least for a straightforward two-state case.

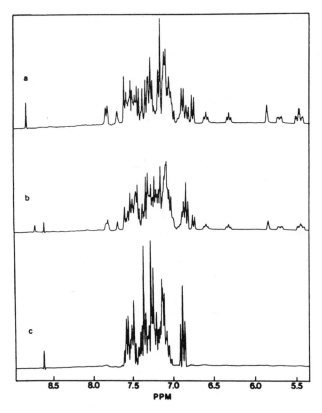

Figure 1. Changes in the ^{1}H NMR spectrum of lysozyme with temperature. Part of the 300 MHz ^{1}H NMR spectrum of the protein showing resonances of aromatic residues. The top spectrum, at 65°C, is of native lysozyme and the bottom, at 82°C, is of denatured lysozyme. The middle spectrum, at 77°C which is close to the mid-point of denaturation, shows the superposition of the spectra of both native and denatured states. *From Ref. 26.*

The Ca^{2+}-dependent nuclease of *Staphylococcal aureus* is a small protein of 149 amino acid residues which contains no disulphide bridges and which, like lysozyme, undergoes reversible thermal unfolding.[28] The protein has four histidine residues and four $H^{\varepsilon 1}$ resonances in the NMR spectrum of the native form can be observed clearly, Fig. 4, despite the rather broad nature of one of them.[29] These resonances are, however, accompanied by several smaller resonances. Each minor peak has a chemical shift comparable to that of one of the major peaks throughout the pH range over which the resonances titrate. From this it was concluded that the minor peaks arise from one or more folded forms which are native-like but distinct from the major form. At higher temperatures resonances of the unfolded form can be seen in the spectrum, those for the histidine $H^{\varepsilon 1}$ lying between 8.3 and 8.4 ppm, see Fig. 4. This shows that, as with lysozyme, there is slow exchange between folded and unfolded states. In this case, however, three sets of peaks, corresponding to two folded and one unfolded state, can be clearly resolved.

Saturation of individual resonances in the spectrum of Staphylococcal nuclease demonstrates that magnetization transfer effects comparable in magnitude to those

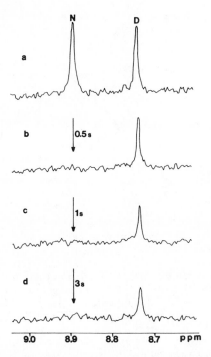

Figure 2. Magnetization transfer between native and denatured lysozyme. (a) Resonances of His 15 $H^{\varepsilon l}$ in the native (N) and denatured (D) states at 77°C. The effects of saturation at the resonant frequency of the N state for 0.5, 1.0 and 3.0s prior to recording the spectra are shown in **(b)**, **(c)** and **(d)**. The decrease in the intensity of the resonance of the D state results from transfer of the saturation resulting from interconversion of the two states of the protein.

seen with lysozyme can be obtained; Fig. 5 shows one such experiment.[29] Saturation of one of the minor resonances, labelled H4* in Fig. 5, leads to changes in the intensity of a resonance of the unfolded state, labelled U, revealed here by difference spectroscopy. It also leads, however, to a decrease in the intensity of the corresponding resonance H4 of the major folded form. This experiment shows directly that all three states of the protein are interconverting under these conditions. This eliminates the possibility that the minor and major folded species are different chemical species and reveals that they are different conformations of the same molecule.

The saturation experiments described above were not able to demonstrate unambiguously that all of the minor resonances showed behavior similar to that of H4, because of the limited selectivity of such experiments. They were not able, therefore, to suggest whether or not the various minor resonances arise from a common folded form. In order to overcome this problem, a 2-D experiment was carried out.[30] This exchange experiment involves, in fact, the same pulse sequence that is routinely employed to observe NOEs between pairs of proteins.[3] Cross-peaks arising from exchange, however, represent correlations between the frequencies of resonances of the different interconverting states. An example of such an experiment is shown in Fig. 6. In this experiment three of the major histidine resonances,

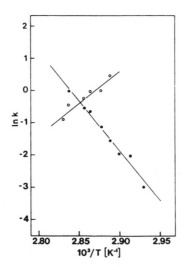

Figure 3. Temperature dependence of the unfolding (●) and folding (○) rates of lysozyme derived from a study of the time developments of magnetization transfer between the $H^{\varepsilon 1}$ resonances of His 15 in the folded and unfolded states. Closely similar values were obtained from experiments involving resonances of Cys 64 and Met 105, which are situated in different regions of the protein structure. The ratio of the two rate constants at each temperature agrees closely with the equilibrium constant for the unfolding process measured independently from resonance intensities.

H1, H3 and H4, have cross-peaks to a minor resonance; the resolution is not, however, sufficient to identify a similar peak for H2. This shows clearly that three of the histidine residues at least each give rise to two resonances corresponding to interconverting major and minor folded forms. Moreover, each of the major peaks and the corresponding minor peaks show a cross-peak to a resonance from the unfolded state. This similarity of behavior strongly implies that a common minor folded species gives rise to these minor resonances. This state will be designated N*, whilst N and U are used to represent the major folded species and the unfolded state respectively.

Examination of Fig. 6 indicates that the intensities of the cross-peaks between N* and U are in all cases greater than the intensities of the cross-peaks between N and U despite the lower intensities of the resonances from N*. Similarly, in 1-D experiments saturation of the resonance of U was observed to cause a proportionately much larger effect on H4* than on H4.[28] Because there is no significant difference between the relaxation rates of H4 and H4*, this result implies that the equilibrium unfolding rate of the minor species N* is much greater than that of the major folded form N.

It is important to note at this point that neither the single-saturation 1-D experiments nor the 2-D experiment described above have revealed the complete pathways of interconversion in this system. All we know from the experiments is that the three states N, N* and U interconvert. We do not know whether the interconversion is in all cases direct rather than via the third form. In other words the rate constants for one of the three interconversions below could be zero provided both the other two are non-zero.

This raises several interesting questions. Thus, for example, do N and U interconvert directly? If not, then N* is a required intermediate on the folding pathway between the denatured states and the major folded form of the protein. Do N and N* interconvert directly? If they do, a conformational rearrangement must be possible within the folded protein. In order to answer these questions experiments have been carried out in which pairs of resonances are simultaneously saturated. One such experiment showed that the intensity of the resonance H4 of the major folded state, N, is decreased to a greater extent when resonances of the corresponding proton of N* and U are saturated together than when the resonance of N* alone is saturated.[29] This indicated that magnetization can be transferred directly from U to N, without the need to pass through N*; if the equilibrium were D $\Leftrightarrow$ N* $\Leftrightarrow$ N no additional perturbation of N could be caused by the double saturation. In other

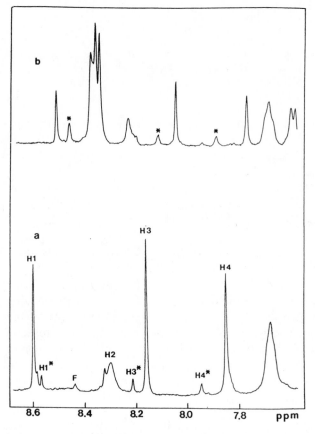

Figure 4. Part of the 500 MHz ^{1}H NMR spectrum of Staphylococcal nuclease at 37°C (**a**) and 55°C (**b**). Resonances H1 to H4 arise from histidines in the major folded state and the starred resonances arise from three residues in the minor folded state. 55°C is close to the mid-point of the unfolding transition and resonances from the histidines in the unfolded state are seen at about 8.4 ppm in spectrum (**b**). *From Ref. 29.*

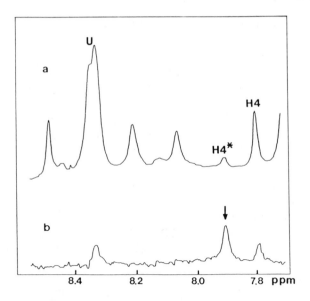

Figure 5. A magnetization transfer experiment on Staphylococcal nuclease, carried out at 300 MHz. The top spectrum is of the protein at 55°C (see Fig. 4). The bottom spectrum is a difference spectrum obtained by subtracting spectra recorded without and with a 3s saturation pulse applied at the H4* resonant frequency. *From Ref. 29.*

words, the experiment shows that N* is not an essential intermediate on the folding pathway from U to N. A series of such double-saturation experiments has been carried out and demonstrates that in fact all three of the states of the protein inter-convert directly with each other.[29] Thus, both N and N* can fold and unfold directly, and N and N* can interconvert without passing through the unfolded state. More detailed magnetization transfer experiments are in progress to obtain values for at least some of the rate constants involved in these processes.

At this stage the difference between the two native states has not been established. One suggestion that would be consistent with the fact that direct interconversion of the two states can take place, and that there is a difference in the folding kinetics of the two states, is that the difference lies in *cis-trans* isomerism of a proline peptide bond.[29] It is known that in ribonuclease A a metastable form contains a non-native proline isomer.[30] The possibility of *cis-trans* isomerism also requires that the possibility of spectroscopically unresolved heterogeneity in the folded state must be taken into consideration in the detailed analysis of the kinetics of interconversion.[31] These and other possibilities are being investigated by magnetization transfer NMR in conjunction with single-site mutations.

Concluding Remarks

Magnetization transfer techniques are one of the several approaches that are now being used to explore by means of NMR spectroscopy aspects of the folding and

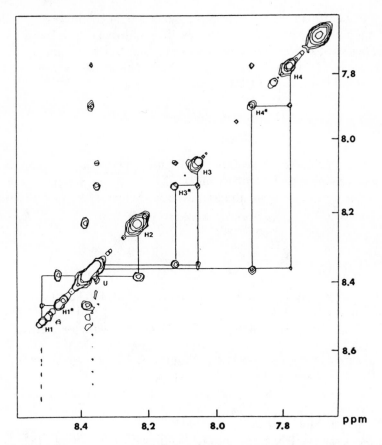

Figure 6. Contour map of part of the 500 MHz ^{1}H two-dimensional chemical exchange spectrum of Staphylococcal nuclease recorded at 55°C. The diagonal corresponds to the one-dimensional NMR spectrum. Cross-peaks indicate that exchange of magnetization takes place between the pairs of resonances at the corresponding frequencies.

unfolding of proteins.[18] In this article we have tried to emphasize that quantitative kinetic data can be obtained from analysis of experimental data, at least in favorable circumstances, and that pathways of interconversion between different folded and unfolded states of a protein at equilibrium can be identified. This appears to offer unique possibilities for studying folding and, for example, to explore the effects on such pathways of single-site mutations.

Magnetization transfer NMR experiments also offer an exciting approach for studying the interconversion of different conformational states of a folded protein at equilibrium. In cases such as Staphylococcal nuclease, where multiple folded conformations exist, it is possible to use the techniques to follow interconversion of such states directly. If the exact nature of the conformational difference can be established then, as with the folding pathways, there will be opportunities to study the effect of perturbations to the protein, for example by single-site mutations, or by binding molecules or ions, on a specific conformational transition. Similar opportunities exist for the study of conformational transitions induced by binding provided conditions where resonances of both bound and unbound species are present at equilibrium can be established.

Acknowledgments

This is a contribution from the Oxford Enzyme Group which is supported by the Science and Engineering Research Council. The work on Staphylococcal nuclease has been carried out in conjunction with Robert O. Fox who also contributed many of the ideas discussed in this article.

References

1. O. Jardetzky and G.C.K. Roberts, NMR in Molecular Biology, Academic Press, New York (1981).
2. J.H. Noggle and R.E. Schirmer, The Nuclear Overhauser Effect, Academic Press, New York (1971).
3. A. Bax, Two Dimensional NMR in Liquids, Reidel, London (1982).
4. V.F. Bystrov, Prog. NMR Spectrosc., **10**, 41 (1976).
5. I.D. Campbell, C.M. Dobson and R.J.P. Williams, Biochem. J., **231**, 1 (1985).
6. R. Kaptein, E.R.P. Zuiderweg, R.M. Scheek, R. Boelans and W.F. van Gunsteren, J. Mol. Biol., **182**, 179 (1985).
7. M.P. Williamson, T.F. Havel and K. Wuthrich, J. Mol. Biol., **182**, 295 (1985).
8. A. Brunger, G.M. Clore, A.M. Gronenborn and M. Karplus, Proc. Natl. Acad. Sci. USA, **83**, 3801 (1986).
9. F.M. Poulsen, J.C. Hoch and C.M. Dobson, Biochemistry, **19**, 2597 (1980).
10. G. Wagner, Q. Rev. Biophys., **16**, 1 (1983).
11. G.R. Moore, M.N. Robinson, G. Williams and R.J.P. Williams, J. Mol. Biol., **183**, 429 (1985).
12. R. Porter, M. O'Connor and J. Whelan (eds.), Ciba Found. Symp., **93**, 1 (1983).
13. J.A. McCammon and M. Karplus, Acc. of Chem. Res., **16**, 187 (1983).
14. C.M. Dobson and M. Karplus, Methods Enzymol., in press (1986).
15. C.M. Dobson in "Structure and Dynamics: Nucleic Acids and Proteins," E. Clementi and R.H. Sarma, eds., Adenine Press, New York, p. 451 (1983).
16. C.K. Woodward, I. Simon and E. Tuchsen, Mol. Cell. Biochem., **48**, 135 (1982).
17. S.W. Englander and N.R. Kallenbach, Q. Rev. Biophys., **16**, 521 (1981).
18. C.M. Dobson, P.A. Evans and R.O. Fox, in "Structure and Motion: Membranes, Nucleic Acids and Proteins," E. Clementi and R.H. Sarma, eds., Adenine Press, New York, p. 265 (1985).
19. I.D. Campbell, C.M. Dobson, R.G. Ratcliffe and R.J.P. Williams, J. Magn. Reson., **29**, 397 (1978).

20. I.D. Campbell, C.M. Dobson, G.R. Moore, S.J. Perkins and R.J.P. Williams, FEBS Lett., **70**, 96 (1976).

21. A.G. Redfield and R.J. Gupta, Cold Spring Harbor Symp. Quant. Biol., **36**, 405 (1971).

22. J. Boyd, G.R. Moore and G. Williams, J. Magn. Reson., **58**, 511 (1984).

23. M. Delepierre, C.M. Dobson, M.A. Howarth and F.M. Poulsen, Eur. J. Biochem., **145**, 389 (1984).

24. C.M. Dobson and C. Redfield, to be published.

25. R.E. Wedin, M. Delepierre, C.M. Dobson and F.M. Poulsen, Biochemistry, **21**, 1098 (1982).

26. C.M. Dobson and P.A. Evans, Biochemistry, **23**, 4267 (1984).

27. C.M. Dobson, P.A. Evans and K.L. Williamson, FEBS Lett., **168**, 331 (1984).

28. C.B. Anfinsen, Science, **181**, 223 (1973).

29. R.O. Fox, P.A. Evans and C.M. Dobson, Nature, **320**, 6058 (1986).

30. C.M. Dobson, P.A. Evans and R.O. Fox, to be published.

31. K.H. Cook, F.X. Schmid and R.L. Baldwin, Proc. Natl. Acad. Sci. USA, **76**, 6157 (1979).

Distributions and Fluctuations of Protein Structures Investigated by X-Ray Analysis and Mössbauer Spectroscopy

F. Parak, M. Fischer and E. Graffweg
Institut für Physikalische Chemie der Westfälischen
Wilhelms Universität Münster
Schloßplatz 4-7
4400 Münster, Fed. Rep. Germany

H. Formanek
Botanisches Institut der Universität München
Menzinger Straße 67
8000 München 19

Abstract

The X-ray structure determination of myoglobin at five temperatures allows an estimation of the structural order at 0 K. Extrapolation to 0 K of the $< x^2 >^x$-values determined between 80 K and 300 K shows that this protein has a large zero point disorder indicating a distribution of slightly different structures. At temperatures above 200 K, Mössbauer spectroscopy measures structural fluctuations of the molecules at a time scale between 10^{-7} to 10^{-9} s occurring within the distribution of structures obtained by X-ray methods. These fluctuations can be described as a diffusive motion of segments of the molecule within a limited space. RSMR-experiments show that the characteristic size of these segments is larger than 5Å but smaller than the diameter of the molecule.

Investigations of model compounds prove that forming and breaking of hydrogen bridges plays an important role for protein specific motions on the time scale between 10^{-7} and 10^{-9} s. Water serves as a plasticizer but also dry myoglobin reveals protein dynamics. Mössbauer spectroscopy on membrane proteins gives $< x^2 >^\gamma$-values similar to those of myoglobin crystals. The onset of protein specific motions occurs, however, at lower temperatures.

Introduction

Chemical composition, three-dimensional structure and the state of aggregation determine the function of a molecule. In inorganic and organic molecules with relative low molecular mass the structure is well defined, normally. Dynamics is mainly determined by the ionic or covalent binding of the atoms. In a good approximation one can assume that the structure of molecules of the same type is identical or allows only a few well defined conformations.

In biomolecules an additional principle has to be considered: "distributions." Frauenfelder and coworkers[1] have found that the rebinding kinetics of CO flashed by laser pulses from myoglobin cannot be understood by a well defined potential barrier. Instead, a distribution of activation energies has to be introduced indicating that each individual molecule of the sample under investigation has a slightly different structure. The conformation of the molecules is determined as an ensemble average. The ensemble contains molecules with structures distributed around the average structure. It is now common to say that each individual molecule can be in one conformational substate. X-ray structure investigations favor this picture.[2,3] In this contribution a rather direct proof of structure distributions will be given.

From the concept of distributions of structures it is reasonable to assume that one biomolecule can fluctuate between different conformational substates if the thermal energy is large enough to overcome the energy barrier between these states. Time sensitive spectroscopic methods like fluorescence are very suitable to investigate such fluctuations. In this contribution, results obtained by another method are described. If a ^{57}Fe atom is incorporated into a biomolecule this atom can be used as a marker for fluctuations. Mössbauer spectroscopy allows the simultaneous determination of the average fluctuation amplitude and of the characteristic time of motion. Since crystals are not necessary, it is easy to study the influence of the hydration shell on protein dynamics; also membrane proteins can be investigated. In the following we discuss Mössbauer experiments on myoglobin crystals and on dry myoglobin. A comparison with membrane proteins is performed. It should be emphasized, however, that Mössbauer spectroscopy on ^{57}Fe is sensitive only for motions with a characteristic time faster 10^{-7} s. Slower fluctuations are not discussed in this contribution. Protein specific motions are discussed mainly in the time window 10^{-7} to 10^{-9} s.

Conformational Substates Determined by X-ray Structure Analysis

Let us assume that a molecule has one well defined structure which is determined by a deep minimum in energy. At 0 K all molecules of a crystal will be in this structure. Disorder occurs only due to zero point vibrations. The picture changes if atoms or groups of atoms can occupy different positions which differ not too much in energy. The individual molecules of the crystal are then frozen in a distribution of slightly different structures which increases the disorder above the value of zero point vibrations.

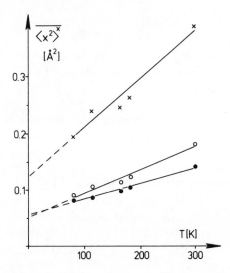

Figure 1. Mean square displacements, $< x^2 >^x$, obtained from X-ray structure analysis of metmyoglobin at different temperatures. Closed circles: average values over all backbone atoms; open circles: average values over all side chain atoms. Crosses: average values over 160 water molecules found in X-ray structure analysis.

X-ray structure analysis measures disorder by the Debye Waller factor which depends on the mean square displacement, $< x^2 >^x$, of each atom from its average position. A structure determination at temperatures as close as possible to 0 K would give the zero point distribution of the molecule structure. Such a measurement is, however, difficult. Moreover, it is probable that many molecules will be frozen in a metastable structure yielding an unrealistically high disorder. A better estimation of the structural distribution at 0 K can be obtained from the temper-

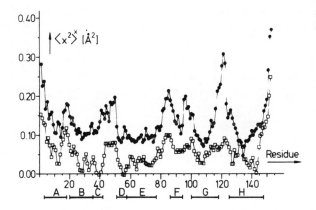

Figure 2. Mean square displacements, $< x^2 >^x$, of the average values of the backbone atoms of the residues in myoglobin.[4,5] A linear temperature dependence of $< x^2 >^x$ is assumed between 80 K and 300 K. Full circles give values at 300 K and squares give values extrapolated to 0 K from a linear regression of the experimental data.

ature dependence of the $< x^2 >^x$-values. For myoglobin the X-ray structure was determined at 300 K, 185 K, 165 K, 115 K and 80 K.[4,5] Figure 1 shows the average mean square displacements, $< x^2 >^x$ of all backbone atoms (full circles) and all side chain atoms (open circles), respectively. In both cases the temperature dependence is linear. Extrapolation to T = 0 K yields a large disorder. A detailed analysis shows that for the individual residues the temperature dependence of the $< x^2 >^x$-values of 147 backbone averages and 134 side chain averages (total number of residues: 153) becomes linear using $\pm \sigma = 0.01 + 0.1 < x^2 >^x$ as error bars. Linearly with temperature decreasing $< x^2 >^x$-values are a good indication that the sample is in thermal equilibrium at each measurement. A linear extrapolation to T = 0 K gives the lower limit for the zero point distribution of structures, neglecting any zero point vibrations. In Fig. 2 the average values of the backbone atoms of the 153 residues are given at 0 K and at 300 K. The values are obtained from a linear regression using the experimental data at five temperatures. The high temperature values represent the structural disorder as a function of the position of the residue as discussed by Frauenfelder et al.[2] Our results are in good agreement with these data. At low temperature the disorder is strongly reduced. Nevertheless, extrapolation to 0 K reveals still a distribution of structures. It is clearly seen that the tendency of disorder along the chain is very similar at 300 K and 0 K. Nonhelical regions show in both cases the largest $< x^2 >^x$-values. It is interesting to note that the GH corner which has a very large disorder at 300 K becomes relatively well defined at 0 K.

The results of the X-ray structure investigation clearly show that myoglobin cannot be frozen into one unique structure. However, the question whether the zero point structural distribution is nearly continuous or results from a small number of slightly different but well defined structures cannot be answered.

Segmental Motions Investigated by Mössbauer Spectroscopy

Mössbauer spectroscopy labels protein dynamics at the position of a ^{57}Fe nucleus yielding mean square displacements, $< x^2 >^?$, which come from motions with a characteristic time, τ_c, faster than 10^{-7} s. In myoglobin, protein specific motions become important at temperatures above 200 K. Additional broad lines occur at this temperature indicating motions in a characteristic time of about 10^{-9} s at 300 K. A discussion of the results is given in the literature.[6-13] X-ray structure investigations and Mössbauer spectroscopy on myoglobin give a consistent picture.[14,15] This is, however, no proof that there are not alternative interpretations.

The Mössbauer spectrum is determined by the cross section, σ_a, for absorption:

$$\sigma_a(E) = C \int_{-\infty}^{+\infty} e^{i(E_a - E)t/\hbar - \Gamma_a |t|/2} I(\vec{k_o}, t) \, dt \tag{1}$$

where C is a constant containing nuclear properties and $\hbar$ is Planck's constant divided by 2π. E_a is the resonance energy of the absorbing ^{57}Fe nucleus and Γ_a is essentially equal to τ_N^{-1}, the reciprocal of the lifetime of the Mössbauer level. (τ_N = 1.4 10^{-7} s for ^{57}Fe). Setting $I(k_o, t)$ equal to 1 the integration over time yields the well known Lorentzian energy dependence of $\sigma_a(E)$. All information on dynamics of the absorbing ^{57}Fe nucleus is contained in the intermediate scattering function $I(k_o, t)$, depending on the wave vector $\vec{k}_o$ of the incoming radiation ($|\vec{k}_o| = 2\pi/\lambda$; λ = 0.86Å) and on the time t. It is impossible to determine $I(k_o, t)$ from the measured Mössbauer spectrum in an unambiguous way. Instead, it is necessary to propose a model which yields the intermediate scattering function. Then, one can calculate the Mössbauer spectrum. Disagreement with the experiments disproves the model while agreement only tells that $I(k_o, t)$ represents a possibility.

As already mentioned, Mössbauer spectra of ^{57}Fe in myoglobin crystals show line broadening and additional broad lines above 200 K. This is a clear indication for diffusive processes. Since free diffusion is impossible in crystals one has to assume a diffusive motion of molecular segments which is limited in space. Restoring forces make sure that the segments diffuse only around their average position in the molecule as determined by X-ray analysis.

Brownian motion in limited space is described by the Brownian oscillator. For the interpretation of Mössbauer spectra the case of overdamping was successfully applied. Here, the frequency ω of the restoring force is small compared to the friction constant β. The Mössbauer spectra become a superposition of an infinite number of Lorentzians with a weight $[k^2 < x^2 >^{\gamma}]^N/N!$ and a width $N\hbar\alpha$ of the N-th line. The characteristic time of the motion is given by $\alpha^{-1} = 2\beta/\omega^2$. In this model each myoglobin molecule of the crystal performs essentially the same motion.[16]

Recently another diffusion model was proposed by Frauenfelder.[17,18] In this model each myoglobin molecule is frozen longer than 10^{-7} s in one set of conformational substates. Each set can be characterized by traps with a distance d and separated by potential barriers with the height ε. The distribution of motions of the molecules can be obtained from the distribution of the energies ε which are taken from CO flash experiments.[19] Taking the probability 1 for 12 kJ/mol $\leq \varepsilon \leq$ 41 kJ/mol and 0 for all other energies, this model gives the correct shape of the Mössbauer spectrum. In the picture of the hierarchy of protein dynamics,[19] Mössbauer spectroscopy is mainly influenced by the equilibrium fluctuations in tier 2. The two models differ in the explanation of the temperature dependence of $< x^2 >^{\gamma}$. They predict differences in RSMR experiments (Rayleigh Scattering of Mössbauer Radiation). A further investigation is in progress.

In the RSMR technique 14.4 keV radiation of a Mössbauer source is Rayleigh scattered into the angle 2θ by the electrons of the sample. An energy analysis of the scattered radiation can be performed by a Mössbauer ^{57}Fe absorber yielding the fraction of elastic and inelastic Rayleigh scattering. This allows the determination of a mean square displacement, $< x^2 >^R$, averaged over all atoms of the sample. As in the case of Mössbauer absorption spectroscopy only motions with a characteristic time faster than 10^{-7} s contribute. The RSMR-technique has recently been used to investigate the size of the segments performing the protein specific motions. $< x^2 >^R$ values of the myoglobin should be independent from the angle 2θ at which the scattering experiment is performed. Independently vibrating atoms of the myo-globin molecules give an unrealistic angular dependence of $\overline{< x^2 >^R}$. Physically rea-sonable results are obtained if one assumes collective motions of segments with a characteristic dimension larger than 5Å.[20] Experiments which characterize the segment size more accurately are in progress. It is, however, already shown that the diffusive motions in limited space are performed by segments and not by the mole-cule as an entity. RSMR experiments on reflexions of a single crystal of myoglobin show that there are nearly no contributions from acoustic lattice vibrations to the mean square displacements.[16]

Forces Determining Protein Specific Motions

Mössbauer spectroscopy reveals protein specific motions with a characteristic time between 10^{-7} and 10^{-9} s. The increase of the $< x^2 >^\gamma$-values above 200 K is not found in iron complexes of low molecular mass. In these complexes the structure is determined by covalent and ionic bonds. In proteins, the primary structure is formed by covalent bonds too. However, the three-dimensional architecture of the molecules is held together by weak forces. Hydrogen bridges, entropy effects like the hydrophobic interaction, van der Waals forces and Coulomb interaction of charged side groups form a balanced system where the net energy stabilizing the tertiary structure is relatively small. It is reasonable to assume that these weak forces make segmental motions possible.

The influence of hydrogen bridges on disorder in proteins was analyzed by Gavish.[21] Mean square displacements of the backbone atoms of the residues as obtained by X-ray structure analysis were correlated with the number of possible

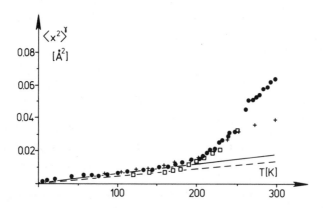

Figure 3. Temperature dependence of the $< x^2 >^\gamma$-value of iron as determined by Mössbauer spectroscopy. Full circles: Deoxy myoglobin crystals[10] (for comparison); crosses: masked Asp with ^{57}Fe; squares: reaction centers of photosynthetic bacterials.[25]

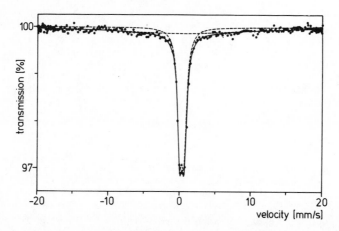

Figure 4. Mössbauer spectrum of a DPPC/Fe absorber at 270 K. The fit of the experimental data is performed with a quadrupole splitted narrow and broad Lorentzian.

hydrogen bridges. For myoglobin and lysozyme it was found that the density of hydrogen bridges is nearly inversely proportional to the mean square displacement of the residues. This proves the important role of H-bonds.

From the consideration above it seems probable that one can build model complexes where the Lamb-Mössbauer factor of the iron shows a similar temperature dependence as found in myoglobin. For a first investigation we used a Fe-complex of aspartic acid masked at the basic and the acid end (N-tBoc-L-Asp-α-benzylester). The sample was prepared by dissolving the masked Asp together with ^{57}FeSO in H_2O and metoxyethanol and concentrating it slowly at 60°C. The compound has imino, carbonyl and carboxyl functions which can act as donor and acceptor in the formation of hydrogen bridges, respectively. In the absence of bulk water the carboxyl group can bind the iron. Figure 3 shows the mean square displacement $< x^2 >^\gamma$ of the iron as obtained from the Lamb-Mössbauer factor determined on this sample. The increase of $< x^2 >^\gamma$ above 200 K is similar as in myoglobin. It should be noted that a kind of saturation is reached at 300 K. Figure 4 shows the Mössbauer spectrum of another model compound. A sample containing dipalmitoyl-phosphatidylcholin instead of aspartic acid was prepared in the same way as the sample before. It is obvious from the Mössbauer spectrum that broad lines are necessary in addition to the narrow lines in order to fit theory to the experimental data. The broad line indicates motions with a characteristic time of about 10^{-9} s at 270 K. The mean square displacement of the iron again increases drastically above 200 K.

The investigation of model compounds demonstrates that quasidiffusive motions of molecular segments and weakly bound molecules is rather similar. It also emphasizes the important role of hydrogen bridges for protein dynamics.

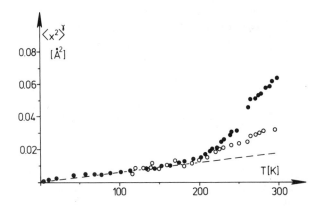

Figure 5. Temperature dependence of the $< x^2 >^\gamma$-value of the iron in myoglobin as determined by Mössbauer spectroscopy. Closed circles: Deoxy myoglobin crystals.[10] Open circles: freeze dried myoglobin (met).

The Influence of Water

The mobility of a protein is strongly correlated with the mobility of the surrounding water. For a discussion we refer for instance to Ref. 22. In the case of myoglobin 160 water molecules per protein molecule could be found by X-ray structure analysis.[23] Estimates show that a crystal of myoglobin contains not more than 450 waters/myoglobin. This means that more than 1/3 of the waters have strong interaction with the protein surface. Figure 1 gives the temperature dependence of the $< x^2 >^x$-values of the water averaged over the 160 molecules. It can be understood by a linear temperature dependence. The decrease with temperature is, however, stronger as for atoms of the protein. This shows once more that the mobility of the water is larger than the mobility in the protein. Extrapolation to T = 0 K gives a large distribution around the average position.

Earlier investigations of myoglobin have indicated that freeze dried myoglobin exhibits practically no protein specific dynamics.[6,24] In a new investigation a myoglobin powder kept for one week over phosphorpentoxyde has been investigated by Mössbauer spectroscopy. First results are shown in Fig. 5. In comparison to myoglobin crystals protein specific dynamics is strongly reduced in the dry sample but nevertheless clearly visible. It is important to note that the protein specific increase in the $< x^2 >^\gamma$-value occurs at about the same temperature in dry myoglobin and in myoglobin crystals. Water serves as a plasticizer for the protein. The mechanism is, however, not completely clear. Structural rearrangement in the presence of water, shielding of charged groups and reduction of direct contacts between molecules may be of importance. Viscosity changes in the protein environment can also play a role. In this context a comparison with the dynamics of membrane bound proteins is interesting. Results on the chromatophores of Rhodospirillum Rubrum[25] are shown in Fig. 3. In comparison to myoglobin the protein specific increase of $< x^2 >^\gamma$ occurs at lower temperatures. Similar results were recently obtained in the investigations of Fe containing Bacterio Rhodopsin.[26]

Acknowledgments

This work was supported by the Deutsche Forschungsgemeinschaft and the Bundesministerium für Forschung and Technologie.

References

1. R.H. Austin, K.W. Beeson, L. Eisenstein, H. Frauenfelder and I.C. Gunsalus, Biochem., **14**, 5355 (1975).
2. H. Frauenfelder, G.A. Petsko and D. Tsernoglou, Nature (London), **280**, 558 (1979).
3. H. Hartmann, F. Parak, W. Steigemann, G.A. Petsko, D. Ringe Ponzi and H. Frauenfelder, Proc. Natl. Acad. Sci. USA, **79**, 4967 (1982).
4. F. Parak, H. Hartmann and G.U. Nienhaus, Proceeds of the Conference on Protein Structure: Electronic and Molecular Reactivity, ed. by B. Chance and B. Austin, in press (1986).
5. F. Parak, H. Hartmann, K.D. Aumann, H. Reuscher, G. Rennekamp, H. Bartunik and W. Steigemann (to be published).
6. F. Parak and H. Formanek, Acta Chryst., **A27**, 573 (1979).
7. H. Keller and P.G. Debrunner, Phys. Rev. Lett., **45**, 68 (1980).
8. F. Parak, E.N. Frolov, R.L. Mössbauer and V.I. Goldanskii, J. Mol. Biol., **145**, 825 (1981).
9. S.G. Cohen, E.R. Bauminger, I. Nowik and S. Ofer, Phys. Rev. Lett., **46**, 1244 (1981).
10. F. Parak, E.W. Knapp and D. Kucheida, J. Mol. Biol., **161**, 177 (1982).
11. E.W. Knapp, S.F. Fischer and F. Parak, J. Phys. Chem., **86**, 5042 (1982).
12. E.R. Bauminger, S.G. Cohen, I. Nowik, S. Ofer and J. Yariv, Proc. Natl. Acad. Sci. USA, **80**, 736 (1983).
13. I. Nowik, E.R. Bauminger, S.G. Cohen and S. Ofen, Phys. Rev., **A31**, 2291 (1985).
14. F. Parak, in "Structure and Motion: Membranes, Nucleic Acids and Proteins," eds. E. Clementi, G. Corongiu, M.H. Sarma and R.H. Sarma, Academic Press, p. 243 (1985).
15. F. Parak and E.W. Knapp, Proc. Natl. Acad. Sci. USA, **81**, 7088 (1984).
16. F. Parak, H. Hartmann, G.U. Nienhaus and J. Heidemeier, in "Proceed. of the First EBSA Workshop: Structure, Dynamics and Function of Biomolecules," eds. A. Ehrenberg and R. Rigler, in press.
17. H. Frauenfelder, in "Structure and Motion: Membranes, Nucleic Acids and Proteins," eds. E. Clementi, G. Corongiu, M.H. Sarma and R.H. Sarma, Academic Press, p. 205 (1985).
18. H. Frauenfelder and F. Parak, in preparation.
19. A. Ansari, J. Berendzen, S.F. Bowne, H. Frauenfelder, I.E.T. Iben, T.B. Sauke, E. Shyamsunder and R.D. Young, Proc. Natl. Acad. Sci. USA, **82**, 5000 (1985).
20. G.U. Nienhaus and F. Parak, Hyperfine Interactions, **29**, 1451 (1986).
21. B. Gavish, Biophys. Struct. Mech., **10**, 31 (1983).

22. F. Parak in "Methods in Enzymology," **Vol.** **127**, ed. L. Packer, Academic Press, p. 196 (1986).
23. H. Hartmann, W. Steigemann, R. Reuscher and F. Parak, submitted to Eur. Biophys. J. (1986).
24. Yu. F. Krupyanskii, F. Parak, V.I. Goldanskii, R.L. Mössbauer, E. Gaubmann, H. Engelmann and I.P. Suzdalev, Z. Naturforschg., **37c**, 57 (1982).
25. F. Parak, E.N. Frolov, A.A. Kononenko, R.L. Mössbauer, V.I. Goldanskii and A.B. Rubin, FEBS Lett., **117**, 368 (1980).
26. F. Parak et al., to be published.

Rotational Motions of Tryptophan and Tyrosine Residues in Proteins

Enrico Gratton, J. Ricardo Alcala and Gerard Marriott
Departments of Physics and Biochemistry
University of Illinois at Urbana-Champaign
Urbana, IL 61801 USA

Proteins are flexible structures; their flexibility is crucial for biological function. The physical origin of protein flexibility has been discussed and it arises from the intrinsic flexibility of the polypeptide chain.[1,2] The time scale of protein motions extends from picoseconds to several seconds depending on the process involved.[3] The physiological importance of protein flexibility is well established but the functional importance of the dynamics in a given time range must still be proven. Several classes of motions can be distinguished on the basis of the element of the protein involved and of the time scale. It is likely that all of these motions can have some functional significance. It is our goal to characterize some of the motions and in particular the motions of residues. It has been proposed that correlated motions of residues occurring in the nanosecond time range can be relevant to biologically important events such as enzyme catalysis.[3] However, theoretical investigations of protein motions based on molecular dynamics calculations have suggested that motions of residues occur in picoseconds.[4] For example, in the molecular dynamics calculations one of the tryptophan residues in lysozyme displayed high amplitude fluctuations in few picoseconds.[5] Also calculations of the rotational motions of the tyrosine residues of bovine pancreatic trypsin inhibitor (BPTI) have shown high amplitude motions in few picoseconds.[6] Simulations of the motions of several other proteins have given similar results.

The decay of the emission anisotropy provides a direct measurement of the rotational correlation time of fluorescent molecules. We have shown that using frequency domain fluorometry a time resolution of a few picoseconds can be achieved for the measurement of rotational correlation times of small molecules in solution.[7] Such high resolution is obtained by the application of differential methods. The rotational motion of a residue in a protein is quite complex due to the presence of several factors including the anisotropic restricted rotation of the residue with respect to the protein and the tumbling of the entire protein molecule. Furthermore, if the protein is non-spherical, then the rotational motion of the entire molecule requires several exponential terms to be properly described. We will assume that the rotational motion of the entire protein can be described by an average rotational correlation time, i.e., we are assuming that the protein under study is spherical. Instead we will describe the internal motion of the residue as a restricted motion in a cone. The equations describing the decay of the emission anisotropy have been derived for this model. Using some reasonable approximations, these equations reduce to the sum of two exponentials whose characteristic times and relative amplitudes are a function of the physical parameters of the model.[8,9] If the rotational motion of the internal residue is much faster than the rotation of the entire protein, then the two exponentials reflect the motions of the residue and of the protein, respectively. The pre-exponential factors instead depend on the cone aperture and upon the time zero anisotropy. In principle, for a dipole transition, the maximum value of the time zero anisotropy in a randomly oriented system is 0.4. This value occurs if the excitation and emission dipoles are non-colinear. Since in indole it is possible to excite at least two electronic transitions, the value of the zero time anisotropy is strongly dependent upon the excitation wavelength. It has been shown that for excitation at 300 nm this value is approximately 0.31,[10] which corresponds to an angle of about 22° between excitation and emission transition moments. It is important to determine the exact value of the time zero anisotropy, since a smaller value than 0.31 should be an indirect evidence for a rotational motion which is too fast to be resolved. A measuring artifact can be present when the value of the anisotropy is extrapolated at time zero. This effect arises from the non-exponential behavior of the decay of the emission anisotropy. Although the anisotropy can be described by a sum of exponentials, the measured anisotropy which is constructed from the parallel and perpendicular (with respect to the excitation) emission decay curves can contain other terms originating from the time decay of the intensity. The intensity decay is invariably non-exponential. Lifetime heterogeneity is an indication of the coexistence of different molecular configurations which can also have different rotational behavior. Since in frequency domain fluorometry the value of the phase delay and modulation ratio between the parallel and the perpendicular components is directly measured, the lifetime information must be obtained in a different experiment. Then the problem arises of how to associate the observed lifetime values with the rotational species present. By proper association, the above artifact can be avoided.

We have performed a series of lifetime and anisotropy decay measurements on several single tryptophan proteins, in lysozyme and BPTI. The experiments were carried out in solvents of different viscosity and at different temperatures to better discriminate between internal and external motions. We report here on the results of the later two proteins since a direct comparison of our experiments with molecular dynamics calculations can be performed. The measurement of the rotational correlation times of two fluorescent tryptophan residues in lysozyme and tyrosines in BPTI showed no evidence for fast rotations of relatively high amplitude as predicted by molecular dynamics calculations. In contradistinction, two rotational motions were observed which can be identified with the rotation of the entire protein and with the internal rotations by the tryptophan and tyrosine residues, respectively. The assignment was possible due to the different sensitivity of the tumbling of the molecule to the external viscosity which followed a Stokes-Einstein relationship while the internal motion was less sensitive to the external viscosity. Our approach differs from the usual analysis of the anisotropy decay. We introduced the lifetime heterogeneity observed for many single and multiple tryptophan proteins,[11] and in particular in lysozyme and BPTI, as an essential ingredient in the analysis of the anisotropy decay. We propose that the physical origin of lifetime heterogeneity reflects also in the rotational motions of the tryptophan. Consequently, a distribution of rotational rates must be considered. In this work we simplified the distribution of rates using only two components. Using this approach we recovered the correct value for the time-zero anisotropy. Otherwise, using a single species, a very fast rotational component must be assumed. Since we have not observed this fast motion, this hypothetical component must have a rotational correlation time shorter than 10 psec, and it must disappear, without shifting to longer times, as the temperature is decreased and the viscosity increased.

Acknowledgment

Financial support for this work was provided by a grant from the National Science Foundation PCM84-03107.

References

1. G. Careri, P. Fasella and E. Gratton, Crit. Rev. Biochem., **3**, 141 (1975).
2. H. Frauenfelder and E. Gratton, in "Methods in Enzymology," in press.
3. G. Careri, P. Fasella and E. Gratton, Ann. Rev. Biophys. Bioeng., **8**, 69 (1979).
4. M. Karplus and J.A. McCammon, Ann. Rev. Biochem., **53**, 263 (1983)
5. T. Ichiye and M. Karplus, Biochemistry, **22**, 2884 (1983).
6. J.A. McCammon, B.R. Gelin and M. Karplus, Nature, **267**, 585 (1977).
7. J.R. Alcala, E. Gratton and D.M. Jameson, Analytical Instrum., **14**, 225 (1985).
8. K. Kinoshita, Jr., S. Kawato, and A. Ikagami, Biophys. J., **20**, 289 (1977).
9. A. Szabo, J. Chem. Phys., **81**, 150 (1984).
10. B. Valeur and G. Weber, Photochem. Photobiol., **25**, 441 (1977).
11. E. Gratton, J. Alcala, G. Marriott and F. Prendergast, in "Progress and Challenges in Natural and Synthetic Polymer Research," Japan, July (1985).

Protein Fluctuations and Hemeprotein Affinity for Ligand

Bernard Alpert
Laboratoire de Biologie Physico-Chimique
Université Paris VII
2 place Jussieu 75251 Paris

Abstract

Binding processes in hemeproteins involve the iron atom and the protein properties: structural and dynamics. Crystallographic data of Perutz have given an appreciate popularity to a correlation between the displacement of the iron out of the porphyrin plane and the affinity of the hemeprotein for ligand. However, the dynamics character of the protein matrix may explain the ligand binding properties in hemeproteins. The internal motions of the apoprotein could be "the structural parameter" involved in the mechanism regulating the ligand binding.

Introduction

Many works have been performed to understand the different affinities of hemeproteins for oxygen. The structural work of Perutz et al[1] on the liganded and unliganded forms of hemoglobin has allowed to develop a model[2] correlating the iron atom position and its reactivity for ligand.

However, recently we demonstrated that the ligation process in hemoglobin cannot be controlled by an iron electronic change in the binding site.[3] So, the iron movement seems not to be accompanied by an iron reactivity change for ligand.

Many data support the idea that the apoprotein plays an important conformational and dynamic role in the hemeprotein-ligand association.[1,4-9] Thus, in an effort to obtain a new vision on the processes controlling the affinity of hemeproteins for ligands, the hemeprotein ligand binding was analyzed in terms of mobility of the polypeptide chains of the apoprotein.

Hemoglobin Iron Reactivity

X-ray absorption near edge structure spectroscopy (XANES) of porphyrin-iron reflects the electronic repartition[10] and the chemical reactivity[3] of the iron. Thus XANES investigation on hemoglobin having different ligand affinity allows to understand the relations connecting the protein affinity changes with the iron electronic organizations. This study was performed on human[3] (Fig. 1) and carp hemo-globin[11] (Fig. 2). The latest presents the largest affinity changes ($\simeq$ 200 fold) ever observed with pH and allosteric effector.[12,13] The results obtained demonstrate that a change in protein quaternary structure, or the breaking of salt bridges associated with the ionization of some amino-acids (Bohr effect) do not affect the electronic structure of the iron. Thus, the chemical reactivity of the iron cannot be responsible for the difference in hemoglobin affinities. The mechanisms which control the affinity changes must be investigated in other parts of the macromolecule or in the dynamic character of the binding process.

In agreement with the fact that the quaternary structural state and the ligand binding reaction in hemoglobin are correlated,[4,5,7] we should consider that changes in the ligand affinity may come from significant variations in the internal motions of the protein.

Myoglobin Fluctuations

The binding of oxygen to the iron in hemeprotein is dependent on the iron-ligand interaction and on the oxygen diffusion process through the protein.[14] The

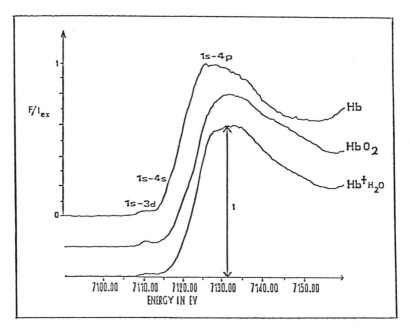

Figure 1. Iron K-edge spectra of human hemoglobin in different forms. X-ray absorption near edge structure (XANES) represents the ratio of the iron fluorescence intensity (F) to the intensity of the incident beam (I_{ex}) as a function of X-ray energy. The figure exhibits the principal iron transitions using the atomic description. Heterotropic effectors do not affect the shape and the energy positions of the K-edge maxima. Differences in the XANES spectra come only from the iron ligand.

migration of the oxygen in the protein interior can be revealed by the fluorescence quenching rate of a fluorophore inside the protein.[15-16]

For mammalian hemeproteins, an increase in temperature of 10°C decreases its oxygen affinity by about 2 fold.[17-18] Since horse myoglobin is significantly more stable than other species, temperature effects on the oxygen diffusion constant have been measured on this protein. Oxygen quenching experiments were performed on Mb^{desFe} (myoglobin without iron). In absence of oxygen the lifetime τ_0 of the metal-free porphyrin in the heme-pocket is almost constant in the range of 15 to 35°C, τ_0 decreases from 17.8 ns at 20°C to 17.5 ns at 35°C. The apparent bimolecular fluorescence quenching rate constant k was obtained from the Stern-Volmer relation:

$$F_0/F = 1 + k\tau_0[O_2]$$

F_0 and F are the porphyrin fluorescence intensities in absence and in presence of oxygen and $[O_2]$ is the concentration of oxygen in the solution.

At low oxygen concentration limit (Fig. 3) we found:[19]

$$k = 8.4 \times 10^8 M^{-1} s^{-1} \text{ at } 20°C, \text{ and}$$

$$k = 13 \times 10^8 M^{-1} s^{-1} \text{ at } 35°C.$$

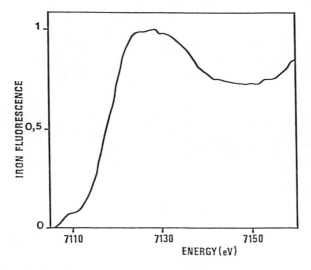

Figure 2. Normalized XANES spectra for carp deoxyhemoglobin. Experiments were performed at pH 6 and 9, in presence and absence of IHP effector. All spectra obtained are identical to each other.

So, oxygen displacements inside the protein matrix are temperature dependent, and certainly follow the rapid fluctuations of the amino-acid residues of the protein. This seems to indicate that the protein mobility could be correlated with the ligand binding process. Thus if the protein fluctuations govern the motion of the ligand inside the apoprotein they should have an important ligand binding implication, particularly on the geminate recombination process.

CO Geminate Recombination

The choice of nanosecond CO cage recombination for this approach was suggested by the well-established fact that the CO geminate molecules which rebind the iron are inside the apoprotein interior around the heme-pocket.[9,20] Geminate CO binding process is simultaneously controlled by the interaction of the iron with the CO and the movement of this molecule through the protein. Therefore, the temperature of the CO geminate recombination becomes a powerful tool to analyze the respective contribution of the chemical iron attraction and the diffusional modulation by the protein matrix in the ligand-protein interaction.

The geminate recombination was monitored at the isobestic points of structural changes (419 and 438 nm) so that the conformational relaxation of the hemeprotein was not observed. The CO rebinding was observed after complete photodissociation in the purpose of following the pure geminate religation process without any

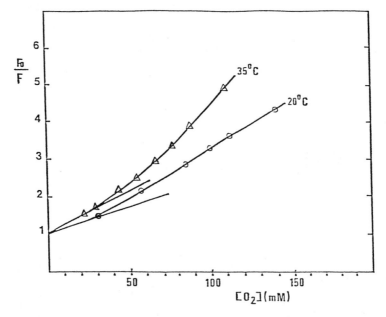

Figure 3. Stern-Volmer plots of Mb^{desFe} quenching by oxygen at two temperatures. The $k\tau_0$ are obtained from the initial slopes of the plots. All experiments were realized at pH 7.

Table I. Temperature dependence on the percentage of iron-CO pairs recombined at the end of geminate process. All experiments were performed at pH 7.

Temperature (°C)	7	15	25	35
iron-recombined fraction	0.77	0.70	0.62	0.55

contamination of back reaction from some CO molecules trapped inside the protein medium. As expected for a diffusion process, the real geminate iron-CO pair recombination follows a $t^{1/2}$ kinetics.[20,21]

However, in the picosecond time range the ligand molecules stay positioned in the heme-pocket, and the geminate phenomenon follows a single exponential kinetic law. This was attributed to the chemical reactivity of the hemoglobin iron without any diffusion.[14] This geminate regime was not used in our ligand migration study. The temperature dependence of the nanosecond CO geminate recombination has been investigated. Table I presents the percentage of iron-CO pairs in function of temperature obtained at the end of geminate process (500 ns after photodissociation pulse). One can see that in contradiction to the chemical kinetics law of ligand binding, increasing temperature decreases the geminate ligand binding yield. This phenomenon indicates that nanosecond geminate recombination is more controlled by diffusion process inside the protein moiety than by the iron chemical reaction.

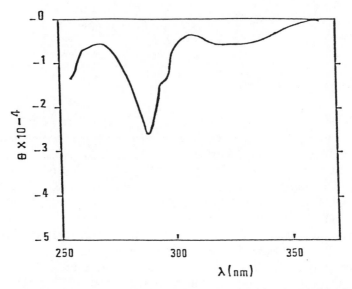

Figure 4. Circular Dichroism spectra of carp deoxyhemoglobin recorded in the U.V. wavelength region. A same spectrum was obtained for the stripped hemoglobin at pH 6 and 9.

In the geminate regime, the iron and ligand are coupled together in a spatial cage with dimensions inferior to those of the apoprotein.[9,20] After the photodissociation, a fraction of the liberated CO molecules encounters the geminate iron atoms during their random walk and recombines with the iron, whereas the rest of ligand molecules diffuses in the surrounding medium. A rise in temperature increases the kinetics of the chemical iron-CO binding, and also the protein fluctuations. Thermal motions of the protein increase more the iron-CO distance than the chemical iron activity. Thus, the result of these two opposite actions favors more the random process, consequently the recombination exhibits a diffusional character. So, the nanosecond geminate process observed in the room temperature range reflects the continuous rearrangements of the amino-acid residues and the chains mobility of the apoprotein relatively close to the heme. The ligand binding process is in close relation with the ligand penetration mechanism through the apoprotein.

Temperature Effect on Protein Conformation

The strong variation of ligand binding properties with temperature could be the result of a protein conformational change. Circular dichroism (C.D.) was used in the past to follow changes in the structure of hemeprotein.[22-24] In this work C.D. spectra were performed between 5 to 35°C for human hemoglobin. Thermal energy does not produce any modification in the C.D. spectrum. Thus, the large temperature effect on the position of the ligand binding curve of mammalian hemoglobin is not observed in the protein structure.

Then, we must accept the fact that the protein conformation is not directly correlated with the affinity. The protein contribution expected in the binding variation could reside in the dynamics of the macromolecule, and the vibrational energy of the protein could be correlated with the differences in affinity. With this point of view, internal motions of the apoprotein appear to be the "structural parameter" involved in the regulation of the ligand binding.

Affinity and Protein Mobility

In contradiction to previous suggestions by Perutz,[2] it is now clear that hemeprotein affinity for ligand is not controlled by some subtle changes in the properties of the iron atom.[3,11] Furthermore, although the binding affinity of carp hemoglobin to oxygen is considerably altered by pH ($\simeq$ 200 times), no structural variations can be detected both by C.D. spectra (Fig. 4) and Resonance Raman of the ν_{Fe-His} band.[25] Conformational changes between the constrained (T) and relaxed (R) states seem not to be the origin of the affinity change. This possibility has already been demonstrated on another allosteric enzyme: aspartate transcarbamylase.[26] However, a change in the mobility of the polypeptide chains of the apoprotein could be produced without affecting the protein structure. NMR spectra show an exchangeable proton resonance (9.1 ppm) at low pH which disappears at high pH.[27] From these data it must be concluded that the protein flexibility varies with the pH and conserves the same quaternary structure.

Similarly temperature variation does not result in any modification in the protein structure, but modifies the structural fluctuations of the protein. We do not yet have an idea concerning the distribution of local fluctuations in the protein, but it is obvious that protein fluctuations and protein structure organization are strongly connected. We can imagine that all modifications in the distribution of different local fluctuations of the protein will necessarily correlate with conformational changes in other regions of the protein. So a correlation should exist between conformational structure modifications and local motions changes.

The implication of this model is that the free energy of the affinity changes in hemeprotein for ligand is never localized at the iron site, but must be distributed throughout all the protein.[4,6,7] With this vision, the protein matrix fluctuations and the ligand diffusion inside the different domains of the protein become important problems to investigate.

Acknowledgments

This review is a synthesis of parts of works realized in our laboratory by J. Albani, S. Pin and C. Zentz, with different collaborators from other different laboratories.

References

1. M.F. Perutz, H. Muirhead, J.M. Cox and L.C.G. Goaman, Nature, **219**, 131 (1968).
2. M.F. Perutz, Nature, **228**, 726 (1970).
3. S. Pin, P. Valat, R. Cortes, A. Michalowicz and B. Alpert, Biophys. J., **48**, 997 (1985).
4. J. Wyman, Advan. Protein. Chem., **4**, 407 (1948).
5. J. Monod, J. Wyman and J.P. Changeux, J. Mol. Biol., **12**, 88 (1965).
6. G. Weber, Biochemistry, **11**, 864 (1972).
7. J.J. Hopfield, J. Mol. Biol., **77**, 207 (1973).
8. D.A. Case and M.J. Karplus, J. Mol. Biol., **132**, 343 (1979).
9. B. Alpert, S. El Mohsni, L. Lindqvist and F. Tfibel, Chem. Phys. Letters, **64**, 11 (1979).
10. A. Bianconi, Appl. Surf. Sci., **6**, 392 (1980).
11. S. Pin, R. Cortes and B. Alpert, Febs Letters, in press.
12. R.W. Noble, L.J. Parkhurst and Q.H. Gibson, J. Biol. Chem., **245**, 6628 (1970).
13. A.L. Tan, A. De Young and R.W. Noble, J. Biol. Chem., **247**, 2493 (1972).
14. P. Valat and B. Alpert, Laser Chem., **4**, 173 (1985).
15. J.R. Lakowicz and G. Weber, Biochemistry, **12**, 4161 (1973).
16. E. Gratton, D.M. Jameson, G. Weber and B. Alpert, Biophys. J., **45**, 789 (1984).
17. F.J.W. Roughton, A.B. Otis and R.L.J. Lyster, Proc. Roy. Soc., **144**, 29 (1955).
18. A. Rossi Fanelli and E. Antonini, Arch. Biochem. Biophys., **77**, 478 (1958).

19. J. Albani, B. Alpert and D.M. Jameson, Eur. J. Biochem., submitted 1986.

20. S. Pin, P. Valat, H. Tourbez and B. Alpert, Chem. Phys. Letters, **128**, 79 (1986).

21. L. Lindqvist, S. El Mohsni, F. Tfibel, B. Alpert and J.C. Andre, Chem. Phys. Letters, **79**, 525 (1981).

22. S. Beychoc, I. Tyuna, R.W. Benesch and R. Benesch, J. Biol. Chem., **242**, 2460 (1967).

23. S.R. Simon and C.R. Cantor, PNAS, **63**, 205 (1969).

24. M.F. Perutz, J.E. Ladner, S.R. Simon and C. Ho, Biochemistry, **10**, 2163 (1974).

25. M. Coppey, S. Dasgupta and T.G. Spiro, Biochemistry, **25**, 1940 (1986).

26. G. Hervé, M.F. Moody, P. Tauc, P. Vachette and P.T. Jones, J. Mol. Biol., **185**, 189 (1985).

27. D. Dalvit, S. Muira, A. De Young, R.W. Noble, M. Cerdonio and C. Ho. Eur. J. Biochem., **141**, 255 (1984).

Cytochrome Oxidase in Energy Transduction

M. Brunori* and P. Sarti
Dept. of Biochemical Sciences and CNR Center of Molecular Biology
University of Rome La Sapienza, Rome, Italy

G. Antonini and F. Malatesta
Dept. of Experimental Medicine and Biochemical Sciences
University of Rome Tor Vergata, Rome, Italy

M.T. Wilson
Dept. of Chemistry
University of Essex, Colchester, UK

Summary

Cytochrome-c-oxidase, an integral membrane protein of the inner mitrochondrial membrane, is the terminal enzyme of the respiratory chain. This paper summarizes the role of this enzyme in energy conservation, and presents a possible regulatory model based on $\Delta\mu H^+$ linked conformational changes of the macromolecule.

Introduction

In recent years, a wealth of novel information on the structure and function of membrane proteins has appeared in the literature. Among others, proteins involved in energy transduction have been actively and successfully investigated. The interesting information obtained on the structure and topology of bacteriorodopsin and cytochrome-c-oxidase (by electron microscopy and image reconstruction)[1,2] has been followed by the resolution of the tridimensional structure of the photosynthetic reaction center by x-ray crystallography.[3]

A general feature which has emerged from biochemical and biophysical studies of energy transducing membrane proteins is related to their vectorial organization across the biological membrane. In the case of cytochrome oxidase, it was found[2,4] that the protein assembly spanning the membrane is very asymmetric with reference to the plane passing through the center of the bilayer, a feature which is immediately associated to a vectorial function of this enzyme.[5-7]

---

* Fogarty Scholar in residence, F.I.C., N.I.H. (Bethesda, MD 20205, USA).

The interaction of the protein moiety with the membrane plays an important role in the overall modulation of activity. Binding of specific components of the mitochondrial membrane (such as cardiolipin) to cytochrome oxidase is considered to be crucial to catalysis,[8] although other phospholipids may play a similar role. But over and above specific complex formation, the membrane is crucial to energy transduction because it provides an ion-impermeable physical separation between two compartments. Full appreciation of the significance of this role of a biological membrane has come from acceptance of Mitchell's chemiosmotic theory.[9,10]

In what follows, we shall first summarize some general properties of a molecular transducer, and then present an analysis of the functional properties of cytochrome oxidase with special reference to its role in energy conservation.

Some General Features of a Molecular Transducer

Any molecular transducer has to fulfill some basic requirements to achieve the purpose of transporting a small molecule against the diffusion gradient. These may be summarized as follows, with reference to the scheme depicted in Fig. 1:

a) The transducer has to be the site of linkage between two chemical reactions, involving two ligands X and Y, one of which provides the free energy for the uphill transport of the other.[11,12] The linkage may be based on an allosteric type of mechanism and, in the case of proton pumps, we are dealing with heterotropic effects which demand the coupling of a ligand X (such as oxygen or electron) to a proton (in this case indicated by Y).[13]

b) The transport has to be exerted between two physically separate compartments (A and B in Fig. 1) which, in principle, should be completely impermeable to the ligands X and Y. The two compartments may be close in space with the transducer spanning the membrane (the more common case considered below), or may be removed from one another and be located in different parts of the body (as exemplified by the blood in the case of fish hemoglobins).[11,14]

c) The transducer should contain a mobile element, which is an essential part of the process of transport insofar as it makes translocation of ligand Y between the two compartments A and B irreversible. This mobile element may be represented by a different mechanical system external to the transducer, in the special case of blood.[11]

When we restrict our interest to proton pumps, the linkage is analogous to the classical Bohr effect, which in hemoglobin involves the interaction between the oxygen binding sites (the hemes) and proton binding sites represented by amino acid side chains of the protein moiety.[13,15] Although it is not the subject of this article, it may be worthwhile to recall that the Bohr effect in hemoglobin has been accounted for on the basis of a classical allosteric model[16,17] in which the relative population of the two functionally different states of the macromolecule is affected by pH.

Cytochrome-c-oxidase (EC 1.9.3.1) As A Molecular Transducer

Cytochrome oxidase is the mitochondrial enzyme (fundamental to aerobic life) which catalyzes the electron transfer from reduced cytochrome c to dioxygen (which is reduced to water), according to

$$4c(+2) \; + \; 4H(+) \; + \; O_2 \; \xrightarrow{\text{Oxidase}} \; 4c(+3) \; + \; 2\,H_2O \qquad (1)$$

This reaction, the last step of the redox respiratory chain, is an exoergonic process whose excess free energy is coupled to ATP synthesis (site III). According to Mitchell's chemiosmotic theory,[9] this process is driven by the proton-motive force (electrochemical gradient, $\Delta\mu H^+$) maintained by the redox reactions along the components of the respiratory chain.

Cytochrome oxidase contributes to energy conservation in two ways: (i) the reduction of dioxygen to water consumes protons from the mitochondrial matrix in a scalar reaction (scalar protons, with the stoichiometry indicated by Eq. 1); and (ii) the net translocation of protons from the matrix to the cytosolic side (vectorial protons), coupled to the redox reaction. The latter function of the enzyme has been actively debated in the past, but is now largely accepted and it has been measured under a variety of experimental conditions both in intact mitochondria[18] and in proteoliposomes (i.e., small unilamellar phospholipid vesicles containing purified cytochrome oxidase, called below COV).[19,20]

Several interesting questions concerning the proton-pumping activity of cytochrome oxidase have been raised and addressed by different scientists in the last years, such as: (i) What is the ratio between electrons transferred to dioxygen and protons translocated (the so-called pump's stoichiometry), and is this a fixed or a variable

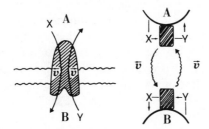

Figure 1. Schematic drawing of a molecular transducer with its basic components. *Left:* The more familiar case of a transducer spanning the membrane which separates two compartments, A and B. *Right:* The transducer is transported from one compartment to another by a different "mobile element" (the circulating blood in the case of hemoglobin). X and Y indicate the two coupled ligands: v is the velocity of the "mobile element" of the pump.

quantity? (ii) What is the structural basis of the pump and the role of the various subunits of the enzyme in controlling translocation? (iii) What is the mechanism of transduction, in terms of trigger and control processes of the pump?

Some of these questions are being actively investigated taking advantage of the increasing experimental information available on this enzyme, in spite of the limited resolution on its tridimensional structure. Application of transient kinetic methods to the study of the reaction of cytochrome oxidase (and to the effects of the electrochemical gradient on activity) has provided novel and relevant information, some of which is reported below.

A Model for the Control of Cytochrome Oxidase Activity

The rate of cytochrome c oxidation and/or oxygen consumption catalyzed by cytochrome oxidase reconstituted into small unilamellar liposomes is affected by the presence of ionophores and/or uncoupling molecules, similarly to what has been observed in tightly coupled mitochondria (respiratory control[5,10,21]).

As shown in Fig. 2, the apparent rate constant for the oxidation of reduced cytochrome c in air increases by a factor of approximately 10-fold when the reaction is carried out in the presence of valinomycin (a specific K^+ carrier) and FCCP (a proton carrier).[10] Thus the structural environment which the oxidase experiences in the mitochondrion is artificially reproduced by making use of artificial phospholipid vesicles in which the enzyme is vectorially inserted in the bilayer with the cytochrome c binding site facing the bulk. This preparation provides an excellent experimental system to test the effect of microcompartmentation on the redox and proton pumping activities of the enzyme, also in view of the fact that the scalar protons (see Eq. 1) are consumed in the internal space of the liposome.[22,23]

Analysis of transient and steady-state kinetic experiments carried out with COV following either the time course of oxidation of cytochrome c added externally, or the time course of proton ejection in the bulk, both in the absence and in the presence of ionophores, have led[24,25] us to propose a model for the control of cytochrome oxidase activity. This model emphasizes the linkage between the onset of the electrochemical gradient across the membrane and the conformational state (and thus the catalytic efficiency) of the enzyme. According to this model, cytochrome oxidase reconstituted into liposomes exists in two distinct states, which are in rapid equilibrium, according to the scheme reported in Fig. 3.

This model is based on the following features:

a. Both states of the enzyme, P and S, are catalytically competent, i.e., they are both capable of catalyzing the reaction depicted in Eq. 1 with proper stoichiometry;

b. The two states have different turnover number (indicated by k(P) and k(S)), due to a different rate constant for the intramolecular kinetic step assigned to the electron transfer from cytochrome a-Cua to the oxygen binding binuclear center, cytochrome a3-Cua3, which is rate limiting under these conditions;[26]

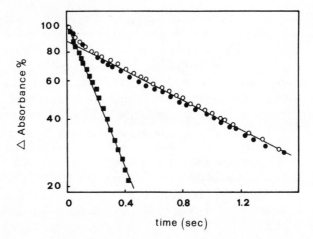

Figure 2. Time course of the oxidation of cytochrome c by cytochrome oxidase reconstituted into small unilamellar vesicles, in the presence of oxygen. Open circles: coupled COV; filled circles: the same in the presence of nigericin (an electroneutral proton carrier); filled squares: the same in the presence of valinomycin and FCCP. Analysis according to first order kinetics.

c. Proton translocation (with an apparent stoichiometry of 0.9 H$^+$/e$-$) is a unique property of the P state (P=pumping), while the S state, though competent in electron transfer, is failing to pump protons. Therefore this has been designated as S (S=slipping), following the proposal of other authors[27] that the mitochondrial system may be "burning" redox energy without proton translocation;

d. The population ratio (P/S) is coupled to the value of the electrochemical gradient across the bilayer; thus the P state is the prevailing one in the absence of the gradient (with k(P) $\gg$ k(S)), while the onset of the gradient above a critical threshold value stabilizes the S state.

Quantitative analysis of the time course of cytochrome c oxidation by COV leads to the conclusion that the effect of $\Delta\mu H^+$ is fully developed after approximately the first turnover, leading to a stabilization of the S state (as indicated by the inhibition of the redox reaction rate, and the parallel loss of acidification in the bulk).

Figure 3. Schematic representation of the model (see text). P stands for Pumping; S for Slipping. k(P) and k(S) represent the catalytic rate constants for the two states of the oxidase.

Moreover it was found that the electrical component of the gradient[25] is the dominant potential in controlling the equilibrium constant between states. An important point which emerged from the analysis is the independence of the first order rate constant of cytochrome c oxidation on the increase in the value of the electrochemical gradient above a critical threshold, which implies that the phenomenon is not proportional to the gradient but acquires a "saturation" value. Given some basic information on the geometry of the system, its electrical properties[28] and the number of oxidase molecules present per liposome,[29,30] an electrical potential difference of $150-200$ mV is sufficient to shift the enzyme from the P to the S state. It is of some interest that the corresponding free energy gap is not very different from that characteristic of the two allosteric states in hemoglobin ($\Delta G° \simeq 4$ kcal/mole).[17] It is not difficult to see that this effect discovered for oxidase is indeed reminiscent of the behavior of voltage-gated ion channels, whose conductance is linked to the value of the electrical potential across the axonal membrane.[31]

Concluding Remarks

The kinetic data outlined above have led us to propose a simple and general model for the control of cytochrome oxidase activity by the electrochemical potential across the (ion-impermeable) bilayer of the liposomal membrane. This model is based on a linkage between the value of $\Delta\mu H^+$ and the conformational state of the enzyme, which in turn controls the rate of the electron transfer reaction and the efficiency of coupling to the proton pump. The transition between the two functional states (albeit still derived only from kinetic data) is reminiscent of other regulatory effects in functioning macromolecules, such as the allosteric transition in hemoglobin involving quaternary structural changes of the protein. In the latter case the "state function" is controlled by the number of oxygen molecules bound to the tetramer and/or by allosteric effectors (such as organic phosphates and protons); in the case of cytochrome oxidase, the electrochemical gradient across the membrane acts similarly to an allosteric ligand, controlling the structural state, catalytic turnover number and energy transducing efficiency of the enzyme. This model has all the elements of regulation and control, and is susceptible to experimental tests, which are in progress. Its applicability and significance in the control of mitochondrial energy transduction remains, however, an open interesting question.

Acknowledgments

Work partially supported by a Grant from the Ministero della Pubblica Istruzione of Italy and from the European Community (CEE Stimulation Action 086-J-C.CD). The authors express their thanks to Mr. E. D'Itri for skillful technical assistance.

References

1. R. Henderson and P.N.T. Unwin, Nature, **257**, 28 (1975).
2. J.F. Deatherage, R.Henderson and R.A. Capaldi, J. Mol. Biol., **158**, 487 (1982).
3. J. Deisenhofer, O. Epp, K. Miki, R. Huber and H. Michel, J. Mol. Biol., **180**, 385 (1984).
4. R.A. Capaldi, F. Malatesta and V. Darley-Usmar, Biochim. Biophys. Acta, **726**, 135 (1983).
5. M. Wikström, K. Krab and M. Saraste, "Cythochrome oxidase. A synthesis," Academic Press (1981).
6. M. Brunori and M.T. Wilson, Trends Biochem. Sci., **7**, 295 (1982).
7. A. Azzi, Biochim. Biophis. Acta, **594**, 231 (1980).
8. S.B. Vik, G. Georgevich and R.A. Capaldi, Proc. Nat. Acad. Sci., **78**, 1456 (1981).
9. P. Mitchell, Biol. Rev., **41**, 445 (1966).
10. D.G. Nichols, "Bioenergetics," Academic Press, London (1982).
11. J. Wyman, Acad. Naz. Lincei, Ser VIII, **64**, 409 (1978).
12. T.L. Hill, Trends Biochem. Sci., **2**, 204 (1977).
13. J. Wyman, Adv. Protein Chem., **19**, 223 (1964).
14. M. Brunori, M. Coletta, B. Giardina and J. Wyman, Proc. Nat. Acad. Sci., **75**, 4310 (1978).
15. E. Antonini and M. Brunori, "Hemoglobin and Myoglobin in their reactions with ligands," North Holland, Amsterdam, 21 (1971).
16. M.F. Perutz, Nature, **228**, 726 (1970).
17. J. Monod, J. Wyman and J.P. Changeux, J. Mol. Biol., **12**, 88 (1965).
18. M. Wikström, Nature, **266**, 271 (1977).
19. E. Sigel and E. Carafoli, Eur. J. Biochem., **89**, 119 (1978).
20. R.P. Casey, M. Thelen and A. Azzi, J. Biol. Chem., **255**, 3994 (1980).
21. P. Hinkle, J.J. Kim and E. Racker, J. Biol. Chem., **247**, 1338 (1979).
22. J.M. Wrigglesworth, Proc. FEBS Meet. 11th, **45**, 95 (1978).
23. P. Sarti, A. Colosimo, M. Brunori, M.T. Wilson and E. Antonini, Biochem. J., **209**, 81 (1983).
24. P. Sarti, M.G. Jones, G. Antonini, F. Malatesta, A. Colosimo, M.T. Wilson and M. Brunori, Proc. Nat. Acad. Sci., **82**, 4876 (1985).
25. M. Brunori, P. Sarti, A. Colosimo, G. Antonini, F. Malatesta, M.G. Jones and M.T. Wilson, EMBO J., **4**, 2365 (1985).
26. M.T. Wilson, J. Peterson, E. Antonini, M. Brunori, A. Colosimo and J. Wyman, Proc. Nat. Acad. Sci., **78**, 7115 (1981).
27. D. Pietrobon, G.F. Azzone and D. Walz, Eur. J. Biochem., **117**, 389 (1981).
28. J.M. Wrigglesworth, J. Inorg. Biochem., **23**, 311 (1985).
29. M. Müller and A. Azzi, J. Bioenerg. Biomembranes, **17**, 385 (1985).
30. Manuscript in preparation.
31. A.L. Hodgkin and A.F. Huxley, J. Physiol., **117**, 500 (1952).

Proteins and Glasses

Hans Frauenfelder
Department of Physics
University of Illinois at Urbana-Champaign
1110 West Green Street
Urbana, IL 61801, USA

Protein Models

Models are intermediate way stations on the road to a complete theory of protein motions and protein dynamics. All roads lead to Rome and many models may contain some of the right ingredients for a successful theory. We can learn from nuclear theory, where apparently contradictory approaches, the liquid drop and the nuclear shell model, were both way stations to a unified description of nuclear dynamics. Here we describe some similarities between proteins and glasses, in particular spin glasses. This similarity does not mean that proteins *are* glasses but that essential physical characteristics are common. Since many more theorists work in the fields of glasses and spin glasses than in proteins, we may be able to borrow from their results or even entice them to join our efforts. On the other hand, proteins have the advantage of 3 1/2 Gy of R&D and there may be many experiments that can be performed more easily and more reliably with proteins than with glasses. The various fields consequently may be able to progress faster together than individually.

In the present paper we sketch the salient experimental facts and describe the most important concepts that connect proteins, glasses, and spin glasses. Details can be found in the references.

Experimental Facts

A number of experimental observations lead to the realization that proteins, glasses, and spin glasses share some characteristics:

(i) Nonexponential Time Dependence. We found in 1973 that the binding of carbon monoxide (CO) to myoglobin (Mb) is nonexponential in time below about 200 K.[1] In a good approximation, the rebinding function $N(t)$ can be written as

$$N(t) = N(0) (1 + t/t_o)^{-n}.$$

(1)

169

Here N(t) is the fraction of Mb molecules that have not rebound a CO at time t after dissociation; t_o and n are temperature-dependent parameters. A similar behavior is found in all heme proteins studied and appears to be a general property of proteins.[2] In glasses and spin glasses, nonexponential time dependence is also observed frequently.[3-6]

(ii) Inhomogeneous Systems. Nonexponential time dependence can arise from the fact that each protein molecule has a different binding rate (inhomogeneous system) or that each molecule is capable of nonexponential binding (homogeneous system). Using repeated flashes ("hole burning in time") we have shown that proteins at low temperatures are inhomogeneous.[7,8]

(iii) Debye-Waller Factor. The protein inhomogeneity can be observed directly by studying the Debye-Waller factor of the individual non-hydrogen atoms in a protein.[9-11] With careful evaluation some of the individual side chains can be seen to assume different positions, thus contributing to large Debye-Waller factors.[12]

(iv) Mössbauer Effect. The Mössbauer effect in heme proteins shows two characteristic features: The recoilless fraction f(T) (Lamb-Mössbauer factor) decreases precipitously above about 180 K,[13,14] and the sharp elastic line is accompanied by a broadened quasielastic one.[15,16] Rayleigh scattering also shows a sharp decrease above a certain temperature.[17] The sharp decrease of f(T) above a critical temperature is also seen in glasses.[18]

(v) Specific Heat. The specific heat of proteins below 1 K is similar to that of glasses and suggests a few tunnel states per molecule.[19,20]

(vi) Proteinquakes.[21,22] Consider a protein reaction, for instance the photodissociation of MbCO, MbCO + light → Mb + CO. The protein structures in the liganded and the unliganded state are slightly different.[12,23] The protein consequently must rearrange its structure after the Fe-CO bond is broken. This rearrangement occurs in a series of fims (functionally important motions) that release the strain at the active site. Such a reaction is similar to an earthquake and we call it a proteinquake. The released strain energy is dissipated in the form of waves or through the propagation of a deformation. In the photodissociation of MbCO, consideration of many different experiments shows that the proteinquake occurs in a series of steps,

$$ MbCO \xrightarrow{\hbar\omega} Mb^*_4 \xrightarrow{fim\ 4} Mb^*_3 \xrightarrow{fim\ 3} Mb^*_2 \xrightarrow{fim\ 2} Mb^*_1 \xrightarrow{fim\ 1} deoxyMb. \qquad (2) $$

Mb^*_4 to Mb^*_1 are intermediate protein states. Fim 4 occurs rapidly even at 3 K, fim 3 takes place near 20 K, fim 2 starts at about 40 K, and fim 1 sets in near 200 K.

Details concerning the proteinquake are given in Ref. 23. For the discussion here, the two last motions, fim 2 and fim 1, are of particular interest. Both relaxations are nonexponential in time. Fim 2, which so far has been studied in greatest detail, can be shown to be complex: not only the ensemble average is nonexponential in time, even the motion in each individual protein must be nonexponential.[22]

Concepts

The experimental facts described above lead to a number of concepts. While these concepts are obtained predominantly from work with heme proteins, most often Mb, it is likely that they apply to essentially all biomolecules. We briefly describe the main concepts here.

(1) *Distributions.* The experimental observations (i) and (ii) together yield some important concepts. The nonexponential time dependence of ligand binding can formally be decomposed by writing

$$N(t) = \int dk \, f(k) \, \exp\{-kt\} \tag{3}$$

Nonexponential binding thus implies a *distribution* of binding rate coefficients. Eq. 3 contains nothing new; it is simply a reformulation of Eq. 1. Together with observation (ii), however, (i) yields something new. Observation (ii) proves that each protein can be characterized by a single rate coefficient k. The observed nonexponential time dependence then must be caused by the fact that different proteins (of exactly the same primary sequence) at low temperatures have different binding rates.[7,8] Assuming further that binding is governed by an Arrhenius transition over a barrier of height H so that

$$k(H) = A \, \exp[-H/RT], \tag{4}$$

the distribution f(k) can be caused by a distribution in A, in H, or in both. The experiments rule out a distribution in A only but can be adequately described by one in H alone. The expansion

$$N(t) = \int dH \, g(H) \, \exp\{-k(H)/RT\}, \tag{5}$$

with a temperature-independent enthalpy distribution g(H) describes the data well from 40 to 160 K. A comparison of the binding of CO and O_2 to various heme proteins shows that binding is dominated by steric and not by electronic factors.[24] An alternate explanation[25] of the nonexponential time dependence, based on the influence of the solvent, is unlikely to be correct because it assumes binding to be governed by the electronic matrix element.

(2) *Conformational Substates (CS).* A simple explanation for the nonexponential ligand binding invokes conformational substates.[7,9] A protein in a given state can

assume a large number of structurally somewhat different substates, each with a different activation enthalpy H for ligand binding. Below a critical temperature transitions among substates are absent, each protein is locked into a particular substate, and binding is nonexponential in time. Above the critical temperature transitions among the substates can occur and binding can become exponential in time. The concept of substates led to the reevaluation of the Debye-Waller factor[9,11] and it is consistent with the data from Mössbauer effect and also with an analysis of optical and Raman information.[26]

Conformational substates imply that the ground state of a protein is highly degenerate; the Gibbs energy of the protein possesses many energy valleys separated by high mountains.

(3) *States and Substates.* The existence of CS implies that we should distinguish between states and substates. A protein in a given state, say MbCO, can assume a large number of CS.

(4) *Equilibrium Fluctuations (EF) and fims.* In a given state at sufficiently high temperature, the protein can fluctuate from substate to substate.[27] As the temperature is lowered, a protein may be frozen into a particular substate. We thus expect that not only the activation enthalpy described above is distributed, but that many protein properties must be described by distributions.

In the transition from one state to another, both fims and EF are involved and the general description of a reaction can become very complex.[22]

EF and fims are not completely independent, but are in favorable cases connected by fluctuation-dissipation theorems.[28-32]

(5) *Hierarchy of Substates.* The observation of a sequence of fims described in Eq. 2 implies that substates are arranged in a hierarchy. The detailed discussion[22] shows that there is evidence for four tiers of substates. The arrangement of substates in MbCO consequently is as sketched in Fig. 1.

Comparison with Glasses and Spin Glasses

It is remarkable that many of the concepts that emerge from studies with proteins have often analogs in amorphous solids. We sketch the important similarities here.

(a) *Disorder or Nonperiodicity.* Amorphous solids, glasses, and spinglasses are not periodic, they are disordered.[33-36] Proteins also are disordered, they are not spatially periodic and possess no obvious symmetries. The arrangement of the amino acids in the primary sequence is not periodic and the folding into the tertiary structure increases the "disorder."

(b) *Frustration.* Consider a system of three atoms with spins as shown schematically in Fig. 2. Assume that the interaction between the atoms is such that antipar-

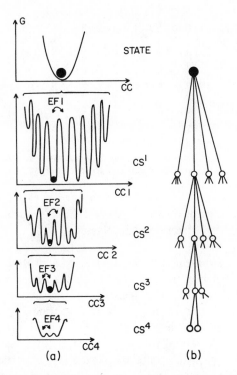

Figure 1. Hierarchical arrangement of the conformational substates in myoglobin. **a)** Schematic arrangement of energy surfaces. **b)** Tree diagram. G is the Gibbs energy, CS denotes conformational substates.

allel spins are favored. The system is then frustrated.[37] If atoms a and b have antiparallel spins, atom c is told to be up by a and down by b. The system has two states of equal energy, separated by a barrier.

In proteins frustration may exist because of steric effects.[38] A particular side chain may be pushed into one position by one adjacent atom, into another position by another neighbor. Proteins consequently may be frustrated in the same sense that the three spins in Fig. 2 are frustrated.

(c) *Energy Valleys.* If a system is both disordered (a) and frustrated (b), it no longer has a unique ground state, its ground state is multiply degenerate. The system possesses many energy valleys that are separated by high mountains.[39-41] In the case of proteins, these valleys can be identified with the conformational substates which were introduced on the basis of the experimental data (i) and (ii).

(d) *Replica Breaking.* In a perfect crystal, the exchange of two atoms leads to a state that is indistinguishable from the initial one. In proteins, however, different substates have different properties, for instance different activation enthalpies. In the language of spin glasses, different substates would be called replicas. The fact

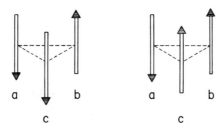

Figure 2. Frustration. A system of three spins, in which two neighbors favor the antiparallel orientation, does not have a unique state of lowest energy.

that different substates are not identical can be called "replica symmetry breaking," a concept well known in the theory of spin glasses.

(e) *Nonergodicity and Timescales.*[40] We consider a protein with substates and assume that a protein can hop from CS to CS with a rate $k_r = 1/\tau_r$, where τ_r is the corresponding relaxation time. The response of the system to an experimental observation depends on τ_r and on the characteristic time t_{obs} of the experiment. If $\tau_r \ll t_{obs}$, the system passes through all substates during the experiment and it is *ergodic*. The CS play no essential role and the multi-valley system can be replaced by a single well. If $\tau_r \gg t_{obs}$, each protein is frozen into a particular substate during the measurement and the system is *nonergodic*; the experiment measures a mixture over all substates. Thus the concept of nonergodicity, or broken ergodicity, depends on the characteristics of the system and the observation. If τ_r depends on temperature, for instance through an Arrhenius relation, the properties of the system and the observation will depend strongly on T. These remarks apply equally well to proteins and amorphous solids.

(f) *Hierarchical Systems.* Experimentally, the evidence for a hierarchical arrangement of the substates in amorphous solids may be less clear than in proteins. Theoretically, however, considerable effort has been devoted to the exploration of hierarchical structures.[40,42,43]

In a hierarchical structure, the question of nonergodicity (e) appears in a new light. In each tier of substates, the relaxation time has to be considered separately. One tier may be ergodic while a higher one may still be frozen and hence nonergodic. Moreover, because of the replica breaking (d) relaxation in some or all tiers may be nonexponential in time, making detailed studies difficult. As compensation, however, the nonexponential character gives information on the underlying distribution(s).

(g) *Relaxation Patterns.* The nonexponential time dependence of relaxation phenomena has a long and distinguished history, dating back to Gauss and Weber in 1835.[44,45] Among many questions, we mention two: (1) Is the time dependence given by a power law as for instance in Eq. 1 or is it described by a stretched exponential of the form

$$N(t) = N(0) \ \exp\{ - (t/t_0)^\beta\}, \quad 0 < \beta \le 1. \tag{6}$$

(2) Is the system homogeneous or inhomogeneous? In the first case, each individual system shows a nonexponential time dependence, in the second each system is exponential but the ensemble is nonexponential. For amorphous solids both questions have been discussed for many years. In proteins we have some answers to both. Binding of small ligands to heme proteins does definitely not follow a stretched exponential but can be approximated by a power law. Moreover, it is definitely an inhomogeneous process, with each individual protein binding essentially exponentially. The process fim 2 in the proteinquake Eq. 2, on the other hand, may be described by a stretched exponential and it is nonexponential even in an individual protein.

(h) *Ultrametricity.* Nonexponential relaxation leads naturally to hierarchical models.[43,46,47] In 1983 Mézard and coworkers found that hierarchical models can be ultrametric.[42,48] Consider a triangle ABC. In a metric space, the lengths of the three sides satisfy the relation

$$AB \le BC + CA. \tag{7}$$

In an ultrametric space, the relation is

$$AB \le BC = CA. \tag{8}$$

Ultrametric triangles are equilateral or isosceles with the unequal side smaller than the two equal ones.

Ultrametricity makes some definite statements about reactions and phenomena on ultrametric lattices have been studied by a number of groups.[47-51]

We do not yet know whether proteins are indeed ultrametric. A look at Fig. 1 suggests this property but a definite conclusion will require more experimental data.

The discussion given here shows that proteins and amorphous solids have many properties in commmon. Exploration of the connections may well lead to a deeper understanding of the aspects of protein structure and function that are connected with the hierarchical arrangement of conformational substates. Before definite conclusions can be drawn, however, the work must be extended to other proteins, many of the experimental data have to be measured with greater accuracy, and many additional tools have to be focused onto the same systems and processes.

Returning to the introduction we can now suggest that we have found another road to Rome: the ultrametric hierarchical model complements other approaches as for instance molecular dynamics. Together the various models provide a more realistic picture of protein dynamics than that given by each individual approach.

Acknowledgments

The work was supported by Grant PCM82-09616 from the National Science Foundation and by Grant PHS GM18051 from the Department of Health and Human Services. I have learned many of the ideas expressed here from discussions with all my collaborators, Anjum Ansari, Sam Bowne, Joel Berendzen, Ben Cowen, Tim Iben, Todd Sauke, Shyam Shyamsunder, Peter Steinbach, and Bob Young. I have also greatly benefitted from discussions with Peter Wolynes.

References

1. R.H. Austin, K. Beeson, L. Eisenstein, H. Frauenfelder, I.C. Gunsalus and V.P. Marshall, Science, **181**, 541 (1973).
2. F. Stetzkowski, R. Banerjee, M.C. Marden, D.K. Beece, S.F. Bowne, W. Doster, L. Eisenstein, H. Frauenfelder, L. Reinisch, E. Shyamsunder and C. Jung, J. Biol. Chem., **260**, 8803 (1985).
3. "Amorphous Solids," W.A. Phillips, ed., Springer, Berlin (1981).
4. F. Mezei, A.P. Murani and J.L. Tholence, Solid State Comm., **45**, 411 (1983).
5. R.V. Chamberlin, G. Mozurkewich and R. Orbach, Phys. Rev. Lett., **52**, 867 (1984).
6. J. Klafter, A. Blumen and G. Zumofen, Phil. Mag., **B53**, L29 (1986).
7. R.H. Austin, K.W. Beeson, L. Eisenstein, H. Frauenfelder and I.C. Gunsalus, Biochemistry, **14**, 5355 (1975).
8. H. Frauenfelder, in "Structure & Dynamics: Nucleic Acids & Proteins," E. Clementi and R.H. Sarma, eds., Adenine Press, Guilderland, New York (1983).
9. H. Frauenfelder, G.A. Petsko and D. Tsernoglou, Nature, **280**, 558 (1979).
10. P.J. Artimyuk, C.C.F. Blake, D.E.P. Grace, S.J. Oatley, D.C. Phillips and M.J.E. Sternberg, Nature, **280**, 563 (1979).
11. G.A. Petsko and D. Ringe, Ann. Rev. Biophys. Bioeng., **13**, 331 (1984).
12. J. Kuriyan, S. Wilz, M. Karplus and G.A. Petsko, J. Mol. Biol., submitted.
13. H. Keller and P.G. Debrunner, Phys. Rev. Lett., **45**, 68 (1980).
14. F. Parak, E.N. Frolov, R.L. Mössbauer and V.I. Goldanskii, J. Mol. Biol., **145**, 825 (1981).
15. F. Parak, E.W. Knapp and D. Kucheida, J. Mol. Biol., **161**, 177 (1982).
16. E.R. Bauminger, S.G. Cohen, I. Nowik, S. Ofer and J. Yariv, Proc. Natl. Acad. Sci. USA, **80**, 736 (1983).
17. Yu. Krupyanski, F. Parak, D. Engelman, R.L. Mössbauer, V.I. Goldanskii and I. Suszcheliev, Z. Naturforsch., **C.37**, 57 (1982).
18. F.J. Litterst, Nuclear Instr. Meth., **199**, 87 (1982).
19. V.I. Goldanskii, Yu. F. Krupyanski and V.N. Flerov, Doklady Akad. Nauk SSSR, **272**, 978 (1983).
20. G.P. Singh, H.J. Schink, H. von Lohneysen, F. Parak and S. Hunklinger, Z. Phys., **B55**, 23 (1984).
21. H. Frauenfelder, "Structure and Motion: Membranes, Nucleic Acids & Proteins," E. Clementi, G. Corongiu, M.H. Sarma and R.H. Sarma, eds., Adenine Press, Guilderland, New York (1985).

22. A. Ansari, J. Berendzen, S.F. Bowne, H. Frauenfelder, I.E.T. Iben, T.B. Sauke, E. Shyamsunder and R.D. Young, Proc. Natl. Acad. Sci. USA, **82**, 5000 (1985).

23. S.E.V. Phillips, J. Mol. Biol., **142**, 531 (1980).

24. H. Frauenfelder and P.G. Wolynes, Science, **229**, 337 (1985).

25. W. Bialek and R.F. Goldstein, Biophys. J., **48**, 1027 (1985).

26. V. Srajer, K.T. Schomacker and P.M. Champion, Phys. Rev. Letters, submitted.

27. A. Cooper, Proc. Natl. Acad. Sci. USA, **73**, 2740 (1976).

28. L. Onsager, Phys. Rev., **37**, 405 (1931).

29. H.B. Callen and T.B. Welton, Phys. Rev., **83**, 34 (1951).

30. L. Onsager and S. Machlup, Phys. Rev., **91**, 1505 (1953).

31. R. Kubo, Progress Phys., **29**, 255 (1966).

32. P. Hänggi, Helv. Phys. Acta, **51**, 202 (1979).

33. J.M. Ziman, "Models of Disorder," Cambridge Univ. Press (1979).

34. R. Zalle, "The Physics of Amorphous Solids," John Wiley, New York (1983).

35. "Amorphous Solids," W.A. Phillips, ed., Springer, Berlin (1981).

36. Heidelberg Colloquium on Spin Glasses. Lecture Notes in Physics 192, J.L. van Hemmen and I. Morgenstern, eds., Springer, Berlin (1983).

37. G. Toulouse, Comm. Physics, **2**, 115 (1977).

38. D. Stein, Proc. Natl. Acad. Sci. USA, **82**, 3670 (1985).

39. S. Kirkpatrick and D. Sherrington, Phys. Rev., **B17**, 4384 (1978).

40. R.G. Palmer, Adv. Phys., **31**, 669 (1982).

41. G. Toulouse, Helv. Phys. Acta, **57**, 459 (1984).

42. M. Mézard, G. Parisi, N. Sourlas, G. Toulouse and M. Virasoro, Phys. Rev. Lett., **52**, 1156 (1984).

43. R.G. Palmer, D.L. Stein, E. Abrahams and P.W. Anderson, Phys. Rev. Lett., **53**, 958 (1984).

44. W. Weber, Annalen der Physik und Chemie (Poggendorf), **34**, 147 (1835).

45. J.T. Bendler, J. Stat. Phys., **36**, 625 (1984).

46. M.F. Shlesinger and E.W. Montroll, Proc. Natl. Acad. Sci. USA, **81**, 1280 (1984).

47. B. Huberman and M. Kerszberg, J. Phys., **A18**, L331 (1985).

48. R. Rammal, G. Toulouse and M.A. Virasoro, Rev. Mod. Phys., in press (1986).

49. A.T. Ogielski and D.L. Stein, Phys. Rev. Lett., **55**, 1634 (1985).

50. S. Teitel and E. Domany, Phys. Rev. Lett., **55**, 2176 (1985).

51. A. Blumen, J. Klafter and G. Zumofen, J. Phys., **A19**, L77 (1986).

Global Ab Initio Simulations: Study of A Liquid As An Example

E. Clementi, G.C. Lie,
L. Hannon, D.C. Rapaport, and M. Wojcik
IBM Corporation
Data Systems Division, Dept. 48B/MS 428
Neighborhood Road
Kingston, New York 12401

National Foundation for Cancer Research
7315 Wisconsin Avenue
Bethesda, Maryland 20814

I. Introduction

By now it is a rather accepted viewpoint that science can be divided into "laboratory-experimental" and "theoretical-computational." Possibly, however, it is not sufficiently realized that computational science has made "enormous" progress in the last few years because of the availability of very fast, high performance computers, generally referred to as supercomputers. In referring to the "tremendous progress," we do not wish to put our emphasis on the size and the magnitude of the computations, but rather on the *viewpoint*. Indeed, now we can think in terms of *"global simulations,"* namely simulations of a full problem, no longer limited to a specific subspecialized field. Let us clarify with an example related to chemistry and physics; when we consider the motions of an ensemble of molecules of water, we can either think in terms of the motions of the atoms within a single molecule or the motion of one molecule of water within the solvation cavity of other molecules of water (namely within the solvation cell), or we can think in terms of the collective motions where many solvation cells are interacting one with another. We can even go one step further and think in terms of very large systems where traditionally one would use fluid dynamics rather than a discrete representation like quantum or statistical mechanics. In other words, because of supercomputers we are able to analyze an ensemble of molecules of water from quantum, to statistical, to fluid dynamics. In this paper we shall demonstrate the feasibility of our *global* approach by presenting a detailed discussion on simulations of liquid water. This is likely the first application of the *global viewpoint* but surely more and more will follow.

As we have previously pointed out,[1] there are a few simple rules which characterize our *global ab initio simulation* approach. The first rule is that the total model for the simulation can be decomposed into an ordered set of submodels, for example, submodels 1,....,i,....,N. The second rule is that the output of submodel (i - 1) contains all that is needed as the input for the submodel i. The third rule—for a global simulation—is that the input to the first submodel should be very simple and minimal. For example, in the particular case of a study of liquid water by global simulation, all we need as input is to know that the hydrogen atom has one electron, that the oxygen atom has eight electrons, and the masses for the nuclei of hydrogen and oxygen. With this knowledge we should be able, in principle and in practice, to characterize a molecule of water, clusters of water, and a large ensemble of molecules, the characteristics of liquid water at different pressure, and temperature, and, also, liquid water as a fluid.

In general, in a global simulation we overlap different dynamics: quantum, statistical and fluid. In the above three different mechanics, the objects of motions are different. At the *quantum mechanical level* the objects of our study are radiations, nuclei and electrons. For *statistical mechanical studies,* the objects are atoms and molecules. Parenthetically, we should have included not only atoms and molecules but also clusters of atoms and molecules. Indeed, considering, for example, liquid water from a statistical mechanical viewpoint, we may ask ourselves what is the object which is in motion and, depending on the property we are considering, we would conclude either the atoms of a single molecule of water or a molecule of water in its solvation cavity or an ensemble of cavities. In other words, the statement that liquid water is made up by *water molecules* might be a bit too naive since, depending on the observation, either electrons or the solvation cavity might be more relevant. Clearly, for fluid dynamics the object in motion is matter either in the liquid or solid or gaseous state; but here matter is considered not as a discrete representation of atoms or nuclei and electrons but rather as a continuous distribution of density.

In solving the equation of motions it is very important to consider the constraint imposed on the problem. Newton's classical equation plays a central role; indeed, it is essentially all that we need to solve in statistical mechanics. In passing from statistical mechanics to quantum mechanics, we impose the constraint of the quantum numbers. In a somewhat similar way, in passing from classical mechanics to fluid dynamics we impose the constraint of the boundary conditions and of the Reynolds number; the latter specifies the fluid dynamical system as the quantum numbers do in quantum mechanics. It is interesting to note that in quantum mechanics we put the constraint onto the particle itself, whereas in fluid dynamics we place the constraint *externally*, via the specifications of the boundary conditions.

The physical dimensions — the space scales — of the systems analyzed via quantum, statistical, or fluid mechanics are vastly different. Indeed we go from the infinitesimally small, to intermediate, to large volumes. The time scales are also drastically different. At a very basic level, to each one of the above representations there corresponds a different viewpoint, namely a different statistics. Thus we are in the Fermi-Dirac statistics for quantum mechanics, whereas for statistical and fluid dynamics we are in classical statistics with Boltzmann type distributions.

The corresponding equations of motions are somewhat different and in the following we shall consider specifically those of Schroedinger, Newton and Navier-Stokes. Before the advent of supercomputers, we could solve the above equations only for relatively simple systems. To be more specific we can treat problems with up to 50 electrons (with good accuracy); we can consider 500 to 1,000 atoms or molecules in statistical mechanics, and we can routinely solve two-dimensional fluid dynamic problems. However, with supercomputers and modern techniques we can expand these limits and solve up to 500 electrons in quantum mechanics (but only with approximations), and up to 20,000 or even 200,000 particles in statistical mechanics and we can start to realistically solve problems in three dimensions in fluid dynamics. Of course, not only supercomputers are needed but also new algorithms capable of exploiting the new machines and new "models." Let us note — parenthetically — that symmetry considerations have been ignored in stating the above limits.

In the *global simulation*, quantum mechanics overlap statistical mechanics and the latter overlaps fluid dynamics. In the rest of this paper we shall stress very much this point. Among the outputs from molecular mechanics is the ability to express the interaction between molecules in an analytical form. The link between quantum mechanics and statistical mechanics is indeed the interaction potentials. We recall that statistical mechanics deals with systems which are either at equilibrium conditions or at non-equilibrium conditions. Non-equilibrium statistical mechanics offers the most natural link to fluid dynamics.

In concluding this introductory section, we would like to propose the notion that our *global ab initio viewpoint* should be seen as an "assembly line" for the efficient production of information. As exemplified by the textile industry, the introduction of assembly lines in the nineteenth century changed the productivity of the economical society of the time, so in an equivalent way we do expect that the use of "global ab initio simulations" will change the productivity of information in our time. A machine, the mechanical loom, was essential to upgrade productivity in the textile industry. Analogically, the supercomputer is the machine essential to changing our productivity in the field of information. As we move from the simulations of relatively simple problems to more and more complex problems, we cannot afford to be constrained to rather simple modeling as has been done too often in the past; the very complexity of the task of modern society imposes a change in our way to model. Recent problems with space exploration, with nuclear reactors, and with safety, for example in chemical manufacturing, have given us a very clear warning that such change is necessary.

In what follows we shall talk about electrons, atoms, molecules, either individually or by clusters; all in the context of liquid water. Of course, this is *only an example*. Other examples could have been analyzed, e.g. the understanding of the conductivity of electrons and holes in materials of interest to electronic components. Here one would start by simulating matter at the solid state physics level, then understanding the statistical mechanics of dislocation, vacancies and imperfections and eventually arrive at a model where the macro-circuits can be simulated via classical electricity. Another example is the design of an efficient airplane, starting from the study of the material up to a full simulation of flight conditions. The "global" character of the approach — the "assembly line technique" to increase productivity and distribution of information — is the aspect we wish to stress.

Below we shall start with our problem — namely the prediction of the properties of a molecular liquid — first at the quantum mechanical and then at the statistical level up to hydrodynamic limit, and we shall conclude showing the feasibility of using molecular dynamics to solve problems of fluid mechanics.

Some of the main concepts of the "global ab initio simulation" approach have been reported earlier.[1,2]

II. Quantum Mechanics as the First Submodel

From quantum mechanical simulation we can obtain the correct structure of a molecule of water, namely a molecule which has the OH bond length and the HÔH bond angle pretty much in agreement with the best experimental data for a single molecule of water.[3] Again, from quantum mechanical simulation we can obtain a binding energy, dipole moment, quadrupole moment,[4] vibrational frequencies,[5] and the excited spectrum in the visible and the ultraviolet; the agreement of those quantities relative to accurate laboratory experiments depends on the choice of the model adopted in the simulation. Broadly speaking, we can talk about two "models": the first model is the *Hartree-Fock model* which represents the electronic structure of a molecule in the form of a determinantal product of molecular orbitals. As known, the molecular orbitals are generally written as a linear combination of basis sets located at the position of the three nuclei of the water molecule. Often, one uses a Gaussian type basis set of functions which can be variationally optimized and which should yield the best possible single determinantal function, called the *"Hartree-Fock limit."* Despite the enormous number of quantum mechanical simulations in the last 20 years, very seldom does one find molecular computation very near to (namely, within a few hundredths of a kcal) the Hartree-Fock limit. In the case of the molecule of water it is well known[4-6] that in order to reach the near Hartree-Fock limit one needs at least two sets of 3d functions and one set of 4f functions located on the oxygen atom; in addition, 2p and possibly 3d functions located on the hydrogen atom are also required. The best Hartree-Fock type computation has an energy of -76.06682 a.u.[7] and has been obtained with a geometrical basis set[8] at the experimental equilibrium geometry.

As is well known, a single determinantal wave function describes the electronic distribution of the single particle (electron) in the average field of the other electrons and nuclei. Differently stated, the Hartree-Fock model does not account for the correlation between the motion of one electron with the motion of the other electrons. This correlation neglect brings about an error known as the *"correlation energy error."* For example, the Hartree-Fock binding energy for water is 161.5 kcal/mol, but the correct value is 232.8 kcal/mol. A number of techniques have been introduced since 1930 to overcome the problem of the neglect of correlation energy. In molecular physics among the most popular techniques are configuration interaction, C.I., the perturbation approaches and, finally, the density functional approximations. In the configuration interaction technique one uses a large number of determinants and variationally selects the weight of each determinant in the linear combination. The determinants differ insofar as the original orbitals of the Hartree-Fock model are promoted into excited orbitals via either single, double, triple, or quadruple excitations. The well known drawback of this technique is that the linear combination of determinants is very slow converging; indeed, we could even say that it is "not convergent." Despite much work in the past 20 years, a problem as simple as the interaction of two molecules of water represents, even today, a notable challenge for the configuration interaction technique. Perturbational techniques are also well known in the field of quantum chemistry and have become somewhat more popular in the last 10 years. However, we should always remember that any perturbational technique exhibits different reliability for two interactive molecules placed at different distances. As known, at very long distances, perturbation techniques tend to be very accurate, but this might not hold true for very short distances especially for the highly repulsive region of the intermolecular potential. Finally, perturbational techniques reach a practical computational limit at around 30 to 40 electrons. Density functional techniques have long been introduced in quantum chemistry and here the name of Wigner,[9] Gombas[10] and Bruckner[11] are worth mentioning. In the last 15 years, density functional techniques have become somewhat more popular due to the work by Kohn and Sham.[12] Whereas in the early days one would attempt to use density functionals as a correction to the Hartree-Fock energy,[13] more and more today one uses the density functional technique attempting to obtain the total energy.

The above preliminary comments are presented in order to appreciate problems which one faces in attempting to use quantum mechanics to obtain the interaction energy of two molecules of water or, differently stated, to obtain the two-body interaction potential.

In the early 1970s Clementi and coworkers[14] obtained an interaction potential for two molecules of water by fitting the computed energy of many dimers of molecules of water at different orientations and positions. These computations were at a level near the Hartree-Fock limit. The potential was used in Monte Carlo simulations of liquid water: the obtained enthalpy and the pair correlation functions were crude, but unmistakably those for water in the condense phase. This marked the beginning of "ab initio" potentials in statistical mechanics. Shortly thereafter there were a few attempts to introduce as much of the correlation energy correction as possible.[15,16] The simulated oxygen-oxygen pair correlation functions obtained with the improved potentials and Monte Carlo simulations were in reasonable agreement with experimental data. In one such work, Matsuoka, Clementi and Yoshimine[17] computed the energy of interaction between two molecules of water using the configuration interaction technique for 66 different geometries of the dimers. The model selected for fitting an analytical potential to the 66 interaction energies was a three-point charge model; two positive charges were placed at the hydrogen position and a balancing negative charge was placed along the axis bisecting the oxygen atom in the molecular plane. The value of the point charges as well as the position of the negative charge along the bisecting axis were considered fitting parameters. In addition to coulombic terms, other exponential terms were added involving couplings between the hydrogen atoms, the oxygen atoms, and hydrogen with oxygen atoms. Two potentials were derived and one of the two is the so-called MCY potential. Since the configuration interaction points were relatively few and since the overall shape of the interaction potential was assumed to remain close to the one given by the Hartree-Fock potential, Matsuoka, Clementi and Yoshimine kept the same model as in Ref. 14 and added to it new terms which are to be considered mainly as correction terms. Monte Carlo simulations were computed with the MCY potential[18] and yielded most reasonable pair correlation functions, and good agreement with X-ray and neutron beam scattering intensities. Clearly, "ab initio potentials" could be used competitively with empirical potentials. Quantum mechanics was, *de facto*, the submodel which provided the input to statistical mechanics. There were — as expected — disappointing aspects. Among the major sources of error remaining in the MCY computation, we note the following. First, even if the basis set selected was relatively large and with polarization functions, still it had a sizable superposition error. The second error source is related to the configuration interaction technique limited to single and double excitations. As is known, such computations are not "size-consistent"; namely, the correction which is computed at a given internuclear distance might not be uniquely related to the correction at a very different internuclear distance. This problem is particularly sensitive for configurations where the two molecules of water are very near to each other and are in a repulsive part of the potential. It was estimated[17] that the two errors combined bring about an uncertainty of approximately 0.4 kcal, an error which is approximately 8% of the total binding energy of the dimer of water, more than sufficient to account for most of the MCY drawbacks. This potential was later revisited[19,20] with additions to the 66 configuration interaction energies previously mentioned.

Recently we have started a new effort at improving the MCY two-body potential. To start with, Drs. Dupuis and Huang have computed approximately 382 different configurations for the dimer of water using fourth order perturbation theory with a relatively large basis set (about the same as for MCY). The larger samplings of the potential energy surface will hopefully yield a more reliable two-body potential. We note that the perturbational approach eliminates the size-consistency problem. However, the selected basis set has still a sizable basis set superposition error; for this reason Dr. Corongiu has corrected the 382 computations above referenced by subtracting the superposition error. We expect to have, shortly, a notably improved two-body potential[21] (an intermediate potential has already yielded a very reasonable pressure).

As is known, if we consider three molecules of water their interaction is not simply the sum of the three pairs of different dimers composing the trimer of water. The three-body correction can in itself be partitioned into a Hartree-Fock contribution and a correlation correction contribution. The Hartree-Fock contribution can be obtained by considering a large number of trimers of molecules of water with different positions and orientations, one from the other, thus scanning the three-body interaction energy hypersurface. The correlation correction can be obtained equivalently by performing configuration interaction type computation on the trimers. Clearly, if we have already found so many problems in computing dimers of water, we can expect only to be in a much worse situation when considering trimers. Fortunately one should expect that the correlation energy correction to the three-body forces be very small, *nearly zero*. Indeed, the major effect in the three-body correction is the induction energy which is a classical electrostatic effect and therefore well represented by the Hartree-Fock model. To verify this point a small number of trimers of water simulations were performed, both at the Hartree-Fock level and at the configuration interaction level. For the configuration analyzed, the correlation energy correction to the three-body effect was found indeed to be essentially zero.[22]

The form for the three-body correction being essentially an induction energy is the one representing the interaction of bond dipoles placed along the OH bonds in each water molecule; these are induced by the electric field of all the other point charges located on the atoms of the other water molecules.[23] The explicit form we have selected for the three-body correction is reported elsewhere together with the explicit correction corresponding to the four-body forces.[24] The latter correction represents mainly the induced dipole-induced dipole interaction, and therefore can be obtained from Hartree-Fock computations of four water molecules, once the two- and three-body effects are subtracted. In practice it turns out that this needs to be done only as a verification, since the parameters determined for the three-body correction are all one needs to write the expression for the four-body correction.

Let us now improve our two-body model by allowing the molecule of water to vibrate. A rather straightforward way to achieve the goal is simply to consider the potential energy between the two molecules as a sum of two contributions, one arising from the *intermolecular* and the second from the *intramolecular* motions. The parameters expressing the intermolecular interaction should depend also on the intramolecular coordinates to account for a proper coupling between the two molecules. This is, however, an approach which requires some notable care and refitting of the two-body intermolecular potential. For a rough approximation we could neglect the direct coupling between the two molecules; by so doing, however, we could lose possibly important details concerning the hydrogen bonding. An approximate interaction potential of the type above mentioned has been reported by G.C. Lie and E. Clementi rather recently.[25] The intramolecular potential was simply taken over from the *ab initio* computation by Bartlett, Shavitt, and Purvis.[5]

The MCYL potential, as we call the merging of MCY with the Bartlett, Shavitt and Purvis potential,[5] yields a dimer which has a binding energy of 5.94 kcal/mol and an O-O distance of 2.87Å, and α and β angles of about 4 and 37 degrees, respectively. It is possibly not the best dimer one could wish for; indeed its energy is likely from 0.2 to 0.4 kcal too large, the oxygen-oxygen internuclear distance is likely somewhat too short (about 0.05Å), and the β-value too small by about 20 degrees. The MCYL model was used in a molecular dynamics simulation. It is of interest to compare the geometry of the water molecule in the gas phase and in the liquid phase as obtained from experimental data and from the MCYL molecular dynamics simulation.[25] The oxygen-hydrogen separation, R(OH), in the gas phase is 0.9572Å; in the liquid phase, experimentally, it is 0.966 ± 0.006Å and, in our simulation, is 0.975Å. The average hydrogen-hydrogen separation is 1.51 ± 0.03Å from experiment in the liquid phase, and 1.53Å from our simulation. The average variation in the O-H bond in the liquid phase has been determined to be 0.095 ± 0.005Å, but in our simulation this value is 0.023Å, thus bringing in evidence a limitation of the MCYL potential. However, as we shall show in the next section, even if rough, the MCYL can yield interesting results.

Let us now consider the question concerning the verification of all the above modeling. There is not much data to verify all the above quantum mechanical modeling, exception made when we consider a single water molecule. Unfortunately, this is not of much consequence for the interaction potentials of non-vibrating water molecules. Even the water dimer is known only to a rather low accuracy level (the binding no better than within a 5% error, corresponding to about 0.2 kcal/mol) and for higher clusters the structural information is rather meager. (For some computational work on the cluster we refer to an older[26] and a more recent analysis.[27])

An exception to the above statement is the availability of laboratory data for the second and third[28] virial coefficients for steam. But these experimental quantities should be used with more care than has been usual in the past. The prevailing notion asserts that a good two-body potential should yield the second virial *in full agreement* with the experimental values. This view should be "amended" on at least two accounts. First, the second virial is not available for near room-temperature; obviously the steam usually exists at high temperatures (the higher the better), about the opposite for most biologically interesting environments and for most simulations on liquid water. Second, the experimental value of the second virial includes both vibrational and other quantum effects, generally ignored in many potentials but likely contributing from 5% to 10%, or even more.[29]

In Fig. 1 we show the computed[30] and the experimental second virial for the two potentials obtained in Ref. 17 [note that the one called CI(1) in the figure is the MCY], for Stillinger's ST2[31] and for the Hartree-Fock potential.[14] For the third virial coefficient we refer elsewhere (Ref. 32).

Because of the limited experimental data on water clusters, one is compelled to compare the two-body interaction potentials with liquid water, where three- and four-body as well as vibrational corrections are all lumped together into the experimental data.

Not the agreement, but the *controlled deviation from* experimental values should be used to assess the quality of the two-body potential. This is a most difficult task since it requires testing of many aspects characterizing liquid water. We note that the situation is much simpler for an effective two-body potential, since in this case one aims at "agreement" with experimental data on the liquid.

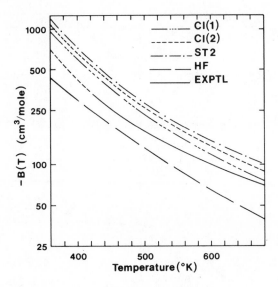

Figure 1. Second virial coefficient from experiments and from computations with empirical (ST2) and ab-initio potentials (Hartree-Fock, MCY and C(2)).

III. Statistical Mechanics as the Second Submodel

The interaction potential obtained via quantum mechanics constitutes the input necessary to obtain a statistical description of many water molecules interacting at a given pressure and temperature.

As is well known, we can consider the ensemble of many molecules of water either at equilibrium conditions or not. To start with, we shall describe our result within the equilibrium constraint, even if we realize that temperature gradients, velocity gradients, density and concentration gradients are characterizations nearly essential to describe anything which is in the liquid state. The traditional methods at equilibrium are Metropolis Monte-Carlo[33] and molecular dynamics.[34] In the Metropolis Monte-Carlo, M.M-C, the temperature is preassigned, and one considers many configurations of water molecules, one related to the next one, in the Markov sense. In the M.M-C algorithm the molecules of water fill in to a desired density, a cubic box which is extended in space through the assumption of periodic boundary conditions. Clearly, the original box has to be large enough to contain a sufficient number of water molecules capable of reproducing all the phenomena with which we are interested. We stress that the assumption of periodicity can introduce artifacts in the statistics for too small boxes since it would be tantamount to introducing an artificial long-range order. Let us note *that most of the simulations on liquid water present in today's literature most likely suffer from this limitation.* A standard way out, to avoid or at least decrease the error associated with a finite size box, is to use long-range corrections; the Ewalds summation[35] and the reaction field[36] are two well-known techniques often used in this context.

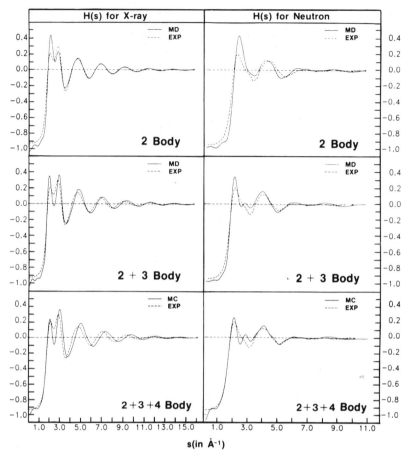

Figure 2. X-ray and neutron beam scattering intensities with MCY, with addition of three- and four-body corrections.

In Fig. 2, we show the x-ray scattering intensity and the neutron beam intensity obtained from simulations of Monte Carlo type and from experimental data.[37] Let us consider the x-ray intensities. The two-body liquid[18] reproduces well the experimental positions of the peaks and the peak-to-peak separation, but not the relative intensity, especially for the initial split peak. In the three-body liquid[23] the intensities show an improvement which is even better in the four-body liquid.[38] In the latter, however, the peaks are somewhat out of phase, an indication that the simulated volume for a single molecule of water is somewhat different from the correct experimental volume. Contrary to tradition, we do not display the pair correlation function for R(O-O), R(O-H) and R(H-H) since these are not direct experimental data and presently are not too accurate.[39-42] For a comparison at the two- and three-body level between Monte Carlo and molecular dynamics, see Ref. 43. All those simulations yield a pressure which is much too large relative to the experimental ones. The high pressure[44] can be removed by a small correction in the MCY potential as we found out recently in the previously quoted work still in progress with Dupuis, Huang, and Corongiu.[21]

From laboratory experiments, the internal energy of liquid water at room temperature is -8.1 kcal/mol. From our simulations, the two-body liquid yields an internal energy of -6.8 kcal/mol. The three-body liquid improves to -7.7 kcal/mol and the four-body liquid brings it to -8.95 kcal/mol. The *quantum correction* decreases the above value by 0.93 kcal/mol, thus yielding a total internal energy of -8.02 kcal/mol, nearly in agreement with the experimental one.

Let us add some detail on the quantum correction, since it affects many measurements, particularly the internal energy and the constant volume heat capacity, C_v. We recall that C_v obtained from temperature fluctuation, following Lebowitz et al.,[45] is 26.5 cal/mol/deg using the MCYL two-body potential.[25] The heat capacity has also been computed in MD simulations using MCY and MCY with three-body correction;[46] the corresponding C_v values are 14.9 kcal/mol and 26.7 kcal/mol, respectively. The latter value does not agree with a Monte Carlo computation using MCY plus three-body, where C_v was computed as 17.32 kcal/mol.[47] We are presently reanalyzing these data in order to rationalize the discrepancies. The experimental value for C_v is 17.9 cal/mol/deg. Let us now return to the quantum correction.

The integral of the Fourier transform of the VACF (see below) corresponds to the number of modes for the atoms of water. Following Berens et al.,[48] all the motions contained in the spectral density (see below) are considered as quantum harmonic oscillators. The intermolecular zero-point energy correction thus found is 15.905 kcal/mol, the vibrational energy 1.703 kcal/mol and the heat capacity 9.050 cal/mol(K). We now treat the vibrations of the isolated water molecules again as quantum vibrators and obtain the zero-point energy, total energy and heat capacity (at 300 K) as 13.56 kcal/mol, 0.0015 kcal/mol and 0.041 cal/(mol K). These corrections bring the MCYL heat capacity to 17.6 cal/(mol K) in notable agreement with experiments. Using MCYL data we have found a quantum correction of 1.342 kcal/mol for the internal energy. One more quantum correction should be added to the internal energy from M.M-C simulation of the MCY liquid, which is due to the intramolecular frequency shift for a water molecule passing from the gas to the liquid phase: these are found to have the value of -0.416 kcal/mol. Thus the total quantum correction to the internal energy obtained from M.M-C simulation of the rigid water model is about 0.93 kcal/mol (namely, $1.342 - 0.416$ kcal/mol) which should be added to the sum of the MCY internal energy corrected by three- and four-body effects. This yields $-8.95 + 0.93 = 8.02$ kcal/mol, to be compared with the experimental value of 8.1 kcal/mol.

Up to now we have considered non-dynamical equilibrium properties, namely time-independent properties; but as we stated above, the richness of a liquid is related to its flow, gradients and dynamics. Let us start now to consider a few dynamical properties of the liquid water.[46] In Figs. 3, 4, 5 and 6 we report the autocorrelation function for the *translational velocity*, $\phi_\alpha(t)$, for the *angular velocity*, $\Omega_\alpha(t)$, and for the *orientational functions*, $C_i^q(t)$, defined below in Eqs. 1, 2 and 3, respectively.

$$\phi_\alpha(t) = \; <v_\alpha(0)v_\alpha(t)> / <v_\alpha^2(0)> \tag{1}$$

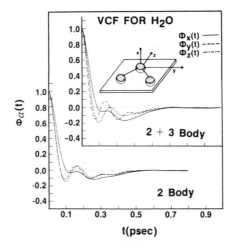

Figure 3. Translational velocity autocorrelation function from two-body MCY with and without three-body corrections.

$$\Omega_\alpha(t) \;=\; <\omega_\alpha(0)\omega_\alpha(t)>\,/<\omega_\alpha^2(0)> \tag{2}$$

Here $v_\alpha(t)$ is the αth component of the molecular translational velocity and $\omega_\alpha(t)$ is the angular velocity about the αth axis at time t. The coordinate system coincides with the principal molecular frame of the molecule at t = 0. In the present convention the molecule lies in the y-z plane with the oxygen lying on the positive z axis and the origin at the center of mass. The angle brackets denote an average over time and over molecules, and the normalization factors $<v_\alpha^2(0)>$ and $\omega_\alpha^2(0)$ are k_BT/m and $k_BT/I_{\alpha\alpha}$, respectively, where m is the molecular mass and $I_{\alpha\alpha}$ is the principal frame $\alpha\alpha$ component of the inertia tensor.

The orientational correlation functions $c_l^\alpha(t)$, is defined as the average of the lth Legendre polynomial of $\cos\theta_\alpha$, where θ_α is the angle between the αth principal axis at some time t = 0 and at a later time t

$$C_l^\alpha(t) \;=\; <P_l(\cos\theta_\alpha)> \tag{3}$$

If the molecule is assumed to move in a harmonic potential well with kinetic energy at the minimum equal to $k_BT/2$ in the x, y and z principal frame directions, the displacement in the αth direction will be

$$r_\alpha(t) \;=\; \frac{k_BT}{\sqrt{<F_\alpha^2>}}\,\sin\sqrt{\frac{<F_\alpha^2>}{mk_BT}}\,t \tag{4}$$

where we have assumed that $r_\alpha(t=0)=0$. Similarly the angular displacement about the αth principal axis will be

$$\theta_\alpha(t) = \frac{k_B T}{\sqrt{<\Gamma_\alpha^2>}} \sin \sqrt{\frac{<\Gamma_\alpha^2>}{I_{\alpha\alpha}k_B T}} \; t \qquad (5)$$

This very simple model provides estimates for the period and amplitude of the oscillatory motions of the molecule within the cage structure in the liquid. The maximum values of θ_x, θ_y and θ_z are found to be 7., 9., 7., respectively, for the two-body liquid, and 6., 9., 6. for the three-body liquid.

Figure 3 shows the *translational velocity autocorrelation functions*, $\phi_\alpha(t)$, for the two-body liquid and the two + three-body liquid, hereafter referred to as the three-body liquid. The $\phi_\alpha(t)$ can be considered to be composed of a slowly varying part (with a period of about 0.6 psec) on top of which is superimposed a high frequency part appearing as the oscillations. In the two-body liquid the ϕ_y and ϕ_z are quite similar to each other and differ from the ϕ_x in the amplitude of the higher frequency oscillations. In the three-body liquid the first zero values in ϕ_α occur sooner, particularly for ϕ_x and ϕ_z, and at longer times the phases of ϕ_y and ϕ_x are more alike.

The self-diffusion coefficient — the integral of $\phi(t)$ over time — is found to be $D = 1.3 \times 10^{-5} \text{cm}^2/\text{sec}$. This is to be compared with the experimental value[49] of 2.3×10^{-5} cm^2/sec and to the value of 2.25×10^{-5} cm^2/sec for the two-body liquid. It could be said that the three-body liquid shows more rigidity in some sense than the two-body liquid. The translational velocity can be considered to be composed of a slowly varying part (with a period of about 0.6 psec.) on top of which is superimposed a high-frequency part appearing as an oscillation. In the two-body liquid the ϕ_y and ϕ_z components are quite similar to each other and differ from the x component in the amplitude of the higher frequency oscillation. In the three-body liquid the first zero value in ϕ_α occurs sooner, particularly for the x and z components, and at longer times the phases of ϕ_x and ϕ_y are more alike.

Let us now interpret the $\phi(\alpha)$ curves in terms of the behavior of a water molecule in its typical environment and let us attempt to understand the changes arising from the three-body forces. As is known, there is little question that the local ordering of water molecules is tetrahedral about a central molecule. The $\phi_x(t)$ is different from the $\phi_y(t)$ and $\phi_z(t)$ because the x motion takes place in a relatively broad potential minimum while there are only half as many such minima for the motion in the y and z directions. As we know, the three-body liquid is more energetic or, equivalently, a molecule of water in a three-body liquid finds itself in a deeper well than in the two-body liquid. The molecule in the deeper well oscillates faster, therefore with higher frequency.

Concerning Fig. 4, *the angular velocity autocorrelation function, $\omega(t)$,* represents oscillations which are much faster than the translational ones, as is clearly seen by looking at the scale of the figure.

In conclusion we obtain the following overall picture: a single molecule of water in a liquid appears to be located within a potential web created by the tetrahedrally oriented neighboring molecules. The single molecule moves about in this well with translational velocity and with angular velocity. The angular process is much faster than the translational process and both are strongly non-isotropic, as can be expected by an overall tetrahedral rather than spherically symmetric arrangement.

Lastly, let us consider the *orientational correlation functions*, $C_l^i(t)$, which are shown in Fig. 5 (inserts a and b) for $l = 1$ and 2. The decay of the rotational order is considerably slower in the three-body liquid, which suggests that the evolution of the cage structure is slower, i.e. that the first shell of neighbors is longer-lived. The y axis continues to reorient more slowly in the three-body liquid with the x and z axis again behaving similarly. The infrared spectrum in the range of vibrational frequency can be related to the Fourier transform of $C_l^i(t)$ through[50]

$$I^{IR}(\omega) \propto x \coth x \int_0^\infty C_1^z(t) \cos \omega t \, dt \qquad (6)$$

where $x = \hbar\omega/2k_B T$.

The *infrared spectrum* is shown in Fig. 6 for the two- and three-body liquids. The three-body forces cause the peak to shift from about 375 cm^{-1} to about 450 cm^{-1}. Experimentally, however, there is a peak at about 700 cm^{-1} so that although the shift is in the right direction the MD results indicate that there are still deficiencies in the potential (the lack of four-body correction and deficiency in the MCY parameterization).

Reorientational relaxation times, τ_{l}^i can be estimated from the assumed exponential decay and those are given in Table I for $l = 1$ and 2. The decay of the orientational correlation is significantly slowed by the three-body forces, and the relaxation times more than doubled in the three-body liquid. The time integral of $C_2^y(t)$ gives an estimate of the NMR relaxation time, τ_{NMR}, associated with intermolecular dipolar coupling. Those are also given in Table I where it can be seen that the three-body forces significantly improve agreement with the experiment.

Let us now consider again the velocity autocorrelation function, this time as obtained from the MCYL potential, and compare it to what we have discussed above for MCY potential with two- or three-body forces. In Fig. 7 we report the velocity autocorrelation function for the oxygen and hydrogen atoms calculated for a temperature of about 300 K. The global shape of the VACF for the oxygen is very similar to what was previously determined for the MCY model. Very notable are the fast oscillations for the hydrogens relative to the oxygen.

The Fourier transform of the VACF, namely the spectral density, are given in Fig. 8. As known, those are related to the infrared spectrum of the liquid water. In Fig. 8 the band centered at about 1,740 cm^{-1} is the intramolecular bending mode while those at 3,648 cm^{-1} and 3,752 cm^{-1} are associated with the intramolecular O-H

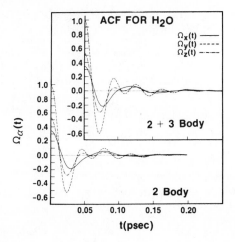

Figure 4. Angular velocity autocorrelation function from two-body MCY with and without three-body corrections.

bond stretches. In going from gas to liquid phase there is an upshift of 55 cm⁻¹ in the bending frequency and downshifts of 198 and 203 cm⁻¹ in the stretching frequencies. These shifts are all in good agreement with experimental IR and Raman results of 50, 167, and 266 cm⁻¹, respectively. Table II compares frequency shifts calculated from various potential models with experimental IR and Raman results. It is clear from the Table that the MCYL results are notably good in reproducing these experimental data. Notice that, since the center of mass of the water molecule is very close to the oxygen atom, the drastic intensity difference between the Fourier transform of hydrogen and oxygen in Fig. 8 allows us to identify immediately that the broad band centered about 500 cm⁻¹ is due mainly to the rotational motion of the molecules, whereas the bands centered around 40 and 190 cm⁻¹ arise from the hindered translational motions. Thermodynamic and transport coefficients from MD simulations with two-body, three-body and vibrational effects are summarized in Table III.

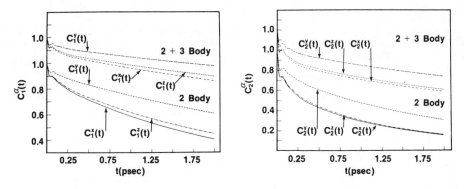

Figure 5. Rotational correlation functions from two-body MCY with and without three-body corrections.

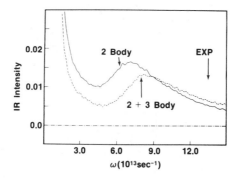

Figure 6. Computed infrared spectrum for vibrational motions from two-body MCY with and without three-body corrections.

IV. Towards the Hydrodynamic Limit: Structure Factors and Sound Dispersion

A non-idle question about liquid water requests the definition of "the particles" existing in a volume filled with a few thousand molecules of water. Obvious, but possibly naive answers identify the particles either with molecules of water or, at a deeper level, with atoms of oxygen and hydrogen or, at an even deeper level, with nuclei and electrons. Well, this may not be correct. Indeed, most likely one should answer that the particles in a volume containing a few thousand molecules of water are *"solvated molecules of water,"* or *"clusters of molecules of water,"* or *"cages of molecules of water."* Considerations of this type bring us to a study of the collective dynamics for liquid water.[25,46b]

The collective properties are most conveniently studied in terms of the spatial Fourier (k) components of the density and particle currents and of the stress and energy fluxes. The time correlation function of those Fourier components detail the decay of density, current, fluctuation on the length scale of the respective $1/k$. Let us recall the definition of microscopic density, $\rho(\vec{r}, t)$, and current, $j(\vec{r}, t)$, at a position $\vec{r}$ of time t, and of the corresponding Fourier components for the density and currents.

Table I. Reorientational relaxation times τ_l^α (in psec) of the $C_l^\alpha(t)$ for two-body and three-body liquid water models. The values in parentheses are $A_2^\alpha \tau_2^\alpha$ and the value of τ_{NMR} is taken from[11].

	τ_1^α		τ_2^α		τ_{NMR}
α	2B	2+3B	2B	2+3B	EXPT
x	2.7	7.2	1.6	3.8	
y	4.9	11.6	2.3 (1.7)	5.5 (4.2)	4.8
z	3.1	8.5	1.6	3.9	

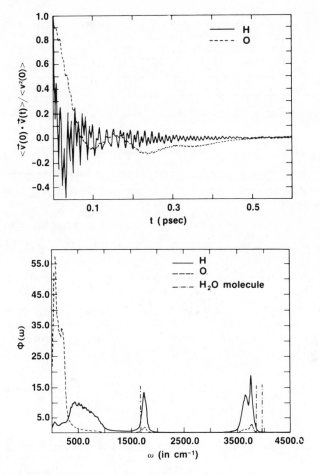

Figure 7. Hydrogen and oxygen velocity autocorrelation function from two-body MCY with vibrations allowed (MCYL), and computed infrared spectrum for intramolecular bending modes and bond stretching.

$$\rho(\vec{r}, t) = \sum_{i=1}^{N} \delta(\vec{r} - \vec{r}_i(t)) \tag{7}$$

$$\vec{j}(\vec{r}, t) = \sum_{i=1}^{N} \vec{v}_i(t)\, \delta(\vec{r} - \vec{r}_i(t)) \tag{8}$$

where $\vec{r}_i$ and $\vec{v}_i$ are the position and velocity of particle i. The Fourier components of the density and current are

$$\rho_{\vec{k}}(t) = \frac{1}{\sqrt{N}} \int_V e^{i\vec{k}\cdot\vec{r}} \rho(\vec{r}, t)\, d\vec{r} = \frac{1}{\sqrt{N}} \sum_{i=1}^{N} e^{i\vec{k}\cdot\vec{r}_i(t)} \tag{9}$$

Table II. Comparison of shifts in intramolecular vibrational frequencies of the water molecules in going from gaseous to liquid phases. All quantities are given in cm-1, negative number indicates down shift in frequency.

Vibrational Modes	Water-Model					Expt'l.e
	CF2a	CCLb	BJHc	WATTS d	MCYL	
μ_1 (symmetric stretching)	307	−118	−322	−152	−198	−167
μ_2 (bending)	224	100	60	91	55	50
μ_3 (asymmetric stretching)	359	−229	−433	−183	−203	−266

aVersion 2 of the central-force model of Stillinger and Raman [J. Chem. Phys., **68**, 666 (1978)]. Results taken from reference given in c.

bA.D. Carney, L.A. Curtiss, and S.R. Langhoff, J. Mol. Spectry, **61**, 371 (1976).

cModified CF2 potential from P. Bopp, G. Jancso, and K. Heinzinger, Chem. Phys. Lett., **98**, 129 (1983).

dR.O. Watts, Chem. Phys., **26**, 367 (1977). Results taken from P.H. Berens, D.H. Mackay, G.M. White, and K.R. Wilson, J. Chem. Phys., **79**, 2375 (1983).

eTaken from D. Eisenberg and W. Kanzmann, "The Structure and Properties of Water" (Oxford University, N.Y., 1969). The stretching frequency assigned to the water molecule in the liquid is taken to be 3490 cm-1, the center of a very broad band in the infrared spectra of liquid water.

$$\vec{j_{\vec{k}}}(t) = \frac{1}{\sqrt{N}} \int_V e^{i\vec{k}\cdot\vec{r}} \vec{j}(\vec{r}, t) \, d\vec{r} = \frac{1}{\sqrt{N}} \sum_{i=1}^{N} \vec{v_i}(t) e^{i\vec{k}\cdot\vec{r_i}(t)} \tag{10}$$

The density and longitudinal and transverse current correlation functions are defined as

$$F(\vec{k}, t) = <\rho_{\vec{k}}(0)\rho_{-\vec{k}}(t)> \tag{11}$$

$$J_l(\vec{k}, t) = <\left[\vec{k}\cdot\vec{j_{\vec{k}}}(0)\right]\left[\vec{k}\cdot\vec{j}_{-\vec{k}}(t)\right]/k^2> \tag{12}$$

$$J_t(\vec{k}, t) = <\left[\vec{k}\times\vec{j_{\vec{k}}}(0)\right]_\alpha\left[\vec{k}\times\vec{j}_{-\vec{k}}(t)\right]_\alpha/k^2> \tag{13}$$

where the angle brackets denote averages over time origins. In the simulation the k vectors are restricted to have the values

$$\vec{k} = (2\pi/L)(l\hat{i} + m\hat{j} + n\hat{k}) \tag{14}$$

where l, m and n are integers, and L = 24.834Å is the side length of the cubical box used in one of our studies, considered in detail below. Specifically, only 19

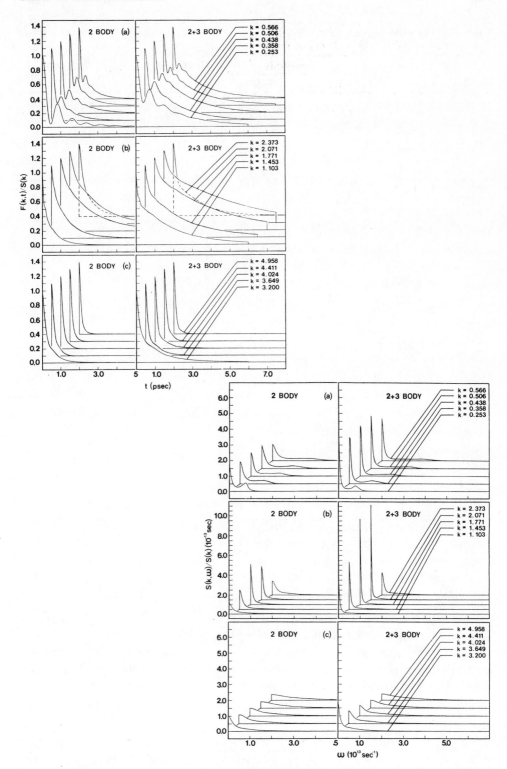

Figure 8. Intermediate scattering function, $F(k,t)$, and dynamic structure factor, $S(k,\omega)$ (left and right), computed from MCY with and without three-body corrections.

Table III. Thermodynamic and transport coefficients for two- and three-body models of liquid water from MD (N = 512) and from experiment.

		Two-Body		Three-Body	Experiment
		MCY	MCYL[a]		
η	kg/m sec	0.6 x 10-3	(0.3 x 10-3)[b]	1.7 x 10 -3	0.9 x 10-3
ϕ	kg/m sec	0.8 x 10-3		1.1 x 10-3	2.5 x 10 -3, 1.2 x 10-3
λ	W/m K	1.35	(0.79)[b]		0.59
G_∞	J/m^3	1.17 x 10^{10}		1.41 x 10 ^{10}	
K_∞	J/m^3	1.13 x 10^{10}		1.17 x 10 ^{10}	
τ_η	sec	0.55 x 10-13		1.18 x 10 -13	
τ_ϕ	sec	1.26 x 10-13		1.57 x 10 -13	
γ_v	bar/K	22.2		30.8	
C_v	J/mol K	70.5	73.6	72.4[c]	75
γ	C_P/C_V	1.08		1.11	1.00
χ_T	/bar	2.15 x 10-5	2.2 x 10-5	2.38 x 10 -5	4.4 x 10-5
χ_s	/bar	1.99 x 10-5		2.14 x 10-5	4.4 x 10 -5
C_T	m/sec	2157	2150	2051	1500
C_s	m/sec	2241		2162	1500
P	bar	8430	7900	8512	200
U	kJ/mol	−28.3	−28.3	−32.7	−33.9
T	K	296	301	304	298
ρ	mol/m^3	(55500)[d]	(55400)[d]	(55500) [d]	55400

[a]From G.C. Lie and E. Clementi, Phys. Rev. A., **33**, 2679 (1986).

[b]From hydrodynamic theory at k = 0.2890A-1.

[c]From Dr. G. Corongiu (private communication) using same potential. The molecular dynamics simulation yielded 111.5 J/mol K.

[d]Input data in the simulations.

values of $|\vec{k}|$ were chosen for study out of all those possible, due to practical considerations. These included all vectors such that $l^2 + m^2 + n^2 = 1 - 6, 8, 10, 19, 33, 49, 67, 88, 120, 160, 208, 253, 304$ and 384, and the number of contributing vectors to each $|\vec{k}|$ varied from 3 to 27. This resulted in a range of $0.253 \leq k \leq 4.958$ Å^{-1}. The wavelength of the phenomenon under study is related to the wavenumber k through $\lambda = 2\pi/k$.

The static structure factor S(k) is given by

$$F(k, t = 0) = <\rho_{\vec{k}}(0)\rho_{-\vec{k}}(0)> = S(k) \qquad (15)$$

which in the $k \to 0$ limit yields the isothermal compressibility (zero frequency bulk modulus) χ_T,

$$\chi_T = -V^{-1}(dV/dP)_T = S(0) / \rho k_B T \tag{16}$$

The frequency spectrum of $F(k,t)$ gives the dynamic structure factor $S(k, \omega)$

$$S(k, \omega) = \frac{1}{2\pi} \int_{-\infty}^{\infty} e^{i\omega t} F(k, t) \, dt \tag{17}$$

which is the quantity most commonly reported in neutron and light scattering experiments, as it is proportional to the measured differential scattering cross-section.

Ordinary hydrodynamic theory[51] gives expressions for $F(k, t)$ and $S(k, \omega)$ that involve the macroscopic transport coefficients, i.e. the shear and bulk viscosities η and ϕ, thermal conductivity λ, the specific heats C_p and C_v, and the adiabatic sound speed c_s.

$$F(k, t) = S(k) \left[\frac{\gamma - 1}{\gamma} e^{-D_T k^2 t} + \frac{1}{\gamma} \cos(c_s k t) e^{-\Gamma k^2 t} \right] \tag{18}$$

$$S(k, \omega) = \frac{1}{2\pi} S(k) \left[\frac{\gamma - 1}{\gamma} \frac{2 D_T k^2}{\omega^2 + (D_T k^2)^2} + \right.$$

$$\left. \frac{1}{\gamma} \left(\frac{\Gamma k^2}{(\omega + c_s k)^2 + (\Gamma k^2)^2} + \frac{\Gamma k^2}{(\omega - c_s k)^2 + (\Gamma k^2)^2} \right) \right] \tag{19}$$

Here, $\gamma = C_p/C_v$, D_T is the thermal diffusivity and Γ is the acoustic attenuation coefficient

$$D_T = \lambda / \rho C_p \tag{20}$$

$$\Gamma = \frac{1}{2} \left[(\gamma - 1) D_T + (\frac{4}{3}\eta + \phi) / \rho m \right] \tag{21}$$

The hydrodynamic equations are valid in the long wavelength limit, that is, at k values corresponding to the wavelength of visible light, i.e. $k \sim 0.001 \text{Å}^1$. There the central (diffusive) Rayleigh peak in $S(k, \omega)$ is well separated from the side (sound) Brillouin peaks. The equations are useful, however, in interpreting the general features in $F(k, t)$ and $S(k, \omega)$ at other k values.

Figure 8 (left) shows $F(k, t)/S(k)$ for the two- and three-body models of water, for the first few wavevectors k = 0.253, 0.358, 0.438, 0.506, 0.566Å^{-1}. In this range of k values both systems exhibit an initial rapid decay followed by a slower oscillatory decay. The rapid decay and oscillations are associated with the compressive elasticity of the fluid while the slower decay at longer times is related to the diffusive mixing of the molecules. The most apparent difference between the two- and three-body liquids is the overall slower decay of spatial order in the latter system. The first minima and maxima in $F(k, t)$ are shifted to somewhat smaller times, indicating that at these wavelengths the three-body liquid is less compressible. The oscillations in $F(k, t)$ are generally more damped in the three-body liquid, which is consistent with the higher viscosities shown in Table III. Thus, sound waves will travel faster and will be more strongly damped than in the two-body liquid.

Figure 8 (right) shows $F(k, t)$ for values of k near the first peak in $S(k)$, i.e. $\lambda \sim 2$ diameters. The rapid and slow decay processes merge into a more gradual decay of quite long duration. This long decay is evidence of the persistence of the first shell of neighbors for the smaller k shown. The even slower decay at larger k indicates that, on average, pairs of molecules remain together for quite long times. The slower decay for the three-body liquid is again explained by the stronger intermolecular binding in that system. At still larger k values the $F(k, t)$ begin to look the same for the two liquids, as is shown in Fig. 8 (left, c). This is as expected because on very small length scales all systems asymptotically exhibit ideal gas behavior, irrespective of the intermolecular potential.

For the purpose of contrast, in Fig. 9 is shown $S(k, \omega)$ for $k = 0.25$ and 0.36Å^{-1} calculated from hydrodynamics (Eq. 19) together with the MD results. It is clear that the MD spectrum decays much more quickly than the Lorentzian given by Eq. 19. The peak intensities are greater and the position of the Brillouin peak has shifted in the MD spectrum, reflecting positive dispersion. At larger k the hydrodynamic $S(k, \omega)$ quickly becomes a low, broad featureless curve. The need for generalizing ordinary hydrodynamics is self-evident from the poor agreement between the calculated and simulation curves in Fig. 9.

A detailed analysis on the sound dispersion brings about again the need to use the hydrodynamics relations only at very, very low k values, and thus the need to use MD for explaining sound dispersion laboratory data. For details on this we refer to a recent paper.[46b]

V. Fluid Dynamical Aspects and Macroscopic Theory

As is well known, fluid dynamics is the study of motion and transport in liquids and gases. It is primarily concerned with macroscopic phenomena in nonequilibrium fluids and covers such behavior as diffusion in quiescent fluids, convection, laminar flows, and fully developed turbulence.

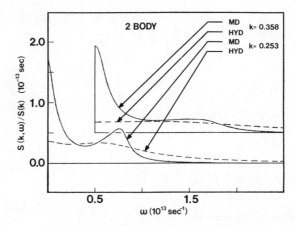

Figure 9. Comparison of $S(k, \omega)$ from hydrodynamics with $S(R, \omega)$ from MD at R = 0.253 and 0.358Å[-1]

Since the phenomena studied in fluid dynamics are macroscopic, the fluid is considered to be a continuous medium, and the theory is not based on the behavior of individual molecules in the fluid but, rather, on their collective motions. Thus, fluid dynamics studies the motion of fluid volume elements which contain a large number of molecules. Such a volume element defines in the continuous medium a point which is small compared to the total system volume, but large when compared to typical intermolecular distances.

We present and discuss results for MD modeling of fluid systems. We restrict our discussion to systems which are in a macroscopically steady state, thus eliminating the added complexity of any temporal behavior. We start with a simple fluid system where the hydrodynamic equations are exactly solvable. We conclude with fluid systems for which the hydrodynamic equations are nonlinear. Solutions for these equations can be obtained only through numerical methods.

All of these experiments were accomplished using Argon atoms interacting through the short range Lennard-Jones or the soft sphere potential. The number of atoms, N, ranges from 10^2 to 10^5, depending on the length scales over which the desired phenomena will appear. The densities and temperatures of the systems are such that the Argon is in liquid state. Shortly we shall consider molecular liquids, water in particular. However, the results reported below are sufficient to prove that we can overlap molecular dynamics with fluid dynamics, thus providing a valid bridge between macroscopic and microscopic theory. Additional evidence in support of this statement are the papers by Tannenbaum, Ciccotti, and Gallico[52] and by Trozzi and Ciccotti.[53]

As a first example (see L. Hannon et al. in Ref. 54) we consider a system bounded periodically in two coordinates and by thermal walls in the other coordinate. The two thermal walls are at rest and maintained at the same temperature, T_W, as shown in Fig. 10. The system is subjected to an acceleration field which gives rise to a net flow in the direction of one of the periodic coordinates. For this system, the hydrodynamic equations reduce to

$$\nabla^2 \vec{v} = \frac{\rho g}{\eta} \hat{k}, \quad v_x = v_y = 0, \quad v_z = v_z(x), \quad \nabla^2 T = \frac{\eta}{2\kappa} \left(\frac{\partial v_z}{\partial x} \right)^2,$$

subject to the boundary conditions

$$v_z(x = -\frac{L}{2}) = v_z(x = +\frac{L}{2}) = 0, \quad T(x = -\frac{L}{2}) = T(x = +\frac{L}{2}) = T_W.$$

The solutions to this set of equations are

$$v_z = \frac{\rho g}{2\eta}[x^2 - (\frac{L}{2})^2], \quad T = T_W - \frac{1}{12} \frac{(\rho g)^2}{2\kappa\eta}[x^4 - (\frac{L}{2})^4].$$

We have modeled this system using MD and both short range as well as soft sphere Lennard-Jones potentials. We found the MD system to exhibit the predicted quadratic profile for the velocity response and, essentially, quartic profile for the temperature response, and these are shown in Fig. 10. In addition, with this system we can check for qualitative agreement with macroscopic theory. Using regression analysis to fit the MD results to the appropriate curves, we can calculate values for the shear viscosity coefficient, η, and the thermal conductivity coefficient, κ. We find the calculated values to be in good agreement with experimental values found for liquid Argon. These results were obtained for acceleration fields as large as 10^{15}cm/sec^2.

We consider now a two-dimensional system bounded periodically in both coordinates. The system is subjected to an acceleration field which gives rise to a net flow. The flow is obstructed in such a way that this system is the two-dimensional analog of three-dimensional flow past an infinitely long grid of infinitely long plates.

In fluid dynamics the behavior in this system is described by the full set of hydrodynamic equations. This behavior can be characterized by the Reynolds number, which is the ratio of characteristic flow scales to viscosity scales. We recall that the Reynolds number is a measure of the dominating terms in the Navier-Stokes equation and, if the Reynolds number is small, linear terms will dominate; if it is large, nonlinear terms will dominate. In this system, the nonlinear term, $(\vec{v} \cdot \vec{\nabla}) \vec{v}$, serves to convert linear momentum into angular momentum. This phenomena is evidenced by the appearance of two counter-rotating vortices or eddies immediately behind the obstacle. Experiments[55,56] and numerical integration of the Navier-Stokes equations predict the formation of these vortices at the length scale of the obstacle. Further, they predict that the distance between the vortex center and the obstacle is proportional to the Reynolds number.

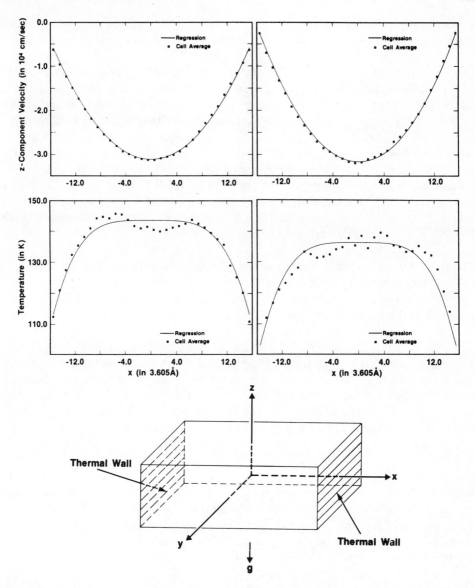

Figure 10. Velocity (top) and temperature (bottom) profiles for the cut-off Lennard-Jones potential (left), and for soft sphere potential (right). The system consists of 1152 atoms enclosed in a box of side length equal to 32:6:6. Bottom: definition of coordinate system used in the simulation.

As we shall show below, modeling this system with MD we find vortex formation even at microscopic scales. The length scale of the vortex at formation is the same as the length scale of the obstacle, as predicted. It is difficult to determine the Reynolds number in the MD fluid because of large variations of the densities and stresses from point to point. However, we do see the vortex center moving away from the obstacle with increased overall fluid velocity (which corresponds to an increased Reynolds number). This result is, at least, in qualitative agreement with macroscopic theory. Let us now analyze the example in more detail.

The results obtained in these studies suggest that the MD approach may prove to be a valuable tool for probing the detailed microscopic flow structures that underlie certain instabilities of continuum hydrodynamics. There are of course severe limitations as to the length and time scales that can be handled, as well as to the opportunities for similar work in 3D, and the exploration of phenomena such as turbulence appears to lie beyond the scope of detailed MD simulation. The gradients present in the MD simulations are much larger than those encountered in normal hydrodynamic situations, and on these grounds it is reasonable to believe that, at best, only qualitative agreement between MD and hydrodynamics might be found. Even at this level of expectation, however, the observed similarity between the flow patterns of MD and real fluids is quite remarkable.

The *first MD system* studied is comprised of two subsystems, a flow system and a bath system, as shown in Fig. 11. The flow system is the system of primary interest. The bath system is a device which acts as a particle source at the inlet and a particle sink at the outlet of the flow.

The flow system is composed of N particles in a two-dimensional box of size L x L. The walls located at $y = -L/2$ and $y = +L/2$ are periodic boundaries; while the walls located at $x = 0$ and $x = L$ are "modified" periodic boundaries, which will be explained shortly. A net flow in the positive x direction is induced by subjecting the particles in the flow system to an acceleration field, g. A thermal plate, located at $x = L/3$ and extending from $y = -L/6$ to $y = +L/6$, serves as an obstruction to the flow.

The thermal plate gives rise to anisotropic behavior in the system. That is, if the length of the considered system is small, the distribution of particle positions and momenta at the tail-end of the flow could be quite different from that found near the head-end of the flow. These anisotropies prevent treatment of the walls at $x = 0$ and $x = L$ as true periodic boundaries. To avoid such problems, a thermal bath system is used as a particle source at $x = 0$ and at the same time also acted as a particle sink for particles leaving the system at $x = L$.

The bath system is composed of N/10 particles in a box of size L/10 x L. The walls located at $y = -L/2$ and $y = +L/2$ are again periodic boundaries; for the particles in the bath, the walls located at $x = -L/10$ (the periodic equivalent of $x = L$) and $x = 0$ are thermal walls. Particles in the bath are not subject to the acceleration field.

Particles in the flow system that are near the wall, $x = L$ ($x = 0$), will interact with particles in the bath system that are near the wall, $x = -L/10$ ($x = 0$). When a particle in the flow system reaches or passes the wall at $x = L$, it is thermalized and becomes part of the bath system. Simultaneously, the particle in the bath system nearest the wall at $x = 0$ becomes a member of the flow system and is

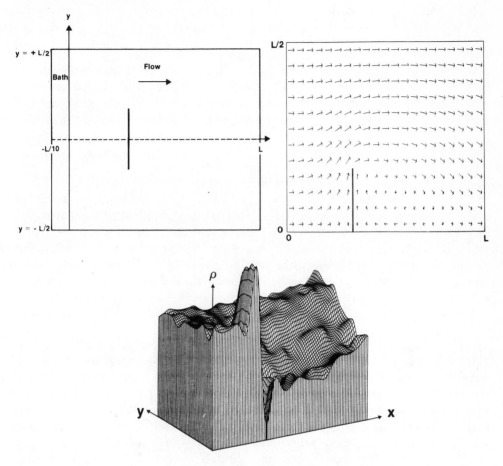

Figure 11. Top left: configuration of obstructed flow system. Top right: velocity field for steady state flow past a plate. The arrow lengths are scaled as the square root of the ratio of individual cell velocity to maximum cell velocity. Bottom: density contour for steady state flow. The approximate location of the plate is marked by shading.

subject to the acceleration field. The bath system thus serves to maintain continuity for particle interactions and to separate source behavior from sink behavior.

The particles interact through the purely repulsive soft sphere potential

$$U(r) = 4\varepsilon \left[\left(\frac{\sigma}{r} \right)^{12} - \left(\frac{\sigma}{r} \right)^{6} + \frac{1}{4} \right]$$

for

$$r = |\vec{r}_i - \vec{r}_j| \leq 2^{1/6}\sigma,$$

$U(r) = 0$, otherwise.

Particles colliding with the thermal walls are re-injected into the appropriate system with new velocities drawn randomly from a Boltzmann distribution characterized by the wall temperature, T_W.[52,53] For particles colliding with the thermal plate obstruction, the case of primary interest, this simulates no-slip boundary conditions.

The above system is set up for n = 10000 particles in a square of side L = 373.22Å using argon parameters for the potential function (σ = 3.405Å, ε = 119.8k_B). The density, ρ = 1.276 gm/cm³,[57] and thermal wall temperature, T_W = 86.5 K, used characterize argon in a two-dimensional liquid state. In the initial system configuration, the atoms are located on a square lattice with velocities drawn randomly from the Maxwell-Boltzmann distribution at temperature T_W. A time step of $\delta t = 10^{-14}$ sec is used to integrate the Newtonian equations of motion for the atoms in an acceleration field via a fourth order Gear's predictor-corrector algorithm.[58] For statistical averages, the flow system is divided into 500 square cells. Snapshots of system behavior are taken every 10 psec by accumulating statistics on density, velocity, and temperature for each of the cells over 1000 time steps.

The simulations are carried out for several values of the acceleration field strength: 0.625, 2.5, 5.0, and 10 × 10¹⁴ cm/sec². For system with g = 2.5 × 10¹⁴ cm/sec², the appearance of two counter-rotating vortices immediately behind the obstructing plate is first observed after 120 psec of simulation time ($< v_x > = 19000$cm/sec). The positions of these vortices increase with increasing velocity until the system reaches a steady state after 200 psec of simulation time ($< v_x > = 21500$cm/sec). The vortex size and location, as well as the average system velocity, remain stable over an additional 300 psec of simulation time. The density contour and velocity field for the steady state system based on statistics taken over the last 100 psec of simulation time are shown, respectively, on the two inserts at the top of Fig. 11. Note that to take advantage of the symmetry, only half of the system is displayed in this figure.

The small size of the system places inherent limitations on the range of parameters that can be simulated. For g < 0.625 × 10¹³ cm/sec², observation of the vortices is hindered by thermal noise (the average system velocity is of the order of thermal noise); for g > 5 × 10¹⁴ cm/sec², the vortex size is influenced by the presence of a boundary at x = L. In the range of velocities studied (0.135c to 0.480c, where c is the sound speed for argon at its triple point, 86400 cm/sec), the fluid is highly compressible. Densities in the system are found to range from 1.422 gm/cm[57] immediately in front of the plate to 1.135 gm/cm[57] immediately behind the plate for the system shown in Fig. 11. This large variation of the density and stress prevents us from making any realistic determination of the Reynolds number and, thus, making any quantitative comparisons with macroscopic theory. Nevertheless, within the above range of simulation parameters, we do observe that the vortex size and the steady state position of the vortex center are roughly proportional to the average mass flux in the system (which itself is proportional to the Reynolds number). This result is, at least, in qualitative agreement with the macroscopic theory. A better determination of the above relation requires simulations with larger systems, which are currently under investigation in our laboratory.

Let us now consider a *second example* of a fluid flow past an obstacle.[59] The initial departure from Stokes flow occurs at Reynolds number $Re \approx 5$, at which point a pair of counter-rotating eddies (or vortices) begin to develop at the downstream boundary of the cylinder.[55,56] The eddies grow in size while at $Re \approx 34$ and an oscillatory wake is seen in the flow downstream of the obstacle. At slightly higher Re, somewhere in the range $55-70$, transverse oscillations begin to occur in the eddy structure accompanied by periodic shedding of rotating fluid regions, and the von Karman vortex street makes its first appearance. Above $Re \approx 100$ the eddy structure immediately behind the cylinder ceases to be visible. There exists a considerable body of photographic documentation of these effects.[60,61] Similar low-Re behavior has also been seen in numerical solution of the equations of continuum hydrodynamics.[62]

This MD simulation uses a purely repulsive short-range pair potential given before. The circular obstacle has diameter D and is positioned at some suitable point in the flow; particles approaching the obstacle experience a repulsion given by $V(r - D/2)$, where $r(> D/2)$ is the distance from the center of the obstacle. The "collisions" with the obstacle are specular; in the language of hydrodynamics this corresponds to a slip boundary.[63]

The boundaries of the region enclosing the fluid are periodic. At the beginning of the simulation the entire system is given the same uniform flow velocity U_0, with thermal motion superimposed. Finally, because collisions with the obstacle dissipate flow momentum, a gravitational field g is imposed in the appropriate direction to maintain the flow; its strength must be determined empirically in order to ensure the desired flow rate.

The particular calculations described here involve $\sim 1.65 \times 10^5$ particles at a density $\rho = 0.83$ (particles/unit area), while the value of D is 74. If the unit of length is given its typical MD value, namely 3.4Å, then the simulation covers a square region of edge 1500Å and the obstacle size is 250Å. The unit of time is 0.31 psec; in terms of scaled units $U_0 = 0.3$ (on this scale the thermal velocity is 0.2) and $g = 2 \times 10^{-4}$. The equations of motion are solved using a third order predictor-corrector method with time step 0.032. The particular run described below extended over 1.2×10^5 steps corresponding to a total of 1.2 nsec; other runs briefly mentioned were of similar size and duration (unless stated otherwise).

One novel aspect of this simulation, apart from the comparatively large (by MD standards) size of the system, is the fact that the computations were carried out using a coupled array of four FPS-264 scientific processors.[63] The processors operate in parallel, each taking responsibility for a separate subregion of the system. A certain amount of data must be exchanged between the processors throughout the course of the calculation to allow for i) motion of particles between adjacent subregions, and ii) the evaluation of interactions for particles lying within a distance r_c of the subregion boundaries.

The results are summarized by introducing a 60×60 grid that covers the entire system and, every 5th step, averaging the properties of all the particles contained in each grid cell at that instant. The results described here represent the mean of either 400 or 1000 such sets of values, but no additional smoothing is applied to the data. The detailed evolution of the flow patterns depends on the initial conditions (a similar lack of reproducibility is observed in real fluids); we describe the principal features seen in one typical simulation run and briefly mention the changes observed in several other runs under altered conditions. At this stage the primary results of the simulations are descriptive in nature.

The uniform flow at time $t = 0$ rapidly changed into Stokes flow around the obstacle. By $t \approx 360$ a pair of counter-rotating eddies had developed adjacent to the downstream edge of the obstacle. The length of the recirculatory region grew steadily, reaching approximately $1.4D$ at $t \approx 850$. The time-averaged flow field at $t = 750$ (an average over 1000 configurations from the preceding interval of length $\delta t = 150$) is shown in Fig. 12; the grid is 30×30 with each cell representing an average of 4 cells of the original grid, while the arrow length is a measure of the averaged local velocity v via the relation $length \propto (1 + v/v_{max})$, and the arrow direction is that of the local flow. A more detailed illustration of the flow pattern appears in Fig. 12 where only the central 30×30 (original) grid cells are shown; the flow separation from the obstacle boundary is clearly visible, as are the flows within the eddies. The backflow velocity amounts to only about 10% of the overall flow speed; the compressible nature of the fluid is manifested most strongly just downstream of the obstacle where a 25% density drop occurs.

The next qualitative change that occurs is the breakdown of the bilateral symmetry of the flow — the counterclockwise (left) eddy begins to grow at the expense of the clockwise (right) eddy, the latter effectively disappearing at $t \approx 1000$. The symmetry is restored by $t \approx 1200$, but at $t \approx 1500$ only the clockwise eddy is seen. This in turn separates from the obstacle and the opposite eddy develops. Alternate eddy formation and shedding is repeated several times and an oscillatory wake develops at $t \approx 2000$. Fig. 12 shows the central portion of the flow field at $t \approx 2400$; the wake — a remnant of previously shed eddies — is clearly visible, as is a counterclockwise eddy still attached to the obstacle. By $t \approx 2700$ the transverse flow component has reversed itself, and the flow pattern is essentially the mirror image of Fig. 12. While the wake oscillations continue until (and presumably beyond) the end of the run ($t = 3600$), the recirculation that characterizes the eddies eventually vanishes and the sinuous wake is found to extend right up to the obstacle.

The Reynolds number[60] is defined as $Re = DU/\nu$, where U is the flow velocity and ν the kinematic viscosity. The average flow velocity is approximately 0.4 (reached at $t \approx 1800$); since $\nu = \mu/\rho$, and the shear viscosity μ is available from earlier 2d simulation,[64] we obtain $Re \approx 25$. This estimate of Re is only a rough one since there is arbitrariness in the choice of U and, in 2d, μ exhibits a complex dependence on both the shear rate and the extent of the sheared region;[61,63] nevertheless it does suggest that the MD fluid is flowing with a value of Re in the range at which eddy formation is observed experimentally. Further simulation over a range of Re would

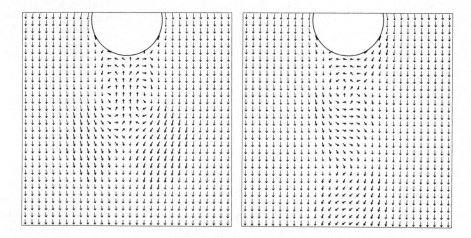

Figure 12. Central region showing eddy pair and location of the circular obstacle at t = 750 (left) and a single eddy and portion of the oscillatory wake at t = 2400 (right). The flow is from top to bottom.

be required to establish whether stationary eddy formation is possible on the length scales accessible to MD.

Behavior qualitatively similar to that just described was observed in another MD run in which the average flow velocity was reduced by 10%. On the other hand, when the flow speed was reduced by 50% there was only the slightest hint of deviation from Stokes flow (run duration 0.7 nsec); this is evidence for the existence of a critical Reynolds number $Re_c > 0$ for the onset of eddy formation. In yet another run, for which the velocity was 25% greater, the eddies stretched almost to the downstream boundary of the system before any oscillation became apparent; clearly the limited system size had a pronounced influence on the flow development in this particular instance.

There has been one other MD study that attempted to observe the appearance of eddies in fluid flow past an obstacle.[64] Eddies were seen shedding from the tips of a thin inclined plate, but since Re_c decreases with increasing curvature of the side of the obstacle, the sharp tip implies $Re_c = 0$,[60] and therefore eddy formation is to be expected even at arbitrarily small Re. The *one key difference* between a gently curved obstacle boundary and one with sharp corners is that, in the former, flow separation occurs at a stagnation point[60] whose position depends on Re, whereas in the latter, separation occurs at the sharp corner itself since the flow cannot follow the abrupt changes in boundary direction.

The results presented are quite remarkable. The theory underlying derivation of the hydrodynamic equations assumes that all gradients and forces acting on the fluid are small. The MD fluids are under the influence of extremely large gradients and forces. Yet, we find results which are in both qualitative and quantitative agreement with macroscopic predictions. The appearance of spatial structure, seen in the last experiment, on such a small scale (10^{-6}cm) provides strong indications that fluid dynamics can be understood from a microscopic viewpoint.

A great deal of research remains to be done in this area. We are currently involved in the study of spatial correlations in these non-equilibrium fluids with the hope of establishing a correspondence between MD and fluctuating hydrodynamic theory. We are also using these systems to study the roles of viscosity and conductivity in fluid behavior under different external constraints. Finally, we plan to continue our research into the formation of spatial structures in fluids.

Conclusion

As stated in the introduction, building up from nuclei and electrons we extend the system to atoms, molecules, molecules in solutions and fluids with eddies and spatial-temporal structures of macroscopic dimensions. We hope to have stimulated the readers in adopting the *global simulation* approach we have advocated.[1,2] Of course, in addition one needs computational facilities, which can either be purchased or assembled on a "do-it-yourself" basis. A "manual" for the latter is given in this volume.[65]

This review is based on a number of IBM-Tech. Reports prepared for a Chaire Francqui series of lectures delivered by one of us (E. Clementi) in the spring of 1986 at the Free University of Bruxelles.

References

1.   E. Clementi, J. Phys. Chem., **89**, 4426 (1985).
2.   E. Clementi, "Computational Aspects for Large Chemical Systems" (Springer-Verlag, 1980), Lecture Notes in Chemistry, Vol. 19.
3.   W.S. Benedict, N. Gailer and E.K. Plyler, J. Phys. Chem., **24**, 1139 (1956).
4.   H.J. Werner and W. Mayer, Mol. Phys., **31**, 885 (1976).
5.   R.J. Bartlett, I. Shavitt and G.D. Purvis, J. Chem. Phys., **71**, 281 (1979).
6.   E. Clementi and H. Popkie, J. Chem. Phys., **57**, 1077 (1972).
7.   G. Corongiu, private communications.
8.   E. Clementi and G. Corongiu, Chem. Phys. Lett., **90**, 359 (1983).
9.   E. Wigner, Phys. Rev., **46**, 1002 (1934).
10. P. Gombas, "Pseudopotentials," Springer-Verlag, New York (1967).
11. K.A. Brueckner and S.K. Ma, Document IRPA 67-150 (June 1967), University of California, La Jolla, California.
12. W. Kohn and L.J. Sham, Phys. Rev., **140A**, 1133 (1965).
13. See for example E. Clementi, IBM J. Res. and Dev., **9**, 2 (1965), for an early proposal on density functionals. E. Clementi, Proc. Nat. Lead. Sci. USA, **69**, 2942 (1972) for early molecular applications. G.C. Lie and E. Clementi, J. Chem. Phys., **60**, 1275 (1974); *ibid.*, **60**, 1288 (1974) for systematic applications to diatomic molecules.
14. H. Popkie, H. Kistenmaker and E. Clementi, J. Chem. Phys., **59**, 1325 (1973).
15. H. Kistenmaker, H. Popkie, E. Clementi and R.O. Watts, J. Chem. Phys., **60**, 4455 (1974).
16. G.C. Lie and E. Clementi, J. Chem. Phys., **62**, 2060 (1975).

17. O. Matsuoka, E. Clementi and M. Yoshimine, J. Chem. Phys., **64**, 1351 (1976).
18. G.C. Lie, E. Clementi and M. Yoshimine, J. Chem. Phys., **64**, 2314 (1976).
19. P. Habitz and E. Clementi, J. Phys. Chem., **87**, 2815 (1983).
20. V. Carravetta and E. Clementi, J. Chem. Phys., **81**, 2646 (1984).
21. E. Clementi, G. Corongiu, M. Dupuis and M.J. Huang (in preparation).
22. P. Habitz, P. Bagus, P. Siegbahn and E. Clementi, Int. J. Quantum Chem., **23**, 1803 (1983).
23. E. Clementi and G. Corongiu, Int. J. Quantum Chem. Symp., **10**, 31 (1983).
24. J. Detrich, G. Corongiu and E. Clementi, Int. J. Quantum Chem., **18**, 701 (1984).
25. G.C. Lie and E. Clementi, Phys. Rev., A33, 2679 (1986).
26. H. Kistenmaker, G.C. Lie, H. Poplie and E. Clementi, J. Chem. Phys., **61**, 546 (1974).
27. E. Clementi, K.S. Kim and M. Dupuis, IBM Research Report KGN-45 (March 27, 1986).
28. J.H. Dymond and E.B. Smith, "The Virial Coefficients of Pure Gases and Mixtures," Oxford Press (1980).
29. K. Refson and G.C. Lie, private communications.
30. G.C. Lie and E. Clementi, J. Chem. Phys., **64**, 5308 (1976).
31. F.H. Stillinger and A. Raman, J. Chem. Phys., **60**, 1545 (1974).
32. G.C. Lie, G. Corongiu and E. Clementi, J. Phys. Chem., **89**, 4131 (1985).
33. N. Metropolis, A.W. Rosenbluth, M.N. Rosenbluth, A.H. Teller and E. Teller, J. Chem. Phys., **21**, 1087 (1953).
34. B.J. Alder, J. Chem. Phys., **40**, 2724 (1964).
35. M.P. Tosi in Solid State Physics, Vol. 16, F. Seitz and B. Turnbull, eds., Academic Press, New York (1964).
36. J.A. Barker and R.O. Watts, Mol. Phys., **26**, 789 (1973); Chem. Phys. Letters, **3**, 144 (1969).
37. A.H. Norten and A.H. Levy, J. Chem. Phys., **55**, 2263 (1971).
38. J. Detrich, G. Corongiu and E. Clementi, Chem. Phys. Letters, **112**, 426 (1984).
39. G.C. Lie, J. Chem. Phys. (in press).
40. J.D. Dore, Faraday Diss. Chem. Soc., **66**, 82 (1978).
41. G. Palinkas, E. Kalman, and P. Kovacs, Mol. Phys., **34**, 525 (1977).
42. A.K. Soper and R.N. Silver, Phys. Rev. Lett., **49**, 471 (1982).
43. M. Wojcik and E. Clementi, J. Chem. Phys., **84**, 5970 (1986).
44. *Note*: See Ref. 25 for comments concerning the pressure values from MCY or MCYL.
45. J.L. Lebowitz, J.K. Percus and L. Verlet, Phys. Rev., **153**, 250 (1967).
46. **a)** M. Wojcik and E. Clementi, J. Chem. Phys. (in press); see also IBM-Tech. Rep. KGN-60 (May 28, 1986); **b)** M. Wojcik and E. Clementi, J. Chem. Phys. (in press); see also IBM-Tech. Rep. KGN-61.
47. G. Corongiu, private communications.
48. P.H. Berens, D.H.J. Mackay, G.M. White and K.R. Wilson, J. Chem. Phys., **79**, 2375 (1983).
49. K. Drynicki, C.D. Grenn and D.W. Sawyer, Disc. Faraday Soc., **66**, 199 (1978).

50. R.W. Impey, P.A. Madden and I.R. McDonald, Molec. Phys., **46**, 513 (1982).
51. J.P. Hansen and I.R. McDonald, "Theory of Simple Liquids," Academic Press, New York (1976).
52. A. Tannenbaum, G. Ciccotti and R. Gallico, Phys. Rev., **A25**, 2278 (1982).
53. C. Trozzi and G. Ciccotti, Phys. Rev., **A29**, 916 (1984).
54. L. Hannon, E. Kestemont, G.C. Lie, M. Marchal, D.C. Rapaport, S. Chin and E. Clementi, IBM-Tech. Rep., "Chaire Francqui" Lecture Series: Part 7 (Feb. 1986).
55. J.H. Gerrard, Phil. Trans. Roy. Soc. London, **A288**, 351 (1978).
56. A.E. Penny, M.S. Chang and T.T. Lim, J. Fluid Mech., **116**, 77 (1982).
57. J.J. Erpenbeck, Physica, **118A**, 144 (1983).
58. C.W. Gear, "Numerical Initial Value Problems in Ordinary Differential Equations," Prentice-Hall, Princeton, N.J. (1971).
59. D.C. Rapaport and E. Clementi, Phys. Rev. Lett., **57**, 695 (1986).
60. G.K. Batchelor, "An Introduction to Fluid Dynamics," Cambridge, U.K. (1967).
61. M. Van Dyke, "An Album of Fluid Motions," Parabolic Press, Stanford, CA (1982).
62. S.K. Jordan and J.E. Fromm, Phys. Fluids, **15**, 371 (1972).
63. D.M. Heyes, G.P. Morriss and D.J. Evans, J. Chem. Phys., **83**, 4760 (1985).
64. E. Meiburg, DFVLR Report FB85-13, Gottingen (1985).
65. See E. Clementi, J. Detrich, S. Chin, G. Corongiu, D. Folsom, D. Logan and R. Caltabiano, A. Carnevali, J. Helin, M. Russo, A. Gnudi, P. Palamidese, "Large Scale Computations on a Scalar, Vector and Parallel 'Super-Computer,' " this volume.

Proton Conductivity of Hydrated Lysozyme Powders, Considered Within the Framework of Percolation Theory

G. Careri
Dipartimento di Fisica
Universita di Roma I
Roma 00185, Italy

John A. Rupley
University Department of Biochemistry
University of Arizona
Tucson, AZ 85721, USA

Hydration Events in Lysozyme Powders

How does water at the surface of proteins enter into or modulate biochemical processes, particularly enzyme catalysis? This and other major questions about protein hydration remain open. We have approached such problems through study of the effect of change in water content on the thermodynamic and dynamic properties of protein powders. The advantage of this system is that the water activity (the hydration level) can be varied from zero (dry protein) to unity (dilute aqueous solution of protein). Thus the changes in chemistry and physics induced by hydration can be examined sequentially by comparison of samples of increasing level of hydration. Unravelling the complexity of hydration then becomes somewhat easier.

Lysozyme is a relatively simple enzyme, for which one has available a rather complete description of the hydration dependence of various properties[1]: IR spectrum, EPR spectrum, heat capacity, other thermodynamic and dynamic properties, and importantly, the enzymatic activity. These data have led to a picture of the hydration process as being step-wise and consisting of three well-defined stages[1]: the first stage (from 0 to about 60 water molecules per lysozyme molecule) is dominated by interaction of water with the charged groups of the protein; ii) the second (from 60 to about 220 waters) is characterized by mobile water clusters centered on the polar surface atoms; iii) the last stage (from 220 to about 300 waters) is where enzymatic activity develops in concert with rapid surface motion and with the condensation of water over the weakly-interacting unfilled regions of the surface.

The above picture and the data that lead to it are in agreement with order-disorder and nucleation theories of statistical physics.[1] With regard to nucleation, it is impossible to condense water on insoluble particles of radius smaller than about 100Å at the equilibrium vapor pressure of water. Therefore a dry globular protein could not become hydrated if its surface comprised only amide backbone and other non-ionizable residues. Soluble elements such as the ionizable residues allow water vapor to condense on the macromolecule, and with increased hydration, clusters of water about these first hydration sites grow until all the surface is involved. With regard to order-disorder transitions, Hill[2] has developed, for the case of interacting adsorbed molecules, a theory of localized unimolecular adsorption on a hetero-geneous surface. Hill's theory[2] predicts two phase transitions: the one at low cov-erage is a two-dimensional surface condensation of dispersed water into clusters, and we identify it with the transition near 60 water molecules per lysozyme mole-cule; the other at high coverage is the condensation that leads to completion of the adsorbed monolayer, and we identify it with the event near 220 waters. For the following discussion of percolation, it is important to note that no phase transition is predicted nor has one been observed for either the adsorbed water or the protein itself, within the range of 60 to to 220 water molecules per macromolecule, namely for the hydration range 0.07 to 0.25 g of water per g of lysozyme.

Detection of a Percolative Transition in the Proton Conductivity

Water molecules bear a strong dipole moment. We have investigated the hydration of lysozyme by studying the MHz-frequency dielectric behavior of lysozyme powders of varied hydration level, with the use of on-line computer analysis of simultaneously acquired data from a high-sensitivity digital a.c. bridge and a digital balance. This technique[3] allows detection of the dielectric properties of a sample capacitor that was constructed so as to have no metal-protein contact, which because of the range of pH investigated, could have produced experimental difficul-ties.

When the dielectric losses were measured for samples of lysozyme that had been isopiestically hydrated in D_2O or H_2O,[3] it was found that the relaxation showed the full theoretically-predicted isotope effect, which implies that the inferred conductivity is dominantly protonic. The pH dependence of the relaxation indicated that ionizable side-chain groups of the protein participate in the protonic con-duction process. Complexation of the active site with substrate significantly decreased the dielectric response; perhaps half the proton flux passes through the active site.

In order to describe carefully the onset of the dielectric response, we have measured the hydration dependence of the capacitance in the low hydration region.[4] The results were found to agree closely with predictions based on the percolation model of a conduction process. This general physical model has been shown applicable to a broad range of processes where spatially random events and topological disorder are of vital importance. A typical application of percolation theory is to the elec-trical conductivity of a network of conducting and non-conducting elements.[5] One of the most appealing aspects of the percolation model is the presence of a percolation transition, where long-range connectivity among the elements of a system suddenly appears at a critical concentration of carriers.

The capacitance of lysozyme powders[4] displays a sharp increase at a threshold water content h_c = 0.150 $\pm$ 0.016 g of water per g of protein. The hydration level for monolayer coverage of the surface is 0.38 $\pm$ 0.04. Thus the percolation threshold, which is the fraction of sites filled by conducting elements when long-range connectivity first appears, is p_c = 0.40 $\pm$ 0.04. This experimental value is in close agreement with the value 0.45 $\pm$ 0.03 predicted by percolation theory for two-dimensional networks.[5] For three-dimensional networks, percolation theory predicts p_c = 0.16 $\pm$ 0.02, which rules out for the protein a conduction process involving the interior of the molecule, where in any case water molecules are known to be sparse.

At the threshold h_c = 0.15 the surface water is mobile[1] and so fulfills the requirement of the percolation model that there be randomness in the arrangement of filled sites. The threshold was found to be constant from pH 3 to pH greater than 8. Thus the local geography of water clusters about ionizable sites of the protein surface is not of primary importance. Apparently, only the number of water molecules acting as interconnected conducting sites is relevant, consistent with the percolation model. Also as expected for this picture, the threshold is the same for both D_2O- and H_2O-hydrated samples. The dielectric behavior at the threshold contrasts with the behavior in the region of saturation of the hydration, where change in either pH or solvent isotope produces a strong effect.[3] Clearly, the chemistry of the system determines the level of the response for full hydration, but the threshold for the response is determined by the connectivity among the sites, as required by the percolation model.

We picture the percolative character of the dielectric properties as reflecting proton transfer along threads of hydrogen-bonded water molecules adsorbed on the protein surface. Because the arrangement of the water molecules undergoes fluctuation, so must the threads. Because the statistical path of the interconnected water molecules is long, the threads act as a short bypassing the local geographical details of the protein surface.

Although the above description was developed for lysozyme, we expect the percolation picture to hold for most proteins: first, the hydration behavior of proteins in general, judged by the sorption isotherms, is closely similar to that of lysozyme; second, percolation theory gives predictions that are, to a first approximation, size independent.

Complexation of lysozyme with substrate increases the percolation threshold h_c to near 0.25 g of water per g of protein. Apparently the presence of a foreign body, where the water bridges may not be favorable for proton transfer, affects long range connectivity on the protein surface. The hydration level h_c $\cong$ 0.25 is so close to the one critical for onset of the enzymatic activity of lysozyme that it implicates protonic percolation in lysozyme catalysis. In this regard, the possible relevance of kinetic coupling among the statistical macrovariables responsible for catalysis (for example, the charges on ionizable side-chain groups) has been proposed and expressed formally in terms of an Onsager matrix.[6] There are believed to be at least three concurrent requirements for lysozyme catalysis[7]: deprotonation of the sidechain carboxylate group of Asp-52; protonation of the carboxylate of Glu-35; and distortion of the substrate within the active site. One can picture coupling among all or several of these processes as being mediated by percolation of protons along the fluctuating threads of water molecules on the surface of the protein. The observation that substrate affects the high hydration proton flow, as well as the threshold behavior, argues that this description based on partially hydrated lysozyme powders is relevant to catalysis in dilute solution.

The example of lysozyme catalysis illustrates how this new level of description, the percolation model, may be useful for understanding catalysis and other functions of proteins. In a broader sense this model is a desirable extension of the statistical picture that has been used for the near-equilibrium Onsager regime, which does not predict threshold behavior and which does not offer a microscopic picture for kinetic coupling among macrovariables.

Percolation vs. Phase Transitions in Proteins and Membranes

Phase transitions are well known to be relevant to biological function: protein folding, allostery, membrane organization, DNA replication, protein aggregation in sickle cell disease, etc. In a phase transition, it is possible to identify regions of the system (phases) that display different chemistry. In the transition, interactions between elements of the system result in changed chemical or physical properties.

The percolation transition, although it exhibits some characteristics of a phase transition, differs importantly in that it reflects change in the connectivity within a system without there being also a change in the chemistry of the elements of the system. A percolative transition can be found even if there is no interaction between elements, as in the case of a mixture of conducting and non-conducting spheres. For this example, one can identify in principle two phases, one consisting of those conducting elements part of the infinite thread, the other of conducting elements not part of it; however, the chemistry of the conducting elements in the two putative phases is clearly indistinguishable. The percolative transition is detected as a discontinuity in a *process*, a property of the entire system, for example conduction, the onset of which reflects the onset of long-range connectivity. This is developed stochastically in a randomly-structured system of many elements, as a result of a change in system composition (in the example, change in the fraction of conducting elements).

In the case of protein hydration, the percolation transition, at h_c = 0.15, occurs between the hydration levels of two phase transitions, at 0.06 and 0.25 g of water per g of macromolecule. The chemistry of the adsorbed water changes sharply at the phase transitions: for the 0.06 hydration level event, where dispersed water condenses into mobile clusters, the heat capacity shows a discontinuity, with change from a value close to that of ice to one greater than that of liquid water.[1] In contrast, the properties of the adsorbed water show no discontinuity at the percolation transition.

The importance of the percolation model for biophysical and biological problems is, first, that it extends the application of transition theory to a wider number of systems, specifically those which exhibit a discontinuity in a connective property but no phase change, and second, that it provides a clear molecular-level picture that can convey a novel view of a process.

In the discussion of the enzymatic activity of lysozyme given in the preceding section, it was surmised that a percolation picture of coupling between events of a catalytic process, suggested by observations on the partially hydrated protein, could be carried over to the dilute solution. The rationale for this was that complexation with substrate affects proton flow at high hydration. It is possible that particular threads of water molecules at the protein surface are more probable, that they pass through the active site, and that they are important for biological function. This constitutes a modification of the fully-random percolation picture. If it is true, protein structural studies should find threads of interconnected water at the surface of fully hydrated lysozyme and possibly other proteins. Diffraction and other structural analyses have focused generally on the description of compact clusters or networks of water at the protein surface. One can suggest that there might be value in the analysis of structural data for the presence of thread arrangements.

Studies on proteins at low hydration have direct application to membranes, the proteins of which, if they are integral to the membrane, have a portion of their surface withdrawn from solvent. A significant amount of water may be bound to the protein surface internal to the membrane. Although purple membrane is hydrated to at least 60 percent of the level of hydration of typical globular proteins,[8] only 25 percent of the surface of the purple membrane protein, bacteriorhodopsin, is exposed at the membrane surface.

The percolation model bears on the transport of ions and small molecules across membranes. This movement is thought to involve, generally, a water channel,[9] although for this discussion it is sufficient to imagine some collection of conducting elements. Several points can be made: (1) The channel or surface need be only partly filled with conducting elements. The percolation threshold, above which there is conduction, is at 0.45 fraction filled for a surface and at 0.16 for a 3-dimensional region. (2) Gating might be associated directly with a simple chemical event, such as binding of a ligand or change in state of ionization of a protein group, without its being mediated by a change in protein conformation. If the system is poised near the critical percolation concentration, a small change in the fraction filled would serve to switch the system. A two percent decrease in water content shifts the lysozyme system from a proton conductive to a non-conductive mode. (3) The central focus of percolation theory is the randomness of the arrangement of conducting elements. A membrane channel or conducting surface that operates through a percolation mechanism would need no structure extending over the full thickness of the hydrocarbon core of the membrane. For example, in the case of the proton pump of the purple membrane there need be no proton wire of hydrogen-bonded groups leading from the membrane surface to the active site. Instead, it is possible that protons diffuse to and from the gate along the threads of a fluctuating random percolation network within the protein-lipid interface. The photoreaction at the active site would serve to give a vectorial kick to the proton movement. In this picture, the requirement for a special non-random structure would be limited to the gate, and the size of the active site would be that found to be typical for other biochemical processes such as enzyme catalysis.

In our opinion the considerations of this section can be extended to include other biological events where a large-scale process can be understood to occur by appropriate correlation of several smaller-scale events.[10] As an example of this reasoning, we note that one may develop an alternative description of the conductive regime of hydrated lysozyme powders, as a series of correlated single proton transfers along a random thread of water molecules. Since each single proton transfer can be considered a small-scale event due to a local fluctuation, the large-scale conductivity results from the correlation of very many of these small-scale fluctuations.

References

1.   J.A. Rupley, E. Gratton and G. Careri, Trends Biochem. Sci., **8**, 18 (1983).
2.   T.L. Hill, J. Chem. Phys., **17**, 762 (1949).
3.   G. Careri, M. Geraci, A. Giansanti and J.A. Rupley, Proc. Nat. Acad. Sci. U.S.A., **82**, 5342 (1985).
4.   G. Careri, A. Giansanti and J.A. Rupley, Proc. Nat. Acad. Sci. U.S.A., in press.
5.   R. Zallen, "The Physics of Amorphous Solids," John Wiley and Sons, New York, N.Y. (1983).
6.   G. Careri, Stud. Nat. Sci. (N.Y.), **4**, [Quantum Stat. Mech. Nat. Sci. (Coral Gables Conf.), 1973], 15 (1974).
7.   T. Imoto, L.N. Johnson, A.C.T. North, D.C. Phillips and J.A. Rupley, "Enzymes," 3rd Ed., Vol. 7, Ed. P.D. Boyer, Academic, New York, 665 (1972).
8.   J.A. Rupley and L. Siemankowski, "Membranes, Metabolism and Dry Organisms," Ed. Carl Leopold, Cornell, Ithaca, N.Y. (1986).
9.   B. Hille, "Ionic Channels in Excitable Membranes," Sinauer, Sunderland, Mass. (1984).
10. G. Careri, "Order and Disorder in Matter," Benjamin/Cummings, Palo Alto, Calif. (1984).

Molecular Dynamics Simulations of Biomolecules in Water

K. N. Swamy and E. Clementi
IBM Corporation
Data Systems Division, Dept. 48B MS 428
Neighborhood Road
Kingston, New York 12401

National Foundation for Cancer Research
7315 Wisconsin Avenue
Bethesda, Maryland 20814

Abstract

Following our previous studies on understanding the structure and dynamics of biomolecules in water we have performed molecular dynamics (MD) simulations for the following systems: i) the ion transport through Gramicidin A (GA) trans-membrane channel, and ii) the hydration structure and dynamics of B and Z-DNA in the presence of counterions (K^+). The results of these simulations are presented both for the structural and dynamical properties. In the case of gramicidin the results are most preliminary and seem to indicate that the waters close to the ion inside the channel point their dipoles towards the ion, while the waters away from the ion are influenced by the bulk waters. In the simulations of B and Z-DNA we find that the waters close to the helix retain their memory of their initial orientations for longer times than those waters away from the helix, in agreement with the relaxation studies of Lindsay et al. on Z-DNA. Further, the velocity auto correlation functions for the counterions (K^+) exhibit a caged behavior unlike that in an ionic solution.

Introduction

Previously, we have reported on the solvation of DNA[1-7] and on the structure of water interacting with ions and Gramicidin.[8-10] In this work we extend the earlier studies by considering not only the structure but also the dynamical characteristics of these systems. To understand the structure and dynamics of biomolecules in water we consider two examples. In the first system we consider the transport of an ion through a Gramicidin A transmembrane channel. The transport of both Na^+ and K^+ ions is considered and the results are analyzed for the structure of water inside the channel. As a second example, we consider the hydration of B and Z-DNA in the presence of counterions (K^+). We analyze the simulation results for the structure of water around the phosphate groups and the counterions as well as information on the dynamical properties of water. The analysis of the dynamical properties is done in different cylindrical subshells where the entire simulation cell is divided into 8 subshells.

A. Gramicidin A channel

The microscopic mechanism of ion transport through membrane channels is important in several areas of biology. Experimentally, ion transport through the Gramicidin channel has attracted attention primarily due to its simple and well defined structure.[11-16] Gramicidin A, a pentadecapeptide with the primary structure HCO - L.Val[1]- Gly[2] - L.Ala[3] - D.Leu[4] - L.Ala[5] - D.Val[6] - L.Val[7]- D.Val[8] - L.Trp[9]- D.Leu[10] - L.Trp[11]- D.Leu[12] - L.Trp[13]- D.Leu[14] - L.Trp[15] - $NHCH_2CH_2OH$ is known to form cation conducting transmembrane channels.[17]

Even though the debate continues, the Urry head-to-head dimerized left handed single stranded helix with about 6.3 residues per turn and with the L-D dipeptide being the repeat unit of the helix is the most accepted predominant structure of the Gramicidin A transmembrane channel.[18-22] In the head-to-head dimerized form the two molecules are held together by six intermolecular hydrogen bonds. Conformational studies of the local energy minima of the Gramicidin structure in the channel performed by Urry and Venkatachalam[23] have concluded that the left handed helix is 2.2 kcal/mol lower in energy than the right handed helix. These calculations have been performed with only selected torsional degrees of freedom of GA and in the absence of water and counterions. One of us (E.C.) and coworkers[8,9] have utilized *ab initio* computations on the ion-GA and the water-GA systems to develop 6-12-1 atom-atom potentials.

From C^{13} NMR experiments it has been shown that there are two binding sites related by two fold symmetry and separated by around 20Å. The entry barrier obtained from rate constant measurements is just over 7 kcal/mol. The rate limiting step for the ion entry into the channel is the partial dehydration step wherein the coordinating water molecules are replaced by carbonyl oxygens. Further, it has been suggested that the rocking or the torsional motion of the carbonyl groups help in solvating the ion. Pullman and coworkers[24] have computed the energy profile of Na^+ interacting with GA where temperature and statistical effects have been neglected; their computed binding energy is far lower than experimentally suggested. Clementi and coworkers[10] have performed Monte Carlo (MC) simulations of ion (Na^+ and K^+)-water-GA system. Brickman and coworkers[25,26] have performed molecular dynamics simulations with a model GA where GA is approximated by a helical arrangement of polar carbonyl groups. This one dimensional model was later replaced by a three dimensional model. Wilson et al.[27] have carried out MD simulations of GA by including 13 water molecules and one ion (Li^+, Na^+, K^+ and Cs^+) placed inside the channel. They analyzed the results for a) the structure of water molecules inside the channel, b) the librational motion of the peptide moiety, and c) the orientational motion of the water molecules. One of the shortcomings of these simulations as pointed out by these authors is the absence of bulk water outside the channel and the use of empirical potential functions.

Previously, Clementi and coworkers[28] have reported the molecular dynamics simulations of GA-ion-water system wherein they included 81 water molecules inside the simulation cell to provide sufficient bulk water characteristics at the channel ends. In these preliminary simulations they considered both Na^+ and K^+ ions and applied periodic boundary conditions along the channel axis. Further, to keep the simulations manageable the channel has been kept rigid and the leap-frog algorithm has been used to integrate both translational and rotational equations of motion. It has been shown by Clementi et al.[29] that the leap-frog algorithm for rotational motion is not a stable algorithm. In the results presented here the molecular dynamics simulation results for the Gramicidin A channel with 145 water molecules and an ion are given. Simulations are performed by placing the ion (Na^+ and K^+) at two different positions inside the channel. The rigidity of the channel has been partly removed by allowing the oxygens of the carbonyls to bend into the pore by including torsional motion. The structural and dynamical properties are then computed from a simulation carried out for 1.2 picoseconds.

B and Z-DNA

It is known experimentally that DNA helical structures are influenced by solvent effects and their stability depends strongly on the presence of counterions in solution. The x-ray studies of hydrated structures are incapable of providing reliable data on the orientation of water molecules. Recently, Clementi and coworkers have performed Monte Carlo simulations of nucleic acids in aqueous solution including up to several thousand water molecules with Li^+, Na^+, K^+ counterions[1,2] considering either A^3, B^4 or Z^5 conformations, single and double helices.[6,7] In these studies they used potential functions obtained from *ab initio* computations explicitly taking into account the hydrogen atoms. These studies have provided detailed information on different aspects of the hydration phenomena. In the earlier simulations the ions were constrained to remain at some fixed positions while the structure of water was optimized with the MC technique. In subsequent simulations this restriction has been removed. The results suggested that some of the counterions are less solvated and the most probable positions of the ions are found to be diffused over a region of space about 1 to 2Å, and that the counterions and water molecules form two slightly different patterns at the two DNA strands, h and h*. From an analysis of the radial distribution functions they concluded that the hydration pattern is directly associated with the observed effect of steric stabilization of the double helix conformation in the presence of counterions.

A final conclusion derived from these simulations is that single and double helices in biological solution should be treated as "super structures" consisting of the traditional DNA, the counterion patterns and the filament of hydrogen bonded water molecules. The electroneutrality of the "super structure" keeps it stabilized.

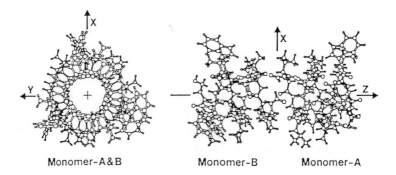

Monomer-A & B   Monomer-B   Monomer-A

Figure 1. Gramicidin A dimer in the x-y plane (left) and x-z plane (right).

Molecular Dynamics Simulations

A. Ion Transport Through Gramicidin A Channel

The molecular dynamics simulations reported in this work are performed with 145 water molecules, one ion (Na$^+$ and K$^+$) and Gramicidin in the Urry head-to-head dimerized form. The simulation box is a hexagonal prism of height 48Å and diameter 20Å. For each ion two simulations are performed, one with the ion in the middle of the channel (z $\sim$ 0.0Å) and the second one with the ion towards the end of the channel (z $\sim$ 14.0Å). The rigidity of the channel is partly removed by allowing the carbonyl oxygens of the peptide moiety to bend towards the pore. The

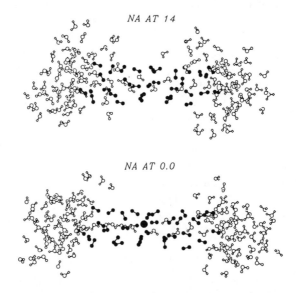

Figure 2. An instantaneous configuration of the ion (Na$^+$), water molecules and the carbonyl groups during the simulation. The bottom inset is for z $\sim$ 0.0 Å and the top inset is for z $\sim$ 14.0 Å Note that there are about 7-8 water molecules inside the channel. The carbonyls and the ion are darkened for easy identification.

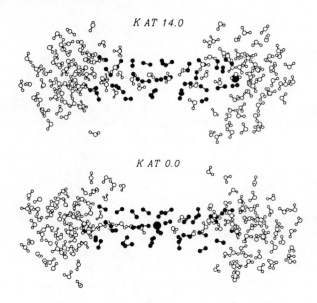

Figure 3. Same as Figure 2 but for K^+ ion.

torsional potentials for this motion is taken from the work of Wilson et al.[27] These potentials are obtained from a least squares fit to the experimental data. The translational equations of motion are solved using fifth order Gear's predictor-corrector algorithm[30] and the rotational equations of motion are solved using second order quarternion method.[31] A time step of $1.0*10^{-16}$ sec. has been used in the simulations. The simulations are performed by equilibrating the ion-water-GA system at $\sim$ 300 K for about 2-3 picoseconds. Following equilibration the simulations are then continued for 1.2 picoseconds.

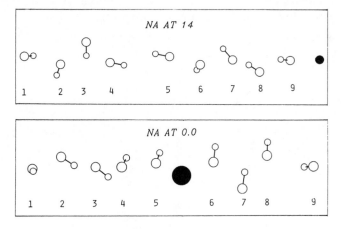

Figure 4. Average dipole moment of the water molecules inside the channel for Na^+. The dipoles are represented as diatomics with big circle representing negative end of the dipole. The ion is also shown as dark circle. The water molecules that are close to the ion point their dipoles towards the ion. Top inset for Na^+ at $z \sim 14.0$ Å and the bottom inset is for $z \sim 0.0$ Å

B. B-DNA

The length of one repeat unit of B-DNA is 33.8Å and consists of 10 base pairs. In this study we consider a G-C and A-T base pair sequence. The helical axis coincides with the z-axis of the simulation cell which also determines the height of the box. The molecular dynamics simulations are performed by taking B-DNA in a parallelopiped of size 44.0Å*44.0Å*33.8Å. An equilibrated MC configuration of 500 water molecules, one turn of B-DNA and 20 counterions (K^+) is added to the box and the box is then filled with an additional 1000 water molecules placed at their lattice sites. The entire system is then equilibrated at $\sim$ 300 K for about 2-3 picoseconds. The translational and rotational equations of motion are then solved using predictor-corrector algorithm as discussed earlier for Gramicidin. The simulations have been performed for 4.0 picoseconds after equilibration with a time step of $0.5*10^{-15}$ sec. Periodic boundary conditions are used to minimize the influence of boundary effects. To speed up the program the water-water interaction potential has been truncated when the oxygen-oxygen distance is greater than 8.0Å. Tests showed that this choice of cut-off results only in small fluctuations in the total energy.

C. Z-DNA

The length of one repeat unit of Z-DNA is 44.56Å and consists of 12 base pairs. We considered a G-C base pair sequence in the present study. The simulation box is a hexagonal prism of height 44.56Å (which also coincides with the helical axis) and of diameter 44.56Å. An initial equilibrated Monte Carlo configuration consisting of 500 water molecules and 24 counterions (K^+) is then added to the system and the rest of the box is then filled with an additional 1351 water molecules to provide additional hydration. The complete system is then equilibrated at $\sim$ 300 K for about 2-3 picoseconds. A time step of $0.5*10^{-15}$ sec is used for integrating the Newton-Euler equations of motion. After equilibration, the simulation is then continued for 3.5 picoseconds.

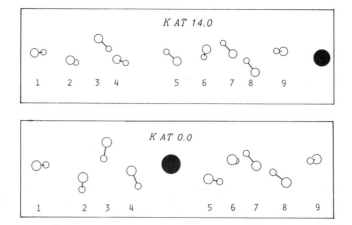

Figure 5. Same as Figure 4 but for K^+ ion.

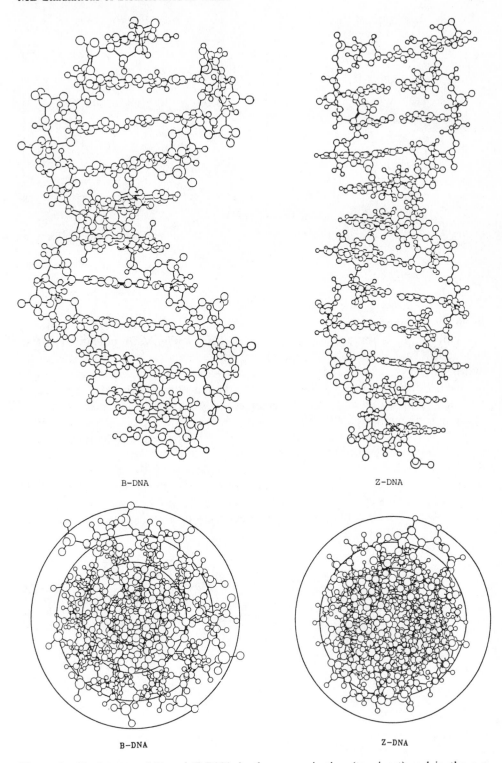

B-DNA   Z-DNA

B-DNA   Z-DNA

Figure 6. Single turn of B and Z-DNA in the x-z projection (top inset) and in the x-y projection (bottom inset). In the bottom inset concentric circles of 3,6,9 and 12Å radii are drawn.

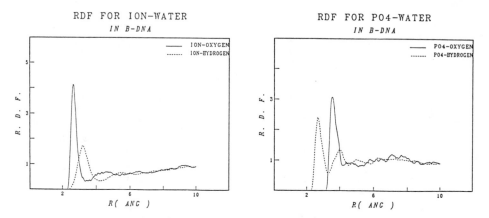

Figure 7. Partial radial distribution functions for phosphorus-water and ion-water in B-DNA.

Potential Functions

In our MD simulations we used the rigid water-water interaction potential of Matsuoka et al.[32] (abbreviated herein after as MCY). The MCY potential is a four site model. Two of the point charges of magnitude $+q$ are placed at the sites of the hydrogen atoms and one charge of magnitude $-2q$ is placed along the C_{2v} axis at a distance R from the oxygen towards the hydrogen atoms. The MCY potential is obtained by fitting *ab initio* binding energies for water dimers obtained at the configuration interaction level. For ion-ion, ion-water, ion-Gramicidin, ion-DNA, water-Gramicidin and water-DNA we used 6-12-1 atom-atom potentials obtained in our laboratory by fitting to *ab initio* data. The interaction potentials for ion-Gramicidin and water-Gramicidin are obtained by fitting the *ab initio* results by considering the interaction of ion and water with individual *residues* of the Gramicidin monomer. For ion-Gramicidin interactions we feel that the least squares fit is too attractive partly due to an insufficiently accurate fit, and partly

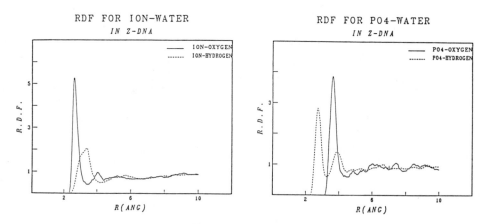

Figure 8. Same as Figure 7 but for Z-DNA.

due to the assumption of additivity of the residues. To remove this unphysical behavior, we are presently refitting the ion-Gramicidin interaction potentials and also computing the *ab initio* interaction energies by considering the *full Gramicidin monomer*. This is a most drastic departure from past studies. We recall that the use of the full Gramicidin monomer implies *ab initio* computations with 276 atoms and about 820 contracted functions (corresponding to 2736 primitive Gaussians) with typically $5*10^8$ non-zero integrals over contracted functions. The results of these

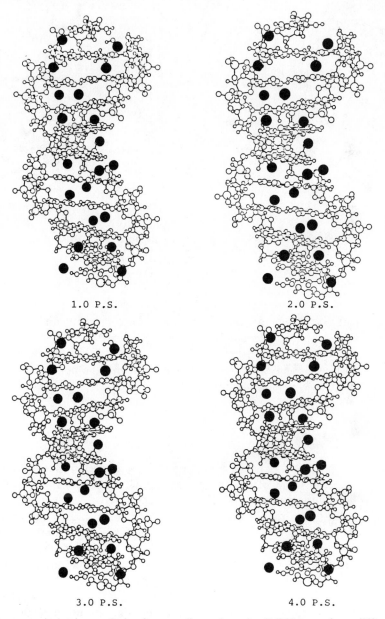

Figure 9. x-z projections of the ion configurations in B-DNA at four different time frames. The ions are darkened for easy identification.

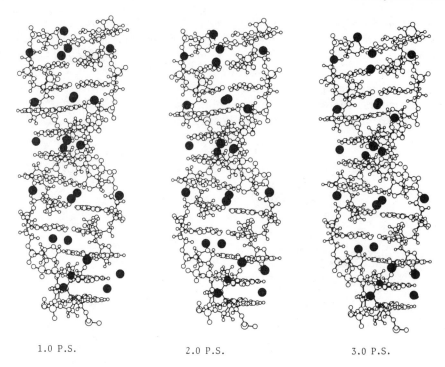

1.0 P.S.   2.0 P.S.   3.0 P.S.

Figure 10. Same as Figure 9 but for ions in Z-DNA.

calculations along with new atom-atom potentials are presently in progress[33] and will be published separately. Hence, the simulation results for the transport of the ion through Gramicidin channel are most preliminary and qualitative, and limited to those results that we feel should not change even with the new and forthcoming potentials.

Parallel Computations

The simulations presented here represent one of the largest computer simulations performed to date, in terms of the number of interacting particles. To give a somewhat rough idea of the computational efforts needed, we recall that in the case of Z-DNA each time step requires approximately 10^7 force and energy calculations for water-water interactions and approximately 10^6 force and energy calculations for water-DNA interactions. Further, the MCY potential for water-water interactions contain time consuming exponential terms. The computations are therefore performed on our experimental parallel computer system lCAP-2 (loosely coupled array processors). The lCAP-2 system consists of IBM 3084 connected to 10 FPS-264 attached processors (APs). A similar configuration called lCAP-1 consisting of IBM 3081 connected to 10 FPS-164 attached processors has been used in the simulation of GA. Multiple FPS attached processors can be used in a single job in parallel.[34-36] We are now working with lCAP-3 where the lCAP-1 and lCAP-2 clusters are front ended by an IBM 3090 supercomputer; this is also the configuration used for the *ab initio* Gramicidin study mentioned above. In the parallel computations the actual evaluation of forces is parcelled out to several independent attached processors by passing the information of the coordinates and orientations of the particles to each AP. Each AP is then assigned only a partial number of force evaluations to perform. The partial forces are then passed from the APs to the IBM host, which collects all the partial forces and passes them to the predictor-corrector algorithm. The new coordinates and orientations are computed sequentially on the IBM host, and then passed to the APs to begin another time step.

Results

A. Gramicidin

Because of the large volume of data and the preliminary nature of the computational results, the analysis for GA are presented graphically, with the emphasis on highlighting the essential features of the structure and dynamics; particularly the structure of water molecules inside the channel.

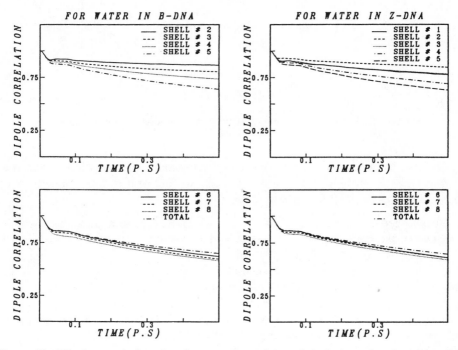

Figure 11. Dipole correlations for the water in various subshells (see text) for waters in B and Z-DNA.

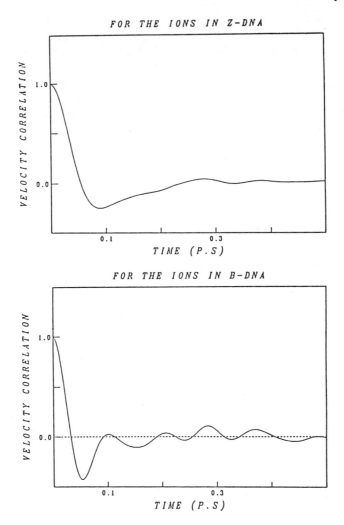

Figure 12. Velocity correlation functions for the ions in B and Z-DNA.

In Fig. 1 we show the Gramicidin dimer in the x-y and x-z projections. The radius of the channel is around 2.0Å. For each ion we performed two simulations by placing the ion at z ∼ 0.0Å and z ∼ 14.0Å.

An instantaneous configuration of the water molecules, ion and the carbonyl groups when the ion is placed at z ∼ 0.0Å and z ∼ 14.0Å is shown in Figs. 2 and 3. In these figures the carbonyls and the ion are darkened for easy identification. As is evident there are around 7-8 water molecules forming a single file inside the channel. This number agrees with the electrokinetic experiments of Levitt.[37] On the other hand, osmotic and diffusion permeability experiments give a single file of 5-6 water molecules inside the channel.[38] The average dipole moment of the water molecules inside the channel during the simulation is presented in Figs. 4 and 5. In these figures the dipole moments of the water molecules are represented as diatomics with the big circles representing the negative ends of the dipoles. Clearly,

Table I. Simulation parameters for the hydration of B- and Z-DNA.

	B-DNA	Z-DNA
Simulation box	Parallelopiped 33.8×44.0×44.0Å	Hexagonal prism 44.56Å Height 44.56Å Diameter
Length of repeat unit	33.8Å	44.56Å
No. of base pairs	10	12
No. of counterions (K$^+$)	20	24
No. of water molecules	1500	1851
Simulation time (picoseconds)	4.0	3.5
Times step ($\times 10^{-15}$ sec)	0.5	0.5

those water molecules that are close to the ion point their dipoles towards the ion whereas the waters away from the ion are influenced by bulk waters. This observation differs from the work of Wilson et al.[27] and is probably due to the absence of bulk water in their simulation. An analysis of the water molecules inside the channel shows that the water molecules are hydrogen bonded with one proton attracted by the nearest carbonyl group and the other to an adjacent water molecule. Such a network of hydrogen bonded chains has been studied by Nagle et al.[39] in connection with proton pumping and conduction.

The ion-GA and water-GA potentials reported by Clementi et al.[8-10] are obtained from a fit of *ab initio* computations by considering the interaction of ion and water with individual amino acids. The potential functions thus derived were used in the MD simulations reported above as well as in the previous Monte Carlo simulations on Gramicidin. The fitting, especially for Na$^+$, is about 10% more attractive than the *ab initio* values. The potentials derived from this fit are deficient because a) the interactions with the ions are long ranged, and b) several residual units will be in near contact with the ion and the water, when the latter are within the GA channel

Table II. Energetics of the water-cation (K$^+$)-DNA system. Interaction energy for water-water (W-W), water-ion (W-I), water-DNA (W- DNA), ion- DNA (I- DNA) and ion-ion (I-I) and temperature of the simulation. Units are kJ/mol-water for W-W, W-I and W- DNA and kJ/mol-cation for I-I and I- DNA. Temperatures are in degrees Kelvin.

Interaction Type	Z-DNA Interaction Energy	B-DNA Interaction Energy
W-W	-27.61 ± 0.95	-27.18 ± 0.84
W-I	-23.90 ± 0.38	-26.71 ± 0.20
I-I	1229.27 ± 19.28	1369.80 ± 2.81
W- -DNA	-29.50 ± 0.28	-23.81 ± 0.17
I- -DNA	-4173.61 ± 28.54	-8800.35 ± 2.13
Tot. PE	-3025.35 ± 30.98	-7508.26 ± 2.17
Temperatures:		
Water (Trans)	346.2 ± 22.9	344.8 ± 18.7
Water (Rot)	345.0 ± 25.7	346.9 ± 27.8
Ion (Trans)	387.9 ± 72.7	342.8 ± 59.6

Table III. Number of water molecules in the first hydration shell for K^+ and PO_4^- at various values of the radii for B- and Z-DNA.

R(Å)	B-DNA (G-C, A-T) K^+		PO_4^-
	MD	MCa	MD
3.0	3.2	4.92	0.0
3.5	3.9	4.92	1.8
4.0	4.9	–	6.2
4.5	6.6	–	8.6
5.0	8.8	–	11.5
5.5	11.1	–	14.9
6.0	14.1	–	19.6

R(Å)	Z-DNA (G-C) K^+		PO_4^-
	MD	MC	MD
3.0	4.2	4.27	0.0
3.5	5.1	4.27	1.8
4.0	6.6	–	7.2
4.5	8.6	–	9.1
5.0	11.1	–	11.6
5.5	14.4	–	14.8
6.0	17.9	–	19.4

(and this fact has been ignored in the derivation of the potentials). These features bring about a large overall fitting error. Indeed, the potential functions yield an energy barrier much larger than the one postulated by Urry et al..[22] This exaggerated attraction of the ion to GA brings about the result that the ion cannot pass through the channel. In an exploratory study of reducing the central barrier we performed two separate simulations with an applied field of 5.0 volts across the channel and considered either only a single ion or an ion and two water molecules. In this very rough study the ion-GA and water-GA interaction potentials were scaled by a chosen fraction of their initial values. In the simulation with only one ion inside the channel, the simulation was carried out for 4.0 picoseconds. During this time the ion did not cross the channel. On the otherhand, in a simulation with one ion and two water molecules the ion moved from an initial configuration of -8.0Å to 16.0Å in 5.0 picoseconds. This leads to the important observation that water is a necessary medium for the transport of ions through the channel. Clearly, the reduction of the interaction potential is most arbitrary but points out the deficiencies of the interaction potentials. Due to these observations, we decided to calibrate the ion-GA interaction potential by performing *ab initio* computations with the entire GA (see above).

B and Z-DNA

In Table I we present the simulation parameters and in Table II the energetics for both B and Z-DNA. The energetics are quoted in kJ/mol-water for water-water, water-ion and water-DNA and in kJ/mol-cation for ion-ion and ion-DNA interactions. In Fig. 6 (top inset) we show a single turn of B and Z-DNA in the x-z

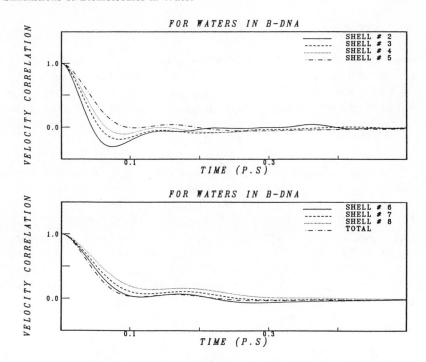

Figure 13. Velocity correlation functions for the waters in various subshells (see text) in B-DNA simulations.

projection. In the bottom inset of the figure x-y projections are given along with concentric circles of 3, 6, 9 and 12Å radius. In Figs. 7 and 8 we show the radial distribution functions for ion-water and phosphate-water systems in B and Z-DNA respectively. In the phosphate-water radial distribution functions the distances are measured from the phosphorus atom. The phosphate-water radial distribution functions indicate a layer of hydrogen atoms in the region between the counterions and the phosphate group. This screening cloud of hydrogen atoms shows up clearly as a peak at r $\simeq$ 2.8Å in the radial distribution functions of hydrogen atoms around the phosphorus atom.[40] For ion-oxygen and ion-hydrogen partial radial distribution functions, the positions of the first maximum and first minimum occurs at 2.7Å and 3.2Å for B-DNA and 2.6Å and 3.4Å for Z-DNA respectively. These results are consistent with the result 2.76Å and 3.35Å for K$^+$ in an ionic solution.[41] In Table III we compare the number of water molecules surrounding an ion and the phosphorus atom for 5 different values of the radii with the results obtained from Monte Carlo simulations. Comparison with previous MC simulations has been somewhat frustrated by the different definition of the solvation shell used in this work. We are now recomputing such data using the same boundaries as in previous work (work in progress).

In Figs. 9 and 10 we plot the ions around the DNA in the x-z projection at four different time frames for B-DNA and three different time frames for Z-DNA. In these figures the ions are darkened for easy identification.

Dynamical properties of B and Z-DNA.

For a proper understanding of the dynamical properties of the water and the ions we divided the simulation cell into several cylindrical subshells of radius 3.0Å. The first shell is a cylinder of radius 3.0Å with origin at the center of the helix. The idea is that the last subshell is away from the helix and contains only water molecules. The dynamical properties are computed in each cylindrical subshell and compared with those for the entire system. In Fig. 11 we plot the dipole orientational correlation functions for the water in several subshells and compare them with those for the total water in the simulation cell. It is seen from these figures that the dipole correlations decay more slowly for those waters close to the helix than those that are farther away from the helix. The behavior of these correlation functions in the subshells 5, 6, 7 and 8 is essentially the same as that for the total water molecules. This observation is in agreement with the relaxation experiments of Lindsay[42] on the hydration of Z-DNA.

In order to obtain some insight into the microscopic dynamics of the ion, we have calculated the velocity auto correlation function, $Z_{ion}(t)$, defined as:

$$Z_{ion}(t) = < \vec{V}(t) \cdot \vec{V}(0) > / < \vec{V}(0) \cdot \vec{V}(0) >$$

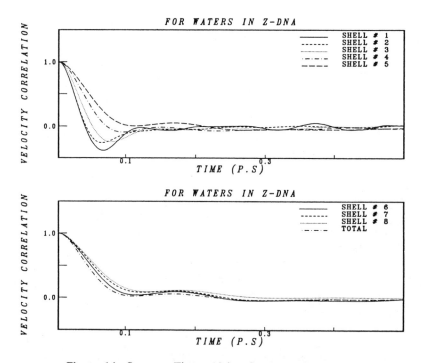

Figure 14. Same as Figure 13 but for waters in Z-DNA.

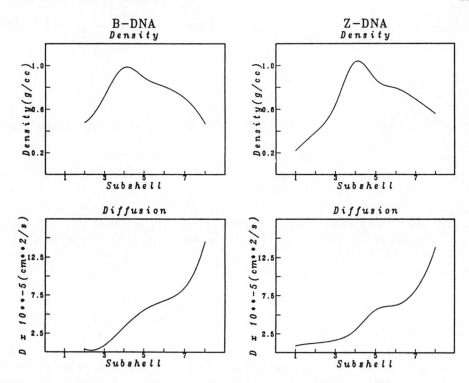

Figure 15. Density and diffusion coefficient variation for water in various subshells for B and Z-DNA.

where $\vec{V}(t)$ is the velocity of the ion at time t and the angular brackets denote the ensemble average. In Fig. 12 we plot the velocity auto correlation functions for the counterions in B and Z-DNA simulations. Because there are only few ions and since most of the ions are close to the helix the velocity auto correlation functions are plotted for the whole system to avoid statistical errors. These correlation functions have the appearance of a damped oscillator, characterized by a single high frequency resembling those for smaller ions, e.g. Li^+ obtained by Impey et al.[41] in their simulation of ionic solutions. This high frequency motion has been ascribed to the "rattling" motion of the ion in a cage.

The velocity correlation functions for the waters in various subshells are plotted in Figs. 13 and 14 for B and Z-DNA respectively.

The analysis of the diffusion coefficients for the waters is complicated due to the fact that the helix has a tendency to pull waters around the helix with subsequent density gradients. But statistically there are no differences in the diffusion coefficients parallel and perpendicular to the helical axis. The diffusion coefficient for the ions in B-DNA is $3.75*10^{-6}$ cm²/sec, much smaller than that found in an ionic solution. This is due to the fact that the ions tend to spend most of the time close to the helix and hence its environment is completely different from that in an ionic solution. The density and diffusion coefficient variation for waters in various sub-shells is plotted in Figure 15.

Conclusions

In this paper we presented the molecular dynamics simulations of a) ion transport through a Gramicidin A transmembrane channel and b) the hydration structure of B and Z-DNA in the presence of counterions. The results of these simulations are analyzed for structural and dynamical properties. In the simulations of Gramicidin A we find that the waters inside the channel form a linear arrangement with about 7-8 water molecules inside the channel. The water molecules that are close neighbors of the ion point their dipoles towards the ion whereas those waters that are away from the ion are influenced by the bulk water molecules. The waters inside the channel are hydrogen bonded via one proton to the carbonyl and the other proton to the nearest water molecule.

In the simulations of the hydration structures of B and Z-DNA we find: i) the water molecules that are close to the helix retain the memory of their initial orientation for longer periods of time than those that are away from the helix, ii) the ions (K^+) in DNA exhibit a caged behavior unlike that in an ionic solution. Finally, we recall that at the time of the MC simulations[1,4] the Manning[43] model of the ionic cloud was still an accepted one. The MC simulations - on the contrary, predicted a counterion structure with well defined positions relative to the phosphate groups. This notable difference in structure was in need of more testing especially by molecular dynamics simulations. The results of the counterions obtained here indicate very limited mobility and a well defined counterion pattern as originally suggested in the MC study.

Acknowledgments

The authors are thankful to Drs. George C. Lie and Giorgina Corongiu for suggestions and to Dr. Giorgina Corongiu for providing the MC data needed to initialize the simulations for B and Z-DNA, and to Dr. Aatto Laaksonen for providing an initial version of the molecular dynamics program.

References

1. E. Clementi and G. Corongiu, J. Biol. Phys., **11**, 33 (1983).
2. E. Clementi and G. Corongiu, J. Chem. Phys., **72**, 3979 (1980).
3. E. Clementi and G. Corongiu, Biopolymers, **18**, 2431 (1979).
4. E. Clementi and G. Corongiu, Int. J. Quantum Chem., **16**, 897 (1979).
5. E. Clementi, G. Corongiu, M. Gratarola, P. Habitz, C. Lupo, P. Otto and D. Vercauteren, Int. J. Quantum. Chem., QCS, **16**, 409 (1982).
6. E. Clementi and G. Corongiu in Biomolecular Stereodynamics, R.H. Sarma, ed., Adenine Press, New York, Vol. 1, p. 209 (1981).

7. E. Clementi and G. Corongiu, Ann. N.Y. Acad. Sci. (USA), **367**, 83 (1981).
8. S.L. Fornili, D.P. Vercauteren and E. Clementi, J. Biomolec. Struct. Dynam., **1**, 1281 (1984).
9. K.S. Kim, D.P. Vercauteren, M. Welti, S. Chin and E. Clementi, Biophys. J., **47**, 327 (1985).
10. K.S. Kim, D.P. Vercauteren, M. Welti, S.L. Fornili and E. Clementi, Biochem. Biophys. Acta., **xx**, xxx, (1985).
11. D.W. Urry in Topics in current chemistry, Vol. 128, Springer-Verlag, Berlin (1985).
12. O.S. Anderson, Ann. Rev. Physio., **46**, 531 (1984).
13. O.S. Anderson, Biophys. J., **41**, 119, 135, 147 (1983).
14. Y.A. Ovchinnikov and V.I. Ivanov, in Conformation in Biology, R. Srinivasan and R. H. Sarma, eds., Adenine Press, New York (1983).
15. G. Eisenman and R. Horn, J. Membr. Biol., **76**, 197 (1983).
16. E. Bamberg and P. Lauger, J. Membr. Biol., **35**, 351 (1977).
17. S.B. Hladky and B. W. Haydon, Biochim. Biophys. Acta., **274**, 294 (1972); G. Eisenman and J.P. Sandblom, Biophys. J., **45**, 88 (1984).
18. D.W. Urry in The enzymes of biological membranes, Vol 1, A.N. Martonosi, ed., Plenum Publishing Company, New York (1985).
19. D.W. Urry, Proc. Nat. Acad. Sci. (USA), **68**, 672 (1971).
20. D.W. Urry, M.C. Goodall, J.D. Glickson and D.F. Mayer, Proc. Nat. Acad. Sci. (USA), **68**, 1907 (1971).
21. D.W. Urry, T.L. Trapane and K.U. Prasad, Int. J. Quantum Chem., Quantum Biology Symp., **9**, 31 (1982).
22. D.W. Urry in Membranes and Transport, Vol. 2, A. Martonosi, ed., Plenum Publishing company, New York (1982).
23. C.M. Venkatachalam and D.W. Urry, J. Comput. Chem., **4**, 461 (1983); C.M. Venkatachalam and D.W. Urry, J. Comput. Chem., **5**, 64 (1984).
24. A. Pullman and C. Etchebest, Fed. Eur. Biochem. Soc., **163**, 199 (1983); C. Etchebest and A. Pullman, Fed. Eur. Biochem. Soc., **170**, 191 (1984); C. Etchebest, S. Ranganathan and A. Pullman, Fed. Eur. Biochem. Soc., **173**, 301 (1984).
25. W. Fischer and J. Brickmann, Biophys. Chem., **17**, 245 (1983); H. Schroder, J. Brickmann and W. Fischer, Mol. Phys., **11**, 1 (1983).
26. E.E. Polymeropoulos and J. Brickmann, Ann. Rev. Biophys. Chem., **14**, 315 (1985).
27. D.H.J. Mackay, P.H. Berens, K.R. Wilson and A.T. Hagler, Biophys. J., **46**, 229 (1984).
28. K.S. Kim, H.L. Nguen, P.K. Swaminathan and E. Clementi, J. Phys. Chem., **89**, 2870 (1985).
29. R. Sonnenschein, A. Laaksonen and E. Clementi, J. Comp. Chem., **7**, 645 (1986).
30. C.W. Gear, Numerical initial value problems in ordinary differential equations, Prentice-Hall, Englewood Cliffs, N. J., 1971.
31. D.C. Rapaport, J. Comp. Phys., **60**, 306 (1985).
32. O. Matsuoka, E. Clementi and M. Yoshimine, J. Chem., Phys., **64**, 1351 (1976).
33. R. Gomperts and E. Clementi, private communication.

34. H.L. Ngyuen, H. Khanmohammadbaigi and E. Clementi, J. Comp. Chem., **6**, 634 (1985).
35. E. Clementi, G. Corongiu, J.H. Detrich, H. Khanmohammadbaigi, S. Chin, L. Domingo, A. Laaksonen and H.L. Ngyuen, Structure & Motion: Membranes, Nucleic acids & Proteins, E. Clementi, G. Corongiu, M.H. Sarma and R.H. Sarma, eds., Adenine Press, New York, p. 49 (1985).
36. G. Corongiu and J.H. Detrich, IBM J. Res., **29**, 422 (1985).
37. D.G. Levitt, Biophys. J., **37**, 575 (1982).
38. P.A. Rosenberg and A. Finkelstein, J. Gen. Physiol., **72**, 341 (1978).
39. J.F. Nagle and S. Tristram-Nagle, J. Membrane Biol., **74**, 1 (1983).
40. W.K. Lee and E.W. Prohofsky, Chem., Phys., Lett., **85**, 98 (1982).
41. R.W. Impey, P.A. Madden and I.R. McDonald, J. Phys. Chem., **87**, 5071 (1983).
42. S.M. Lindsay, private communication.
43. G. S. Manning, Biopolymers, **19**, 37 (1980).

Low Frequency Coherent Vibrations of DNA:
The Role of the
Hydration Shell and Phosphate-Phosphate Interactions

S. M. Lindsay
Physics Department
Arizona State University
Tempe, AZ 85282

Abstract

The vibrational modes of DNA span a range from high frequency localized vibrations, through low frequency collective modes to over-damped Brownian fluctuations. Presumably the most important motions from a biological standpoint are the lowest frequency vibrations (involving the largest units) that are not over-damped by the viscous action of the hydration shell. I describe observations of low frequency collective vibrational modes of DNA which couple to the hydration shell. The dynamics of the hydration shell becomes important in a frequency "window" between the viscoelastic transition of the primary hydration shell (roughly 4 GHz.) and the viscoelastic transition of the secondary shell (roughly 80 GHz.). The role of coupled solvent − DNA dynamics in the A to B and B to Z transition is discussed in terms of the phosphate-phosphate interactions which probably dominate conformational stability. Excitations of coupled modes of the DNA-hydration shell system may also account for the resonant microwave absorption observed in restriction fragments and plasmids.

1. Introduction: The biological importance of gigahertz vibrations and the role of water

Biopolymers are special in their complexity and the involvement of rather large "molecular chunks" in their function. At the moment these components are understood phenomenologically. For example, subtle variations in "static" structure can give clues about the molecular basis of such complex processes as recognition, but a full physical understanding would require knowledge of the movements of the atoms involved, and how these movements (phonons) interact with electronic states to bring about the binding and subsequent reaction. We can say with some confidence, however, that the interesting frequencies of motion must lie between a few GHz. and a few hundred GHz. No sophisticated analysis is needed — the upper end of the frequency range is set by the mass of the smallest "bits" of molecule of biological importance (a basepair, for example) and the value of typical interatomic

239

force constants.[1] The lower end of the time scale is fixed by the effects of the viscous drag of the hydration shell. The time scale of heavily damped motion is characteristic of the viscosity of the damping medium and not details of atomic structure. Thus *the low frequency Brownian fluctuations cannot contribute to biology at the atomic level*. It is my purpose in this chapter to examine this low frequency transition from under- to over-damped motion, and to outline the importance of the hydration shell dynamics and "structure" in conformational transitions and resonant microwave absorption.

I end this introduction by stressing that current theories of biopolymer motion ignore the hydration shell dynamics. Molecular dynamics simulations (which could handle the problem in principle) cannot yet reach gigahertz frequencies. As a final caution, I note that non-linearities can give rise to an important role for fluctuations on time scales many orders of magnitude slower than this in, for example, the nucleation of new phases.[2]

2. The coupled dynamics of DNA and its hydration shell at GHz. frequencies

2.1 Models of water dynamics

The simplest approach to the hydration shell is to treat it as continuous viscous water. The damping of a vibrational mode is then just a consequence of the velocity gradients set up in the surrounding water by DNA motion. In a dilute solution, all modes below a few tens of GHz. are over-damped.[3] This model predicts that acoustic standing wave vibrations at gigahertz frequencies must be overdamped, and cannot give rise to resonant microwave absorption.[4] However, some recent experiments appear to indicate that such modes are far from over-damped.[5,6] Could it be that water is in some way "decoupled" from DNA? We have addressed this question with Brillouin scattering studies of DNA films at various degrees of hydration and as a function of temperature.[10]

2.2 Determination of hydration shell relaxation by Brillouin scattering

In a Brillouin scattering experiment, light is inelastically scattered by an acoustic phonon (sound wave). The wavelength of the scattering phonon depends on the scattering geometry,[7] so the well known linear dispersion of sound waves results in a Brillouin signal shifted in frequency by anything from a GHz. to a few tens of GHz., depending on the speed of sound and the scattering angle (the wavelength of the phonon involved is of the same order as the wavelength of visible light). The damping of these phonons (measured through the linewidth of the Brillouin signal) is both a strong function of the water content of fibers[8,9] and the temperature in a way that can be accounted for by the coupling of the acoustic phonon to two relaxation processes in the surrounding water.[10] We confirm this elementary analysis by independent measurements in which the phonon *frequency* is varied at a fixed temperature, and by theoretical models of the overall spectral profile.[10] A summary of our initial measurements is shown in Fig. 1 where we plot the natural logarithm of the relaxation time (the units are seconds *per radian*) with reciprocal temperature for a sample with "primary" hydration (approximately 0.5 gm of water per gm of DNA) and a sample with "secondary" hydration (approximately 1.5 gm of water per gm of DNA). Overall, the "primary" relaxation strength appears to increase as water is added up to about 1 gm water per gm of DNA.[9] At higher water contents, a fast process associated with the "secondary" shell dominates (though the slow primary process appears to persist).

2.3 *Implications of the measured hydration shell relaxation*

The two most important implications are:

i) GHz. acoustic modes are strongly coupled to the hydration shell.

ii) The structure of the hydration shell leads to specific dynamical excitations associated with that structure — "bound" water cannot be treated as a uniform viscous medium at these frequencies. The "primary" shell undergoes a viscoelastic transition at about 4 GHz. at physiological temperatures. The "secondary" shell undergoes this transition at about 80 GHz.

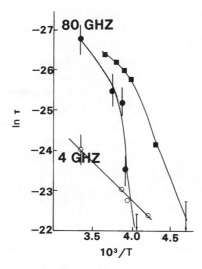

Figure 1. Plot of the natural logarithm of relaxation time with reciprocal temperature for a sample with primary hydration (○), and a sample with secondary hydration (●) as measured by Brillouin scattering. The remaining data (■) are for a sample with secondary hydration and are extracted from the Raman experiments of Tominaga et al.[11] Times are in seconds per radian — corresponding frequencies are marked in GHz.

The simplest possible modification of the "uniform" hydration shell picture is this: Below about 4 GHz. all the water surrounding the double helix is viscously coupled, so, on the whole, modes below this frequency will be over-damped. Above about 80 GHz. all the water is elastically coupled so the water will effectively stiffen (but not damp) motion. In the intermediate region, *there are important extra degrees of freedom associated with the coupled dynamics of the DNA-primary hydration system*. In other words, *in exactly the region of most biological importance, water plays an explicit dynamical role*.

The static structure of the hydration shell is hinted at in several experiments[12,13,21] and also by Monte Carlo calculations.[14] Furthermore the electrolyte surrounding the double helix plays a critical role in conformational stability.[15,16] We therefore expect the coupled motions of the double helix and hydration shell to be of great importance.

3. Water dynamics, conformational transitions and interactions between phosphates

3.1 Phosphate-phosphate interactions and conformational stability

Soumpasis[15,16] has identified the key role played by phosphate-phosphate interactions (mediated by the surrounding electrolyte) in conformational stability. The negative charges are treated as point anions interacting via the electrolyte through an effective pairwise interaction (calculated from a true many-body model based on ion-ion correlation functions). At low salt concentrations, this effective potential is everywhere repulsive, falling off in the familiar Debye-Hückel manner expected for screened Coulomb interactions. At high salt concentrations, however, the interaction is changed dramatically. It is not at all monotonic, and *attractive* in places. This structure reflects the details of the ion-ion correlation functions. Figure 2 shows this effective anion-anion interaction as calculated by Soumpasis[16] for a 1:1 electrolyte of 0.35 and 3.5 M concentration. Note, in particular, the complete reversal near the origin! The length scale is parameterized in terms of the distance of closest approach of the hydrated ions, σ. This parameter may vary between about 3Å and 8Å, but fits to experimental data are obtained with σ close to 5Å.[15,16] I should emphasize that the theory assumes an electrolyte *solution*, so that features in the high salt effective potential do not correspond to *fixed* charge structures. In discussing transitions that occur as solid DNA is diluted, I will outline possible charge structures with the understanding that they may be static, or represent some time average over fluctuating structures.

3.2 Interhelical interactions and conformation transitions

The theory of Soumpasis has proved remarkably accurate in describing the salt dependence of the B-Z transition.[15] It is less satisfactory in describing other transitions.[16] However, as we have shown elsewhere,[17] the A to B transition appears to be driven by solid state interactions. (This does not appear to be the case for the B-Z transition.[18]) We provide a further demonstration of the effect of crystallinity

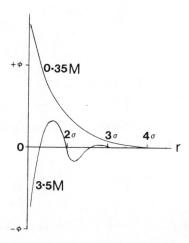

Figure 2. Effective anion-anion interaction potential for low (upper curve) and high (lower curve) salt concentrations. σ is the distance of closest approach of a hydrated ion pair. Note the appearance of new *attractive* regions at high concentrations.

on the A-B transition in Fig. 3 which shows Raman spectra obtained from crystalline and amorphous films as water is displaced by ethanol.[19] The 807 cm[-1] "A-form marker band" appears only in the crystalline sample, demonstrating that aligned strands are needed to form A-DNA. Brillouin scattering experiments show that there are *strong* interactions between adjacent helices in the solid state.[9,17] So strong are these bonds, that highly crystalline films can be nearly insoluble.[20] A-DNA is more closely packed than B-DNA, so the structural fluctuations due to sequence heterogeneity are smaller,[21] and thus A-DNA should be capable of forming a more regular pattern of interhelical contacts in the solid.

The nature of the interhelical contacts in the crystalline regions can be obtained from a study of the unit cell geometry. In A-DNA the double helices pack into a monoclinic cell with (at 54% r.h.) a = 21.1Å, b = 38.9Å, c = 27.3Å, β = 97° and an additional double helix at (1/2, 1/2, 0).[17] Although the closest approach (22.1Å center to center) brings the double helices almost into contact (at their O-2 radius), the three-dimensional packing minimizes O-2 clashes, with the backbones of one double helix lying in the major grooves of its neighbors on the whole. However, geometry does not permit the three-dimensional interlocking of screws of the same handedness and clashes do occur. These are illustrated (for the central molecule only) in Fig. 4 which shows a view down onto the unit cell. The circles correspond to the P-P diameter (nearly 18Å), and the black dots show the location (in the ab plane) of the O-2 atoms in one turn of the central molecule that come within 3.5Å of O-2 atoms on adjacent double helices. There are two such shared clashes per turn (i.e. one per double helix per turn), and it is interesting to note that the "magic" amount of excess NaCl (1% by weight) required to give good crystalline x-ray patterns gives almost exactly one extra Na[+] per double helical turn.

The O-2 interactions fall into two categories: the close approaches at backbone clashes, and the more common interactions as the backbone of one molecule packs

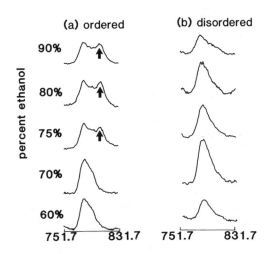

Figure 3. Raman spectra showing (a) the appearance of the 807 cm-1 A form marker band (indicated by an arrow) in crystalline samples as ethanol displaces water (percentages marked next to the spectra) and (b) the lack of the A form marker band as the amorphous samples are dehydrated (abscissa are Raman shifts in cm-1).

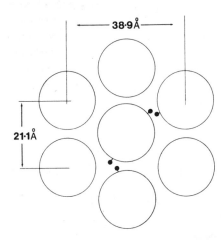

Figure 4. View of the interhelical contacts in A-DNA looking down into the ab plane of the monoclinic unit cell (dimensions are for 54% r.h.). The full circles correspond to the P-P diameter of the double helices and the black dots show the location of the O-2 atoms that come within 3.5Å of each other (for the central molecule only). The breaking of the apparent 6-fold symmetry is a consequence of the incommensurable nature of the 11-fold helix with the nearly 6-fold packing geometry.

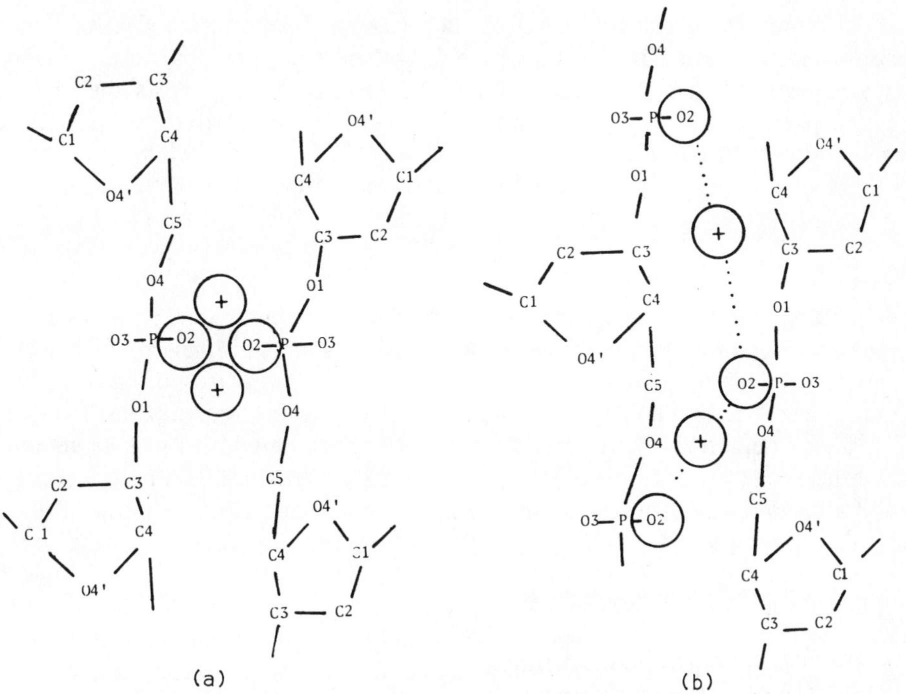

Figure 5. Some possible effective charge distributions for (a) close O-2 interactions at backbone clashes (the O-2 to O-2 separation is typically 3.3Å) and (b) for interactions where the O-2 of one molecule lies in the major groove of its neighbor (separations range from about 8 to 12Å).

into the major groove of another. The structures are sketched in Fig. 5. It is quite possible that many of the O-2 interactions fall into the attractive regions of 1σ and 2σ in the high salt potential (Fig. 2). The strength of the "electrolyte" is indeed in the molar region for fairly dry (A-DNA) fibers, but of course the applicability of a liquid electrolyte model must be somewhat questionable. A system of attractive interactions could be set up with positive charges distributed as indicated in Fig. 5. The closest approaches (5a) do not leave room for a hydrated cation between the oxygens, but nonetheless, if the potential of Fig. 2 applies, these may form the strongest interhelical bonds.

Whatever the detailed structure of these clusters, they presumably form some alternating array of charge held together by the Coulomb interaction. A feature of this interaction is its rapid variation near the origin. In this regard, "cartoons" like Fig. 5 are rather misleading. The simple arrangement of charge in 5(b) is more likely to give an r^{-1} potential than the more complex close packed structure illustrated in 5(a). Note, however, that the high salt potential (Fig. 2) is much more like a hyperbola at 1σ, while the minimum at 2σ is much more like a harmonic (r^2) potential (in reality, the phosphate interactions are probably yet more complicated due to the very large polarizability derivatives associated with the phosphate groups[22]). We have considered a number of charge geometries and screening effects,[9] and conclude that, for *small* displacements of the interacting anions,[17] a Coulomb interaction would almost invariably appear to fall off in a manner close to the classic r^{-1} behavior initially. In consequence, if the O-2 atoms move apart as the center to center separation of neighboring double helices, d, is increased, the interhelical force constants should fall off as

$$c_{ij} = k \, d^{-3}.$$

We measure d by fiber diffraction, so we can test for such Coulomb interactions by comparing Brillouin measurements of the speed of sound with the predictions of a lattice dynamical model in which the interhelical springs are softened as d^{-3}. I should make it clear that the sound speed is a particularly sensitive probe of these long range interactions. Of course these long range interactions must be combined with other short range interactions to form a stable structure, but long wavelength sound waves are not sensitive to short range interactions. Figure 6 shows Brillouin data for the speed of sound perpendicular (a) and parallel (b) to the axis of Na-DNA films as a function of relative humidity (over most of this range the swelling of the crystal is indeed small).[17] The sound speed might also be affected by mass-loading as the water binds, and relaxational softening as water motion becomes more free. Water is relaxationally coupled in this frequency region[10] — for this reason we can exclude mass loading (which would have a small effect anyway). The relaxation strength is not enough to account for much of the softening. On the other hand, if we fit the scale (k) of the interhelical interactions to match the sound speeds at low water contents, the Coulomb softening theory outlined above fits the data remarkably well (curve 3 on Figures 6a and 6b), demonstrating the dominant role of these interactions.

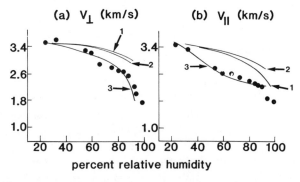

Figure 6. Softening of the sound speed as a function of relative humidity in Na-DNA films (a) perpendicular to the helix axis and (b) along the helix axis. Curve 1 is calculated for mass loading, curve 2 for relaxation and curve 3 for Coulomb softening of the phosphate interactions.

3.3 Collective excitations and conformation transitions

In the above picture of the A-B transition, the mechanism might be understood in terms of a crossover to simple screened Coulomb repulsion as the cation cloud is diluted by water addition. Much more must be known about the structure of the hydration shell before a dynamical pathway for the transition can be constructed. However, it is clear that such a pathway *must involve collective excitations of DNA and its hydration shell*. This point is particularly important for the B-Z transition, for the mechanisms that have been proposed (based on "dry" helix models) involve substantial energy barriers.[23,24] It is possible that a low energy pathway can be constructed with one or more *collective* modes of the polymer and its hydration/electrolyte shell.

4. Microwave resonances

The length dependent microwave absorption reported by Edwards et al.[6] is quite remarkable. The simplest interpretation of the absorption maxima is that they correspond to the frequencies of acoustic standing waves on the DNA. The frequency of such waves will be some multiple of the speed of sound, V, divided by the length of the fragment, L. DNA itself lacks any large dipole moments, but (so the argument goes) the counter ion cloud cannot follow GHz. motions so that the relative compressional displacement of the phosphates sets up an oscillating dipole with respect to the "static" counter ions. The microwave field is essentially homogeneous on a molecular scale, so alternating positive and negative dipoles cancel each other out. An odd number of half wavelengths is needed to absorb energy on a linear fragment. Putting all this together gives the formula that fits the observed absorption peak frequencies quite well:

$$f_n = (n + 1/2)\ V/L.$$

The speed of compressional sound waves, V, obtained by fitting the microwave data, is in close agreement with the value obtained from Brillouin scattering measurements.[25] The remarkable thing about this interpretation is the coherence length it implies for acoustic phonons. These plasmid fragments are a good fraction of a micron in length. In order to build up standing wave amplitude, the vibration must be capable of many passes up and down the polymer. There would be important biological consequences of such vibrational coherence lengths, but it is hard to see how the phonon damping could be so small. Scott has made the novel suggestion that DNA is such a (elastically) non-linear material, that compressional waves travel as Boussinesq solitons.[26,27] These solitons are familiar as shallow water waves (Scott often illustrates his talks with a glass dish full of water). Shallow water is extremely non-linear in the sense that the speed of waves is a strong function of their amplitude (and by "shallow" we mean that the depth is of the same order as the wavelength). Thus crests of waves catch up with troughs, and the water travels in "bunches" rather than being distributed in sinusoidal waveforms. The point about such solitons is that the moving part of the wave becomes localized. The more non-linear and localized the excitation, the more energy goes into the "static" distortion part of the soliton (i.e. the stored spring energy). Consequently, the viscous losses are reduced as the moving area of the polymer in contact with the water is reduced. The key question is whether DNA is as elastically non-linear as shallow water. I share the reservations of Van Zandt.[28] Thermal amplitudes are very small — a tiny fraction of bond lengths, and while non-linearities may make themselves felt, it is hard to see how they could dominate to the extent required to explain the lack of viscous damping.[27] As an aside, it is worth noting that the microwave absorption signal corresponds to less than one pumped quantum of vibration per molecule — thermal energy corresponds to several thousand quanta at these frequencies. The thermal excitation is *incoherent*, so the amplitudes of the vibrations do not add. The displacement is always a small fraction of the bond length.

The dynamical behavior of the hydration shell may offer an alternative explanation of these sharp resonances. As I pointed out at the beginning of this contribution, there are collective excitations of the DNA and the hydration shell in just this frequency region. Excitations of this sort (coupled to compressional phonons on the DNA) could account for the observed resonances.[29]

5. Conclusions and summary

The hydration shell and counterion cloud, interacting with the DNA via the charged phosphates, play an important role in the biologically important dynamics and in the conformational stability of the double helix. Coupled DNA-hydration shell excitations may account for resonant microwave absorption. Theories of the most important vibrational modes of the double helix must include excitations of the coupled DNA-water-ion system as well as explicit phosphate-phosphate interactions mediated by the electrolyte. These interactions have, on the whole, been ignored as dynamical studies focus on the covalent bonds that make up the Watson Crick structure. For many problems, it may be that the only importance of the covalent structure is the role it plays in locating charges in space — the dynamics being dominated by the interactions of this charge system with the electrolyte and, through the electrolyte, with itself.

Acknowledgments

The experimental work referred to was carried out with S.A. Lee, T. Weidlich, N.J. Tao, G. Lewen, J. Powell and C. DeMarco. Support was received from the NSF (PCM8215433), ONR (N00014-84-C-0487) and EPA (68-02-4105 — this chapter was not subject to EPA review and does not necessarily reflect the agency's views).

References

1. S.M. Lindsay, J. Powell, E.W. Prohofsky and K.V. Devi-Prasad, in "Structure and Motion: Membranes, Nucleic Acids and Proteins," eds. E. Clementi, G. Corongiu, M.H. Sarma and R.H. Sarma, Adenine, New York, p. 531 (1985).
2. W.C. Kerr and A.R. Bishop, "The Dynamics of Structural Phase Transitions in Highly Anisotropic Systems," preprint (1986).
3. B.H. Dorfman and L.L. Van Zandt, Biopolymers, **23**, 2639 (1983).
4. M. Kohli and L.L. Van Zandt, Biopolymers, **21**, 1399 (1982).
5. M.L. Swicord, G.S. Edwards, J.L. Sagripanti and C.C. Davis, Biopolymers, **22**, 2513 (1983).
6. G.S. Edwards, C.C. Davis, J.D. Saffer and M.L. Swicord, Phys. Rev. Lett., **53**, 1284 (1984).
7. S.M. Lindsay and J. Powell, in "Structure and Dynamics: Nucleic Acids and Proteins," eds. E. Clementi and R.H. Sarma, Adenine, New York, p. 241 (1983).
8. C. DeMarco, S.M. Lindsay, M. Pokorny, J. Powell and A. Rupprecht, Biopolymers, **24**, 2035 (1985).
9. S.A. Lee, J. Powell, N.J. Tao, G. Lewen, S.M. Lindsay and A. Rupprecht, to be published.
10. N.J. Tao, S.M. Lindsay and A. Rupprecht, Biopolymers, in press (1986).
11. Y. Tominaga, M. Shida, K. Kubota, H. Urabe, Y. Nishimura and M. Tsuboi, J. Chem. Phys., **83**, 5972 (1985).
12. M. Falk, K.A. Hartman and R.C. Lord, J. Am. Chem. Soc., **84**, 3843 (1962). *In this same issue, see also the two papers immediately following that by Falk et al.*
13. M.L. Kopka, A.L. Fratini, H.R. Drew and R.E. Dickerson, J. Mol. Biol., **163**, 129 (1983).
14. E. Clementi and G. Corongiu, Biopolymers, **20**, 351 and 2427 (1981), and Biopolymers, **21**, 763 (1982).
15. D.M. Soumpasis, Proc. Nat. Acad. Sci. (USA), **81**, 5116 (1984).
16. D.M. Soumpasis, J. Wiechen and T.M. Jovin, "Relative Stabilities and Transitions of DNA Conformations in 1:1 Electrolytes: A Theoretical Study," preprint (1985).
17. S.M. Lindsay, "Progress and Challenges in Biological and Synthetic Polymer Research," eds. C. Kawabata and A.R. Bishop, Ohmska, Tokyo (1986).
18. The equilibration of Z-DNA is independent of concentration over six orders of magnitude — F.M. Pohl, A. Ranade and M. Stockburger, Biochim. Biophys. Acta, **335**, 85 (1973).
19. This is an extension of an experiment first performed by Herbeck et al. — R. Herbeck, T.J. Yu and W.L. Peticolas, Biochemistry, **15**, 2656 (1976).
20. G. Lewen, S.M. Lindsay, N.J. Tao, T. Weidlich, R.J. Graham and A. Rupprecht, Biopolymers, **25**, 765 (1986).
21. B.N. Conner, C. Yoon, J.L. Dickerson and R.E. Dickerson, J. Mol. Biol., **174**, 663 (1984).
22. T. Weidlich, S.M. Lindsay and A. Rupprecht, "The Optical Properties of Li- and Na-DNA Films," preprint (1986).
23. S.C. Harvey, Nucleic Acids Res., **11**, 4867 (1983).

24. W.K. Olson, A.R. Srivasan, N.L. Marky and V.N. Balaji, Cold Spring Harbor Symp. Quant. Biol., **47**, 229 (1983).
25. M.B. Hakim, S.M. Lindsay and J. Powell, Biopolymers, **23**, 1185 (1984).
26. A.C. Scott, Phys. Rev., **A31**, 3518 (1985).
27. A.C. Scott and J.H. Jensen, Physics Letters, **109A**, 243 (1985).
28. L.L. Van Zandt, J. Biomol. Str. Dyns., in press (1986).
29. L.L. Van Zandt, "Why Structured Water Causes Sharp Absorption by DNA at Microwave Frequencies," preprint (1986).

Elasticity, Structure and Dynamics of Cell Plasma Membrane and Biological Functions

E. Sackmann, H.P. Duwe, K. Zeman and A. Zilker
Physik Department (E 22, Biophysics Group)
Technische Universität München
D-8046 Garching FRG

Cell plasma membranes can be considered as compound systems of three coupled layers: (1) the glycocalix, a thin macromolecular layer formed primarily by the large head groups of the glycoproteins and glycolipids, (2) the lipid/protein bilayer and (3) the cytoskeleton (a quasi-two-dimensional [gel-like] meshwork of filamentous proteins, spectrin in the case of erythrocytes), which is attached to the cytoplasmic leaflet of the bilayer and which mediates the coupling of the membrane to the meshwork of microfilaments and microtubily of the cell interior. The high flexibility of such compound membranes is manifested in pronounced (thermally excited) surface undulations of the cell envelopes.

In the present contribution shape transformations and shape instabilities of cells are explained in terms of intrinsic bending moments induced by structural transitions in one of the layers: for instance by changes in the state of swelling of highly folded filaments of the cytoskeleton. Experimental evidence for this is provided by analogous shape changes of giant bilayer vesicles made up of phospholipids and macromolecular lipids. Plasma membranes may also exhibit plasticity which is associated with lateral phase separation, the tight binding of macromolecules to one of the leaflets and hysteresis effects.

Introductory Remarks

Our present view of the compound membrane of the red blood cell is shown in Fig. 1, while the major protein composition is summarized in Table I (ignoring biochemically functioning enzymes).

The cytoskeleton is formed by a filamentous protein, the so-called *spectrin* which is primarily interconnected (i) by self-association of its ends, and (ii) by oligomers of actin. The meshwork is coupled to the cytoplasmatic sheath of the bilayer by binding of (part of) the interconnections firstly to the so-called *band III* protein (which functions simultaneously as anion transport system) via the coupling protein *ankyrin* and, secondly, to the major glycoprotein of the cell membrane, the *glyco-*

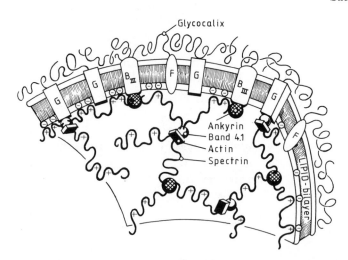

Figure 1. Schematic view of cell plasma membrane (case of erythrocyte) considered as a three layered compound system consisting of (1) the lipid/protein bilayer which contains the functional membrane proteins (not shown), (2) the glycocalix which controls the access of molecules to the membrane and the interaction of the cell with other cells or the walls of the capillaries, and (3) the cytoskeleton which is made up of the quasi-two-dimensional spectrin/actin network and is suspended beneath the inner monolayer by coupling to membrane proteins via coupling proteins (for instance to the anion transport system band III (B_{III}) via ankyrin. (A) and to glycophorin (G) via band 4.1[1], (cf. also Table I). The inner monolayer contains about 20% negatively charged lipids which are supposed to cause additional coupling of spectrin to the lipid/protein bilayer by electrostatic adsorption.

phorin. The latter binding is mediated by the coupling protein *band 4.1* and holds for the interconnections formed by oligomers of *actin.*[1,2,3]

Recent model membrane studies provide evidence that the spectrin molecules are also directly coupled to the lipid bilayer by electrostatic interaction of its positively charged side groups with the acidic phosphatidylserin.

Concerning the fine structure of the spectrin/actin meshwork it is very important to note that spectrin is a highly flexible molecule[4] which in solution exhibits a (slightly ellipsoidal) globular shape of a hydrodynamic radius of 33 nm (Ch. Schmidt, Munich unpublished results). The filaments adsorbed to the inner monolayer are thus expected to exhibit a highly folded structure as ordinary macromolecules.[5]

In erythrocytes the cytoskeleton is closely associated with the membrane thus forming a quasi-two-dimensional gel. In other cells there is a direct transition from a membrane bound cytoskeleton (which may well exhibit a similar structure as that in erythrocytes) to the spider's web of microfilaments and microtubily running through the cell interior.

The lipid/protein bilayer contains all functional enzymes. It exhibits a highly polar structure with respect to a plane separating the two monolayers which is due both to the asymmetry of the integral membrane proteins and to the asymmetric distribution of the four major lipid components: phosphatidylcholine (PC); sphingo-

Table I. Composition of plasma membrane of red blood cell. The membrane contains about 10^8 lipid molecules of which 50% are cholesterol. In the inner monolayer about 20% of the lipid are negatively charged phosphatidylserin. Glycophorin comprises three major species (A: 2×10^5; B: 0.7×10^5; C: 0.35×10^5). Note that band 4.1 is a spectrin-to-glycophorin coupling protein.

Molecule	Number of Species	Molecule	Number of Species
Spectrin	2.2×10^5	Band 4.1	2.3×10^5
Actin	5.1×10^5	Band III	1.2×10^5
Ankyrin	1.1×10^5	Glycophorin	3.1×10^5

myelin; phosphatidylethanolamin (PE); and phosphatidylserin (PS). The first two species reside mainly in the outer and the second two in the inner monolayer.

The bilayer asymmetry is only maintained if the ion pumps of the membrane are active, maintaining the natural asymmetry of the ion-distribution and, in particular, a low intracellular Ca^{++}-level: 10^{-7} Mole ltr^{-1}. It is well possible that the asymmetric distribution of the PS is also maintained by the help of its electrostatic binding to spectrin which is discussed below. Moreover, it has been postulated[6] that the asymmetric lipid distribution is maintained by an ATP-driven lipid pump.

The glycocalix is made up of the very large head groups of the glycoproteins (primarily glycophorin in the case of erythrocytes). The protein head groups are *branched copolymers* of oligo-peptides and oligo-saccharides. They are highly negatively charged since they contain acidic sugar residues (sialic acid). Thus glycophorin contains about 35 such groups.

Model membrane studies provide strong evidence that the glycoprotein head groups have a remarkable tendency to adsorb to the lipid head group region. They form a pancake-like structure at low lateral packing densities but a three-dimensional coil at densities where strong overlapping occurs. The glycocalix is thus expected to form a close protective layer on the cell surface. It is most important to note that the glycocalix is coupled to the cytoskeleton owing to the attachment of the latter to part of the membrane bound glycoproteins.

Molecular Dynamics and Thermally Excited Surface Undulations of Plasma Membranes

Molecular Mobility

From numerous measurements of lateral diffusion coefficients of lipids[7,8,9] and lateral and rotational mobility of proteins[11] it follows that the lipid bilayer moiety of the membrane is in a liquid crystalline state (corresponding to a two-dimensional smectic A phase). The lateral diffusion coefficient of the lipid molecules is about an order of magnitude smaller ($D_1 \approx 10^{-9}$ cm^2/sec) than in corresponding model membranes[7,8] which can be explained in terms of a percolation of the lipid within the mosaic-like arrangement of fluid (protein poor) and rigidified (protein rich) domains of the plasma membrane.[7,10] The band III proteins consist of a mobile (D_1 $\approx 4 \times 10^{-11}$ cm²/sec at 25°C) and an immobile fraction. The former increases from 10% at 25°C to 90% at 37°C suggesting that 10% of the protein are coupled to the spectrin/actin network at physiological temperatures.

The chain lengths of the most abundant hydrocarbon chains of the lipid vary from 16 (20%) to 20 (12%) carbon atoms. Thus, although the bilayer is in a fluid state at 37°C, lateral phase separation due to solidification of part of the lipid arises at decreasing temperature. Experimental evidence for this is provided first by the increase in the fraction of immobile band III,[9] second by the non-random distribution of proteins within the plane of the bilayer (cf. the particle free patches in Fig. 2), and third by an anomalous temperature dependence of the membrane bending elastic moment below about 25°C.[12]

Thermally Excited Membrane Undulations

The high flexibility and dynamics of cell plasma membranes is most spectacularly manifested in the pronounced thermally excited undulations. These lead for instance to the well known "flickering" of erythrocytes: the strong intensity fluctuations visible if the cells are observed by a phase contrast microscope. Figure 3 demonstrates how these surface undulations can be made visible and evaluated by interference contrast microscopy.[12] By using the homemade image processing system BAMBI, the surface of the cell can be reconstructed as a function of time along the Newtonian rings.[12,13] This technique enables the determination of the mean square amplitudes, $u(q)^2$, of the membrane excitations as a function of the wave vector, q. From these the bending elastic modulus, K_c, of the membrane may be determined by application of the equipartition theorem following Brochard and Lennon[14] where S is the area of the cell surface.

$$K_c = \frac{S \times k \times T}{u(q)^2 \times q^4}$$

Very interestingly, the value of the bending elastic modulus of the erythrocyte plasma membrane of $K_c \approx 7 \times 10^{-13}$ erg. as measured by this technique[12] is only by a factor of 3 larger than the bending stiffness of pure lipid bilayers in the smectic A state. This shows that the coupling of the cytoskeleton to the bilayer is rather loose.

On the other side it has been demonstrated recently[12] that the bending elasticity of red blood cells is very sensitively dependent on the structure of the cell membrane, the analysis of the surface undulations (called the "flicker spectroscopy"[12]) provides a very sensitive tool to detect subtle changes of the membrane structure of red blood cells caused by diseases or the application of drugs.[12]

Figure 2. Freeze fracture electron micrograph of erythrocyte membrane showing cytoplasmic monolayer viewed (outer monolayer is removed) from outside of cell. Note the formation of patches which are free of particles (mainly band III proteins).

Shape Transitions of Cells and Instabilities of Plasma Membranes

Shape Transitions

For many years physicists have been fascinated by the shape transformations of red blood cells which can be triggered in various ways, such as by changes in the environmental conditions (pH, temperature, ionic strength), by drugs[12] or by the inhibition of the cell metabolism (ATP depletion). The major stable states are shown in Fig. 4.

In general the transitions between the cullular shapes are reversible. The shape changes exhibit other typical features of phase transitions. Thus if a population of cells is mixed with lipid vesicles of low cholesterol content so that this lipid component is removed from the cell membranes, the cells undergo a simultaneous discocyte-to-stomatocyte transition below a threshold cholesterol concentration.[12]

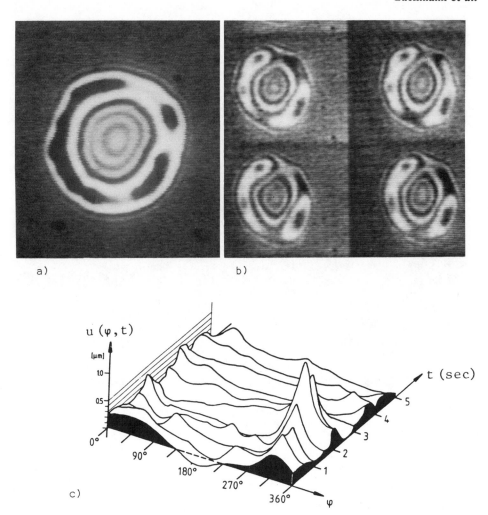

Figure 3. Evaluation of thermally excited undulations of erythrocyte membrane. **a)** Cup-shaped cell observed by reflection interference contrast technique[12] showing the Newtonian rings of constructive and destructive interference which define the contour lines of the cell surface. **b)** Subsequent observation of interference pattern at time intervals of 1/5 sec showing deformation of Newtonian rings due to surface undulation. **c)** Reconstruction of surface along one dark Newtonian ring (cf. Fig. 3a for definition of φ) shown at different times.

An often debated question is whether the stabilization of a certain shape is controlled by the spectrin/actin network or by the lipid protein-bilayer.[1-4,19] However, in view of the compound structure of the plasma membrane the shape is controlled by the cooperative interaction of all three layers. We will discuss below that the shape transformations can be explained in a very general way in terms of the bending elastic properties of the compound membrane.

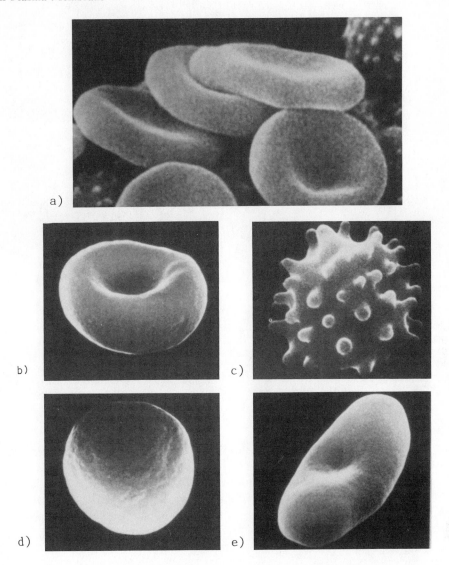

Figure 4. Stable shapes of erythrocytes.[16] **a)** biconcave discocyte which is the "ground state" of a healthy cell under physiological conditions. **b)** cup shaped cell (= stomatocyte) caused by cholesterol depletion. **c)** the echinocyte formed after ATP-depletion. **d)** spherical cell (spherocyte) which may form by osmotic swelling. **e)** shows a cell of a patient with elliptocythosis.

Membrane Instabilities

The membrane coupled cytoskeleton is essential for the global stability of the cells; enabling for instance blood cells to squeeze through the capillaries of the circulatory system or through the pores of the lymphoid tissue with diameters small compared to the size of the cells. On the other side, the cytoskeleton allows for local membrane instabilities associated with the transmembrane transport by vesicles. One example is the endocytosis, that is the uptake of macromolecules by a

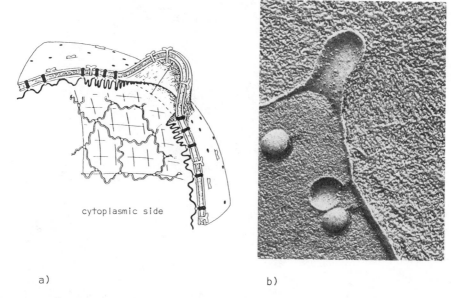

a)   b)

Figure 5. a) Formation of local protrusion (bud-formation). In the case illustrated the event is assumed to be triggered by a change in curvature caused by local decoupling of the cytoskeleton meshwork from the lipid-protein bilayer. **b)** Formation of sharp local protrusions in erythrocyte membrane caused by the (carrier mediated) injection of Ca[++] into cell. The finding that vesicles detached from the sharp protrusions are spectrin free[1] strongly suggests that the spicule formation is associated with a concentration of the spectrin/actin network.

local invagination of the plasma membrane and a subsequent detachment of a vesicle into the cell interior. This is a two-step process consisting of an initial (and *reversible*) step: the local invagination; and an irreversible step: the detachment of the vesicle. Other examples are the budding process shown in Fig. 5 or the fusion of vesicles with cells. The latter event consists also of two steps: the formation of a close contact between the fusing membrane (which is reversible), and the breaking of the diaphragm separating the two compartments.[27]

Modelling of Plasma Membrane Structures and Processes

Partially Polymerized Membranes as Models

In view of the highly complex structure of cell plasma membranes an understanding of the above mentioned processes in terms of structural changes appears impossible without parallel studies of appropriate model systems. The construction of mechanical models of cell envelopes is becoming now possible by the introduction of polymerizable lipids.[17-19] One example of such a model plasma membrane is shown in Fig. 6. It is obtained by preparing giant vesicles of mixtures of conventional lipids and polymerizable amphiphiles which may be linearly connected by photopolymerization, thus forming a two-dimensional macromolecular solution.

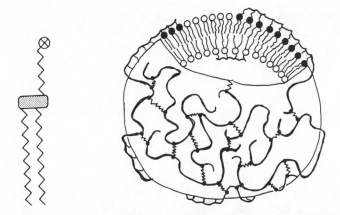

Figure 6. Vesicle composed of solution of macrolipid in phospholipid bilayer. The former is formed by photopolymerization of an amphiphile with a functional group attached to the lipid head via a spacer (cf. insert). The linear macrolipid may be crosslinked by incorporation of a certain fraction of amphiphiles with two polymerizable groups.

The chains of lipids may also be cross-linked by incorporation of a small fraction of amphiphiles with two functional groups. Another model system will be shown below (cf. Fig. 9).

Previously a method has been developed which enables measurements of the bending elastic modules. It is based on the analysis of the thermally excited shape fluctuations of giant quasi-spherical (flaccid) vesicles in terms of spherical harmonics in a similar way as shown in Fig. 3. By measuring the mean square amplitudes $u(q)^2$ of several modes the accuracy of the method is drastically improved. The bending stiffness of fluid bilayers of synthetic phosphatidylcholines is $K_c = (2\pm1) \times 10^{-13}$ ergs. This value is only a factor of 3 smaller than the bending elastic modules cf erythrocytes showing that the spectrin/actin network is a highly flexible system. This agrees with the finding that the K_c-value of a 30% solution of the macrolipid of Fig. 6 in lecithin bilayers is reduced at most by a factor of two as compared with the pure monomer bilayer.

Mimicking of Shape Transitions

Partially polymerized vesicles exhibit typical processes of plasma membranes. One example is the patching and subsequent cap formation triggered by lateral phase separation of the cross-linked lipid[18] quite reminiscent of the capping in the immune response of lymphocytes[28] which shows that the capping process itself may well take place without ATP-consumption.[18] It is thus possible that the patching is the energy driven process in the immunological response of the cells.

More interesting is the finding that mixed vesicles of monomeric and polymerizable lipids may mimic the shape transformation of erythrocytes as will be discussed now. A number of theoretical papers by various workers[20-23] appeared during the last years which postulated that the stabilization of certain cell shapes can be explained in terms of a very simple concept, namely the intrinsic bending moments (spontaneous curvature) of the cell membrane. Following a suggestion by Evans,[27] Singer et al. (cf. 24) explained the effect of drugs on the shape of erythrocytes in terms of an expansion of the inner or the outer membrane leaflet, respectively.

Recently Svetina and Zeks[23] combined this so-called bilayer coupling hypothesis with the bilayer bending elastic theories of Helfrich and Evans and proposed a very attractive model of cell shape transitions which is free of any structural details of the compound membrane with the exception of the introduction of a neutral plane dividing the membrane into an outer (o) and an inner (i) leaflet. The different cell shapes can be explained in terms of two parameters: 1) the reduced volume $v = V/V_S$: the ratio of the actual volume of the cell to that of a sphere, and 2) the relative area difference

$$\alpha = \frac{A_o - A_i}{A_{o,s} - A_{i,s}}$$

where A_o and A_i are the areas of the outer and the inner leaflet of the membrane considered and where $A_{o,s}$ and $A_{i,s}$ are the corresponding values of the sphere. If δ is the distance between the leaflets it is $A_S = 4\delta(\pi A)^{1/2}$. The area difference is related to the spontaneous curvature[21] according to

$$R_o = \frac{1}{4\pi\delta} (A_o - A_i)$$

For a given set of values v and α the cell assumes that shape which minimizes the total bending elastic free energy

$$G_{el} = \iint_S (R_1^{-1} + R_2^{-1} + R_o^{-1})^2 \, dS$$

where R^{-1}_1 and R^{-2}_2 are the principal curvatures and where the integration occurs over the cell surface. Svetina and Zeks calculated phase diagrams of stable cell shapes as a function of v and α. Two important results are: firstly, a non-spherical cell can divide up into two, three or more connected spheres but only two radii can coexist; depending on the size of α, the smaller vesicle can be inside ($\alpha < 1$) or outside ($\alpha > 1$) the larger one; secondly, for a relative volume of $v = 0.6$ corresponding to that of erythrocytes under physiological conditions the discocyte corresponds indeed to a bending energy minimum.

Figure 7 exhibits some typical cell shapes calculated by Svetina and Zeks.[24] Simultaneously it is shown that simple bilayer vesicles may undergo exactly the shape transitions predicted by the Zeks model. The vesicles (composed of positively charged lipids) are kept at a temperature slightly above the chain melting transition

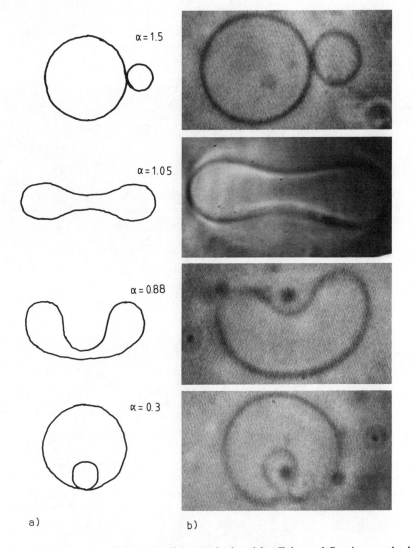

Figure 7. **a)** Shapes of vesicles (or cells) as calculated by Zeks and Svetina on the basis of their bilayer coupling hypothesis model for a reduced volume ($v = V/V_s = 0.5$) and for different values of the reduced area difference $\alpha = \delta A/\delta A_s$ which are given in the Figure. **b)** Shape transformation of giant vesicle of a single lipid component with positively charged (ammonium) head group triggered by temperature changes. The shape transitions are completely reversible.

($T_m = 41°C$). Starting from $T = 42°C$, heating leads to the budding-off of a small vesicle at $T = 43°C$. At subsequent cooling the shape change is completely reversible and an invaginated appears if the vesicle is cooled below the starting temperature, that is to 41°C. This shape change is explained in terms of a difference in the electric surface potential between the inner and the outer monolayer which changes with the temperature due to the dissociation of ions from the lipid/water interface.[15]

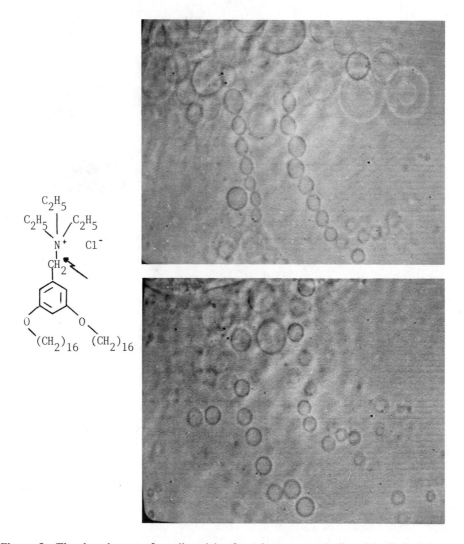

Figure 8. The detachment of small vesicles from larger ones (cell-vesicle fission) is a two-step process: first the strangulation of a vesicle connected with the main "cell body" by a tether (which is a reversible event) and second, the detachment of the vesicles (irreversible step). Experimental evidence for this suggestion is given in the experiment on the right side which shows the spontaneous separation of vesicles (made up of the lipid shown on the left) after the photodissociation of the hydrophilic lipid head group. The hydrophobic molecules form microcrystals owing to lateral phase separation causing destabilization of the vesicle.

Membrane Instabilities

As follows from the above consideration, the formation of a local curvature change may be triggered by a global change in the (chemically) induced bending moment.[15,20] However, in order to detach a vesicle from a cell a local instability at the neck connecting the two spheres must occur. Such instabilities could be effected by a local conformational change caused, for instance, by lateral phase separation. This possibility is demonstrated in the model membrane experiment of Fig. 8. The vesicles are composed of an amphiphile (prepared in the laboratory of Prof. Ringsdorf, Mainz) which can be transformed into a hydrophobic molecule by photodissociation of the charged head group. According to Fig. 8b these lipids can form stable chains of interconnected vesicles. After UV irradiation the chain decays into individual vesicles. This destabilization is caused by the formation of precipitates of microcrystals of the hydrophobic molecule which is formed by photodissociation of the polar head group. The microcrystals form within the bilayer membrane by lateral phase separation. If the crystallites reach a certain size, the membrane at the tether becomes unstable. In biological membranes, the role of the microcrystals could be played by proteins accumulating at the tether.

The Cytoskeleton as a Two-Dimensional Gel

The membrane coupled cytoplasmatic network as depicted in Fig. 1 is a two-dimensional gel. Based on such a picture Stokke et al.[4] showed that the shear elastic properties and also the shape transformations of erythrocytes can be explained by assuming that the spectrin/actin network is an ionic gel. Judged from our model membrane results the shape appears rather determined by the curvature elastic properties. However, in this picture the cytoskeleton plays an essential role for the triggering of chemically induced bending moments. The interesting aspects of flexible spectrin molecules adsorbed to the inner membrane leaflet by electrostatic forces is that it interacts with two solvents: first the aqueous (also hemoglobin-containing) cytoplasmatic fluid and second with the lipid molecules. Both solvents may modulate the entropic elastic properties of the spectrin filaments as well as their resting length. The spontaneous curvature of the whole membrane can thus be modulated either by an expansion (or compression) of the gel or secondly by a change in the coupling of the spectrin to the lipid layer.

It appears that many cellular shape changes can be explained by such a concept. Thus it is known that the inhibition of calmodulin elicits cup-formation of erythrocytes. On the other side calmodulin inhibition leads to an increase in the intracellular Ca^{++}-concentration. Ca^{++}, however, decouples spectrin from PS containing lipid layers which would indeed lead to an expansion of the spectrin filaments and an expansion of the inner leaflet of the membrane with respect to the outer one.

Shape transitions and membrane instabilities could also be caused by a transition of the cytoskeleton from an expanded to a collapsed state as is typical for ionic gels. In fact, such a transition has also been predicted for macromolecular layers

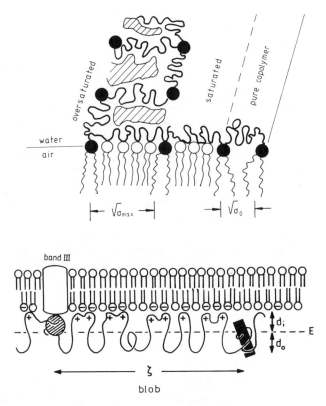

Figure 9. a) Model of membrane coupled cytoskeleton composed of macrolipid consisting of amphiphiles interconnected by hydrophilic chains and monomeric lipids. The latter act as two-dimensional solvent in which the gel swells. The hydrophilic chain is highly folded and adsorbs to the lipid/water interface. The transition from an undersaturated to an oversaturated state of the leaflet is shown. **b)** Possible microstructure of spectrin polymer coupled first to the membrane proteins via ankyrin and second to the negatively charged phosphatidylserine molecules via electrostatic forces. Thus the PS-molecules act as one solvent and the cytoplasm as a second. Therefore the network itself has an intrinsic asymmetry with respect to a neutral plane parallel to the membrane surface.

attached to surfaces.[5] It appears that it has been recently verified experimentally in the model system shown in Fig. 9a. It consists of a macro-lipid composed of amphiphiles interconnected by hydrophilic chains which are swollen by incorporation of monomeric lipids. This system exhibits striking structural similarities to the membrane coupled spectrin meshwork where the hydrophilic chain assumes the role of the spectrin. The model system exhibits two interesting features:

1. The macrolipid readily incorporates monomeric lipid up to a saturation concentration of about 50% monomer. The saturation behavior is determined by the balance of several forces: (i) the elastic energy associated with the expansion of the hydrophilic chain, (ii) the gain in mixing entropy, and (iii) the

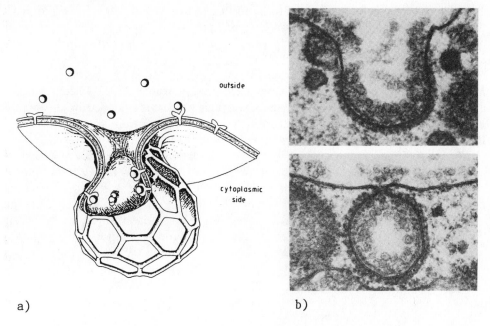

a)   b)

Figure 10. Schematic and electron microscopic view of stabilization of local invagination by adsorption of polymer coat. These so-called coated pits play an essential role for the uptake of macromolecules (such as insulin) or aggregates (such as low density lipoproteins) by the cell. In these, receptors loaded with the macromolecule accumulate until the coated pit is expelled into the cell interior.

change in the adsorption energy caused by the difference in the interaction of the adsorbed hydrophilic chains with the free water surface or the head groups, respectively.

2. The saturated macrolipid exhibits a first order transition from an expanded to a collapsed state.

A more realistic model system of the membrane coupled cytoskeleton could be prepared by replacing the hydrophilic chains by charged ones.

Plasticity of Plasma Membranes and Hysteresis Effects

Very often a local change in spontaneous curvature is stabilized. A most prominent example is the coated pit, a local invagination of the membrane which is stabilized by the adsorption of a coat of a filamentous protein, the so-called clathrin.[28] This and similar types of plasticity are a consequence of a fast biochemical modification of the membrane structure. In synaptic membranes the plasticity seems to play an essential role for the long term memory (Chaneux[29]).

The plasticity of membranes may also be related to memory effects since conformational changes of polyelectrolytes and charged lipid layers caused by the association or dissociation of ions (H^+, Ca^{++}) exhibit pronounced hysteresis effects. An example is the Ca^{++}-induced lateral phase separation in model membranes of charged lipids (phosphatidylserin) and the Ca^{++}-induced fusion of such vesicles. It has been often postulated that the latter effect is responsible for the fusion of synaptic vesicles with the presynaptic membrane. This process is triggered by the influx of Ca^{++} leading to a transient increase in the local concentration (to 10^{-4}M). Now, these ions are rapidly sequestered in order to restore the original low concentration of 10^{-7}M. However, if the avalanche of Ca^{++} -ions would last sufficiently long to cause attachment of the vesicles to the presynaptic membrane the Ca^{++}-mediated contact could well remain if the resting concentration of Ca^{++} is restored. Such a hysteresis effect has indeed been demonstrated in various model membrane studies.[31] Spectrin-like molecules are coupled to the inner leaflet of the presynaptic membrane which could be involved in the fusion process. Since many cellular membrane processes are triggered by a transient increase in the local cytoplasmatic Ca^{++} concentration it is important to look for such hysteresis effects in biological membranes in the future.

Acknowledgments

Financial support by the Fonds der Chemischen Industrie, the Freunde der TU München and the Deutsche Forschungsgemeinschaft (Sa 246-13/3 and Sa 246-16/1) is gratefully acknowledged. Last, but not least, E. Sackmann would like to thank Professor E. Evans for many stimulating and most helpful discussions about membrane elasticity.

References

1.   D. Branton, C.M. Cohen and J. Tyler, Cell, **24**, 24 (1981).
2.   S.B. Shohet and S.E. Lux, Hospital Practise, **19**, 89 (1984).
3.   C.W.M. Heast, Biochim. Biophys. Acta, **694**, 331 (1982).
4.   B.T. Stokke, A. Mikkelsen and A. Elgsaeter, Europ. Biophys. J., **13**, 203 (1986). *Ibid*, **13**, 219 (1986).
5.   a) P.G. de Gennes, J. de Physique, **37**, 1445 (1976) and b) S. Alexander, J. de Physique, **38**, 983 (1971).
6.   M. Seigneuret and Ph. Devaux, Proc. Natl. Acad. Sci., **81**, 3751 (1984).
7.   H.J. Kapitza and E. Sackmann, Biochim. Biophys. Acta, **595**, 56 (1980).
8.   T.M. Jovin and W.L.C. Vaz in "Biomembranes, M.: Biological Transport," Methods in Enzymology.
9.   D. Golan and W. Veatch, Proc. Natl. Acad. Sci., **77**, 2537 (1980).
10. M.J. Saxton, Biophys. J., **9**, 165 (1982).
11. R.J. Cherry, E.A. Nigg and G.S. Beddard, Proc. Natl. Acad. Sci., **77**, 5899 (1980).
12. K. Fricke, R. Laxhuber, K. Wirthensohn and E. Sackmann, Europ. Biophys. J. (to appear 1986).

13. H. Engelhardt, H.P. Duwe and E. Sackmann, J. de Physique, **6**, L395 (1985).

14. F. Brochard and J.F. Lennon, J. de Physique, **36**, 1035 (1975).

15. E. Sackmann, H.P. Duwe and H. Engelhardt, J. Chem. Soci, Faraday Discussion, **81**, 000 (1986).

16. M. Bessis, "Living Blood Cells and Their Ultrastructure," Springer Verlag, Heidelberg, 1973.

17. a) D.S. Johnston, S.A. Sanghem and D. Chapman, Biochim. Biophys. Acta, **602**, 54 (1981), and b) B. Hupfer, H. Rinsdorf and H. Schupp, Makromol. Chemie, **182**, 247 (1981).

18. H. Gaub, H. Büschl, H. Ringsdorf and E. Sackmann, Biophys. J., **45**, 725 (1984).

19. E. Sackmann, P. Eggl, C. Fahn, H. Bader, H. Ringsdorf and M. Schollmeier, Ber. Bunsenges. Phys. Chem., **89**, 1198 (1985).

20. E.A. Evans, Biophys. J., **14**, 923 (1974).

21. H.J. Deuling and W. Helfrich, Biophys. J., **16**, 861 (1976).

22. A.G. Petrov and I. Bivas, Progr. in Surface Science, **Vol. 16**, 389 (1984).

23. S. Svetina and B. Zeks, Biomed. Biochem. Acta, **42**, 86 (1983).

24. M.P. Sheetz and S.J. Singer, Proc. Natl. Acad. Sci. USA, **71**, 4457 (1974).

25. C. Mombers, J. de Gier, R.A. Demel and L.L.M. van Deenen, Biochim. Biophys. Acta, **603**, 52 (1980).

26. T. Tanaka, Phys. Rev. Letters, **40**, 820 (1978).

27. R.P. Rand and V.A. Parsegian, Can. J. Biochem., **62**, 752 (1984).

28. B. Alberts et al., "Molecular Biology of the Cell," Garland Publishing Inc., New York/London.

29. W. Frey, J. Schneider, H. Ringsdorf and E. Sackmann, to be published.

30. J.P. Changeux and A. Danchin, Nature, **Vol. 264** (1976).

31. M.H. Akabas, F.S. Cohen and A. Finkelstein, J. Cell Biol., **98**, 1003 (1984).

Neuronal Signaling. A Simple Thermodynamic Process Involving Complex Membrane Proteins

Franco Conti
Istituto di Cibernetica e Biofisica, CNR
Via Dodecaneso 33
16146 Genova, Italy

Introduction

Since the classical work of Hodgkin and Huxley[1] the complex phenomenology of the transmission of electrical signals along nerve fibers is fully understood in terms of voltage and time dependence of the nerve membrane conductance. Also included in the Hodgkin-Huxley description of nerve excitation was the notion that the time delays characterizing the changes in membrane permeability must arise from simple voltage-dependent reactions involving specific molecular structures within the membrane itself. What could not be foreseen in 1952 was the fact that these structures, nowadays named ion channels, were bound to become one of the most important unifying concepts of membrane biophysics. It is now known that most functions of biological membranes are based on the performance of such specialized proteins, embedded in a common lipid bilayer matrix and characterized by their capability of assuming different conformations with probabilities which are determined, and thereby modulated, by the chemical and physical properties of their environment. Thus, both propagating action potentials (purely electrically driven) and synaptic signals (mediated by chemical agonists) are based on the same type of molecular event involving the opening of an aqueous pore within a membrane-spanning protein. With the advent of the patch-clamp technique pioneered by Neher and Sakmann[2] the unitary electrical events produced by several kinds of ionic channels can be directly observed on the screen of an oscilloscope.[3]

The physical chemistry of the ionic fluxes through the hydrophilic pore provided by an ion channel in its open conformation, as well as the large variety of already characterized channels, and the rich realm of cellular functions which can be regulated through the interplay of different ionic fluxes and ion-dependent intracellular reactions, is already subject matter for textbooks.[4] Far less understood are the molecular mechanisms which underlie the delicate balance between the various possible tridimensional structures of a channel protein. The theoretical notion that very small displacements of hydrophilic groups within the inner pore wall may change drastically the pore permeability to ions, and the fact that the opening of the pore is sensitive to membrane potential changes of few mV may suggest subtle mechanisms involving small regions of the channel proteins acting either as "gates" or as "sensors." However, more critical appraisals of the available electrophysiological data rather hint that the functional conformational transitions of an ion channel may be so drastic as to make such concepts devoid of real meaning.

By cloning and sequencing specific complementary DNAs, the primary structure of an ion channel can nowadays be resolved.[5,6] This extraordinary achievement of molecular biology opens the way to investigations which were not even thinkable a few years ago. Genetically modified "monster" channels can be expressed in oocytes and from their functional alterations important information can be gained about the role played by the specific group or subunit which has been modified.[7] These studies will reduce the complexity of understanding the structure-function relationships in polypeptide chains containing thousands of amino acids.[6] The role of electrophysiology in this context will be that of providing the best quantitative description of the newly designed channels. Such description is obtained by characterizing these microdomains of biological membranes as thermodynamic systems in equilibrium with their heterogeneous environment.

Nerve Membrane Currents as Superposition of Elementary Markov Processes

Three basic hypotheses form the ground of our present understanding of nerve membrane excitability.

1. Although regulating the behavior of a highly nonequilibrium system, where ionic gradients can be of the order of 10^5 M/cm, the physical state of the nerve membrane at any time is by and large independent of the ionic fluxes that flow through it. Apparent deviations from this rule are known to occur in particular conditions,[8,9] but they can be simply explained in terms of indirect couplings due to ion accumulation and/or depletion in the membrane bathing solutions.

2. The total current flowing across any macroscopic membrane area is the sum of quantal contributions from individual ion channels undergoing mutually independent equilibrium thermal fluctuations. The experimental support to this hypothesis is indirect, but quite strong. It relies upon the fair quantitative agreement between the properties of the quantal events recorded from membrane patches which appear to contain only one channel, and those estimated from macroscopic measurements such as toxin-binding,[10] current fluctuations,[11] and gating currents.[12]

3. The functionally important and statistically significant conformational transitions of an ion channel occur within less than 1 μs. Although involving presumably large portions of the channel protein, these molecular events are still so fast as to be practically instantaneous on the time scale of electrophysiological measurements. Even the fine structure of channel fluctuations which has been detected through open-channel current-noise measurements seems to have this same characteristic.[13]

Under the hypothesis stated above the mean membrane current, I, through a membrane containing N identical channels, is given at any time by

$$I = N(i_i + i_g) \tag{1}$$

where i_i and i_g are the mean single channel current contributions arising, respectively, from ionic flows and from the intramembrane charge redistributions (gating currents) which accompany the conformational transitions of the channel-protein. If each channel may assume n metastable conformations (states) with conductances $\gamma_1, \gamma_2, ..., \gamma_n$, the latter quantities are given by

$$i_i = (E_m - E_{rev}) \sum_{j=1}^{n} \gamma_j p_j \tag{2}$$

$$i_g = \sum_{j=1}^{n} q_j \dot{p}_j \tag{3}$$

where E_m is the membrane potential, E_{rev} is the reversal potential at which the net ionic current through an open channel becomes zero, q_j is the equivalent electric charge of state j (defined apart from a constant), p_j is the probability of finding a channel in state j, and $\dot{p}_j$ denotes its time derivative.

For fixed external conditions, the time course of the p_j's is determined by the time-independent Markovian transition-rates α_{jk}, where $\alpha_{jk}\Delta t$ is the transition probability from state j to state k in the time interval δt:[14]

$$\dot{p}_j = \sum_{k=1}^{n} p_k \alpha_{kj} \left(\alpha_{jj} = -\sum_{k \neq j} \alpha_{jk} \right) \tag{4}$$

Accordingly, the membrane current following a step change of E_m (voltage-clamp) is given by the sum of (n−1) decaying exponentials and has an asymptotic value determined by the equilibrium probabilities, $p^{(e)}_j$, of the various channel states under the final external conditions. Both the $p^{(e)}_j$'s and the time constants which characterize the approach to the new equilibrium depend only on the values of the α_{jk}'s for $t > 0$, while the coefficients of the exponentials depend also on the initial conditions.

The expression of the equilibrium probabilities in terms of the rate constants is very simple:

$$p_j^{(e)} = 1 / \left(1 + \sum_{j \neq k} a_{jk}/\alpha_{kj} \right), \tag{5}$$

but there is no convenient expression of the time constants in terms of the α_{jk}'s. In most cases absolute kinetic data are of practical value only in connection with some hypothetical simple scheme of allowed channel-state transitions.

Thermodynamic Description of Single Channel Properties

The above description of the electrical properties of a nerve membrane is still phenomenological. A thermodynamic interpretation of the kinetic parameters is needed in order to establish a correlation between electrophysiological data and physical properties of the channel proteins, such as entropy, volume, and charge distribution.

An ionic channel is a semimicroscopic part of a membrane system. It is in thermal, mechanical and electrical equilibrium with a nonisotropic environment comprising two generally different aqueous solutions and an annulus of lipids structured as a bimolecular leaflet. The externally controlled intensive variables which determine its thermodynamical state are temperature, T, pressure, P, and membrane potential, E_m. By defining the thermodynamic potential appropriate for this situation — the free enthalpy, $G(T,P,E_m)$ — the extensive variables conjugated to T, P, and E_m, are identified as the entropy, S, the volume, V, and the non-linear component of the electric charge, Q_g, drawn from the voltage controlling device in order to change the membrane potential from some reference level to E_m.[15] Experimentally, the most important contribution to Q_g is found to arise from changes in the distribution and/or orientation of polar groups of the channel protein, which take place as the channels redistribute themselves between the possible conformational states.

For a homogeneous membrane containing N independent channel subsystems:

$$G = Ng = -NkT \ln \Delta(T,P,E_m) \tag{6}$$

where k is Boltzmann's constant, and g and $\Delta(T,P,E_m)$ are, respectively, the free enthalpy and the partition function of one channel. The latter quantities allow a correct definition of single channel properties such as the volume, v, the gating charge, q, and the entropy, s, according to basic principles of statistical thermodynamics.

Kinetically observable discrete channel states arise from the existence of particular spatial configurations of the channels yielding deep minima of interaction energy between the channel components. By dividing the single channel configurational space into n non-overlapping domains, each containing such an energy minimum, it is possible to define a partition function, $\Delta_i(T,P,E_m)$, for each domain (or state) i:

$$\Delta(T,P,E_m) = \sum_{i=1}^{n} \Delta_i(T,P,E_m) \tag{7}$$

and any channel state can be viewed as a distinct physical system, with thermo-dynamic properties given by

$$g_i = -kT \ln \Delta_i(T,P,E_m) \tag{8}$$

$$s_i = -(\partial g_i/\partial T) \tag{9}$$

$$v_i = (\partial g_i/\partial P) \tag{10}$$

$$q_i = -(\partial g_i/\partial E_m) \tag{11}$$

Following the above definitions, the equilibrium probability of state i is

$$p_i^{(e)} = \Delta_i/\Delta = \exp\{-(g_i - g)/kT\} \tag{12}$$

and the mean thermodynamic properties of a channel can be written as

$$g = \sum_{i=1}^{n} p_i^{(e)} g_i + kT \sum_{j=1}^{n} p_i^{(e)} \ln p_i^{(e)} \tag{13}$$

$$s = \sum_{i=1}^{n} p_i^{(e)} s_i - k \sum_{i=1}^{n} p_i^{(e)} \ln p_i^{(e)} \tag{14}$$

$$v = \sum_{j=1}^{n} p_i^{(e)} v_i \tag{15}$$

$$q = \sum_{j=1}^{n} p_i^{(e)} q_i \tag{16}$$

Similarly, according to Eyring's absolute rate theory,[16] the transition probability between any two adjacent states are expressed in terms of the free energies, $g_{jk}^{\ddagger}$, of the infinitesimal domains around the saddle-points which characterize the path of minimal effort for the transition itself:

$$\alpha_{jk} = (kT/h) \exp\{ - (g^{\ddagger}_{jk} - g_j)/kT)$$ (17)

where h is Planck's constant.

Since the α_{jk}'s define completely the macroscopic behavior of an ion channel, Eq. 17 provides the most general basis for the interpretation of the influence of the external variables, T, P, and E_m, on nerve membrane currents. Let Z denote any of the intensive variables T, P and E_m and define

$$\zeta = \partial g/\partial Z - (g/T) (\partial T/\partial Z)$$ (18)

It is easily verified that when Z is replaced respectively by P, E_m, or T, ζ becomes equal to v, $-q$, or $-(h/T)$ (h standing here for the thermodynamic enthalpy). Following these definitions, differentiation of Eq. 20 yields

$$- kT(\partial \ln(\alpha_{jk}/T)/\partial Z) = \Delta\zeta^{\ddagger}_{jk}$$ (19)

where $\Delta\zeta^{\ddagger}_{jk}$ is the activation step in ζ needed for the transition from state j to state k.

It is difficult in general to exploit Eq. 19 for estimating enthalpies, volumes, and charge transfers, because the measurable kinetic parameters are complex functions of all the α_{jk}'s. This equation becomes useful only when the value of some relaxation time constant is predominantly influenced by a particular α_{jk} or by a set of α_{jk}'s which have similar activation parameters. The dependence of the steady-state properties of membrane currents on external variables yields more direct information about mean thermodynamic properties of channel conformations.

Let X, x, and x_i be, respectively, the macroscopic mean value, the mean single channel value, and the single channel value in state i of some observable extensive quantity. It can be shown[15] that

$$- kT \, \partial \ln X /\partial Z = \mu_{x\zeta} / x$$ (20)

where

$$\mu_{x\zeta} = \sum_{j=1}^{n} p_i^{(e)} (x_i - x)(\zeta_i - \zeta)$$ (21)

is the cross-correlation between x and ζ, a quantity which depends both on the distribution probability and on the x and ζ values for the various channel states. Useful applications of Eq. 20 are obtained by identifying X with the two most important observables in nerve electrophysiology, namely the ionic conductance, Γ, and the gating charge, Q_g.

Using, respectively, the subscripts o and c to denote mean properties of the "open" subset of channel states, with conductance γ, and those of the "closed" subset, the dependence of the steady-state membrane conductance on T, P, and E_m can be expressed as follows:

$$k\partial(T\ln\Theta)/\partial T = s_o - s_c \tag{22}$$

$$-kT\, \partial\ln\Theta/\partial P = v_o - v_c \tag{23}$$

$$-kT\, \partial\ln\Theta/\partial E_m = q_o - q_c \tag{24}$$

where we have defined, for convenience

$$\Theta = \Gamma/(\bar{\Gamma} - \Gamma) \tag{25}$$

$\bar{\Gamma}$ being the maximal ionic conductance, achievable when all channels are forced to be in the open state.

Equations 22, 23 and 24 relate directly membrane conductance data to differences in the mean thermodynamic properties of the open and closed channels, without any explicit involvement of the multiplicity of channel configurations belonging to the open and closed subsets. However, the quantities $s_o - s_c$, $v_o - v_c$, and $q_o - q_c$, are expected to vary with T, P, and E_m, depending on how the distribution of the various states within each subset varies with the external parameters.

The application of Eq. 20 to the case $X = Q_g$ yields somewhat more complex relationships, because the single channel gating charge cannot be assumed to be a two-valued quantity as the electrical conductance. The sensitivity of Q_g to changes in T and P can be only generally related to the correlation between the gating charge and the entropy, or the volume, of the various single channel conformations. Likewise, the voltage dependence of Q_g is related to the variance of the fluctuations of effective charge distribution of the channel proteins for any fixed set of external conditions.

Major Observable Differences Between Functional States

Even for the best studied ionic channel, i.e., the sodium channel of nerve and muscle membranes, a detailed phenomenological description according to some widely accepted kinetic scheme of possible state transitions is still lacking. A critical review of the large number of models which have been proposed is beyond our present scope. We review here briefly only those experimental observations which allow some general conclusion about the functional ion channel conformations independently of such discussion.

The equilibrium probability for the open state of sodium channels is normally very low, due to an inactivation process[17] which is to a large extent sequentially coupled to the opening transition.[18.] Indeed, two different subsets of non-conducting states can be easily distinguished for this channel: the subset {C} which is highly populated at large membrane polarizations ($E_m \rightarrow -\infty$) and the inactivated subset, {I}, which is most probable for $E_m \rightarrow +\infty$. Operationally, a rough separation between {C} and {I} is obtained by assuming that the transition to {I} states, causing the late decay of sodium currents elicited by step membrane depolarizations, is much slower than the early filling up of the open states. Although being the source of great complications when attempting to find the most accurate kinetic description of the sodium channel, inactivated states have the nice property of bringing within the reach of electrophysiological investigation another set of configurations of the channel proteins which is electrically silent.

Charge redistributions accompanying channel transitions

According to Eq. 24 the E_m dependence of the membrane conductance yields estimates of the gating charge associated with the transition from closed to open states. If several closed and open states are available, these estimates are bound to be smaller than the charge displaced during the transition from the most favored closed state for $E_m \rightarrow -\infty$ to the most favored open state for $E_m \rightarrow +\infty$. For sodium channels in squid axons, the estimates obtained from the analysis of peak currents range from a minimum of 3 electronic charges (e.c.)[19] to a maximum of 6 e.c.[1] Similar values, in the range of 4 to 5 e.c., are derived for the potassium channels. Interpreting sodium inactivation curves as stationary distribution probabilities between the {C;0} subset and the {I} subset, one can estimate that the charge redistribution associated with {C;0} $\rightarrow$ {I} transitions is also in the range of 3 to 6 e.c. per channel.

Gating charge measurements provide more direct information about the charge redistributions which accompany the conformational transitions of an ionic channel. Except for data at relatively large negative E_m values,[20,21] the E_m dependence of Q_g is fairly well fitted by the simple expression which is expected if the elementary gating charge is a two-state variable.[22] This suggests independent contributions to Q_g in this voltage range, from substructures of the sodium channel which can fluctuate between two main conformations. According to Conti et al.,[22] each elementary contribution is about 1.5 e.c., a value consistent with the estimates derived from sodium conductance data if each sodium channel contains 2 to 4 gating substructures.

Entropy differences between functional states

Hodgkin and Huxley[1] described the temperature dependence of the kinetics of sodium and potassium currents using a single temperature coefficient ($Q_{10} \sim 3$) for all the rate constants of their equations. This notion, implying that the equilibrium state probabilities are independent of T, has been corrected only fairly recently by Kimura and Meves,[19] reporting that the voltage dependence of both the activation and the inactivation curve of sodium currents in squid giant axons are appreciably sensitive to temperature variations. From the analysis of these data it can be estimated that the mean entropy of the open conformation of a sodium channel is 26k lower than that of the {C} states, and about 19k lower than that of {I} states.[15] Thus, it appears that the transition {C} → {0} → {I} proceeds first with a sizable increase of structural order, which is then followed by only a slight entropy increase.

Activation and reaction volumes

Pressure has been long known to have a large influence on the physiology of excitable cells,[23] but it has been used only recently as a tool for investigating the structural properties of ionic channels.[22,24,25]

In squid giant axons, the most prominent effect of high pressure is a slowing of the kinetics of the voltage clamp currents roughly equivalent to what is expected if all the most important transitions involved in the opening of both sodium and potassium channels need a positive activation volume, $\Delta V^{\ddagger}$, of about 60 Å[3]. However, a more detailed analysis of the sodium currents[24] suggests that the early activation steps responsible for their delay involve a significantly smaller $\Delta V^{\ddagger}$, while the late decay associated with inactivation is slowed by pressure much more than the early rise. Similar differential effects were found also on the kinetics of the early and late phase of potassium currents,[25] the latter being characterized by a much smaller $\Delta V^{\ddagger}$.

Most interesting is the observation that the pressure dependence of sodium gating currents, I_g, is characterized by apparent activation volumes of only 30 Å[3].[22] This suggests that the major charge redistributions of the channels occur during the initial delay of the sodium currents, whereas the activation volume of 60 Å[3] is associated with some final rate-limiting step in the opening of the channels which yields little contribution to I_g.

Concerning the origin of the relatively large activation volumes observed, it appears fair to exclude simple mechanisms such as changes in the degree of exposure to water of small polar groups. This was tested by measuring the pressure dependence of the rates at which lipophilic ions can cross the hydrophobic core of the nerve membrane matrix.[26] These translocations, which involve the displacement of charged groups from one water-membrane interface through the hydrophobic core of the membrane lipids to the other interface, were shown to involve activation volumes of only about 8 Å[3].

The open-state probability of sodium channels is also influenced by pressure, although not as markedly as the kinetics. Increasing pressure shifts the voltage dependence of the sodium currents towards larger depolarizations, an effect from which one can estimate that the mean volume of the open sodium channel is larger than that of the closed channel by about 30 Å^3.[24] Most interesting is the comparison of this result with the observation that the voltage dependence of the sodium gating charge is little affected by pressure.[22] This implies a low correlation between the volume of the channel and its charge distribution, as if the various closed states had the same volume. The volume change of 30 Å^3 seems to occur during some final step, which is not accompanied by a major charge redistribution. The kinetic data discussed above suggest that such step involves also the largest activation volume.

Conclusions

The description of the nerve physiology in terms of the properties of ion channels has greatly deepened our understanding of the physical basis of nervous excitation. The transmission of the nerve impulse along an axon is now interpreted as a simple thermodynamic process based on the delicate equilibrium between different conformational states of well identified membrane proteins. Unfortunately, when we magnify our view of the nerve membrane and try ideally to focus on a single ion channel, we discover a very complex system indeed. The sodium channel is a polypeptide chain about 2000 amino acids long.[6] It can assume at least three subsets of functionally important configurations with large structural differences which are difficult to envisage as rearrangements of localized short peptide segments. Charge redistributions equivalent to the translocation of six electronic charges across the membrane, entropy differences of the order of 30 k, and volume changes of 30 to 60 Å^3, are likely implying major conformational changes of the whole protein.

It seems legitimate to ask if the complexity of channel proteins is really necessary for optimizing their function. It may be that the sodium channel best fits a number of other conditions, including interactions with other cell components, but it is also conceivable that some of its complexity is merely the result of a random evolution.

References

1. A.L. Hodgkin and A.F. Huxley, A quantitative description of membrane current and its application to conduction and excitation in nerve, J. Physiol. (Lond.), **117**, 500 (1952).
2. E. Neher and B. Sakmann, Single-channel currents recorded from membrane of denervated frog muscle fibres, Nature, **260**, 779 (1976).
3. B. Sakmann and E. Neher, eds., Single channel recording, Plenum Press, New York (1983).
4. B. Hille, Ionic channels of excitable membranes, Sinauer Assoc. Inc., Suderland, Mass. (1984).

5. M. Noda, H. Takahashi, T. Tanabe, M. Toyosato, S. Kikyotani, Y. Furautani, T. Hirose, H. Takashima, W. Inazama, T. Miyata and S. Numa, Structural homology of Torpedo californica acetylcholine receptor subunits, Nature (Lond.), **302a**, 528 (1983).
6. M. Noda, S. Shimizu, T. Tanabe, T. Takai, T. Kayano, T. Ikeda, H. Takahashi, H. Nakayama, Y. Kanaoka, N. Minamino, K. Kangava, H. Matsuo, M.A. Raftery, T. Hirose, S. Inayma, H. Hayashida, T. Miyata and S. Numa, Primary structure of Electrophorus electricus sodium channel deduced from cDNA sequence, Nature (Lond.), **312**, 121 (1984).
7. B. Sakmann, C. Methfessel, M. Mishina, T. Takahashi, T. Takai, M. Kurasaki, K. Fukada and S. Numa, Role of acetylcholine receptor subunits in gating of the channel, Nature (Lond.), **318**, 538 (1985).
8. W. Stühmer, The effect of high extracellular potassium on the kinetics of potassium conductance of the squid axon membrane, Ph.D. Thesis, Technische Universität München, München, RFG (1980).
9. R.P. Swenson and C.M. Armstrong, K^+ channels close more slowly in the presence of external K^+ and Rb^+, Nature (Lond.), **291**, 427 (1981).
10. J.M. Ritchie and R.B. Rogart, The binding of saxitoxin and tetrodotoxin to excitable tissue, Rev. Physiol. Biochem. Pharmacol., **79**, 1 (1977).
11. F. Conti, Noise analysis and single channel recordings, Current Topics in Membrane and Transport, **22**, 371 (1984).
12. W. Almers, Gating currents and charge movements in excitable membranes, Rev. Physiol. Biochem. Pharmacol., **82**, 96 (1978).
13. F.J. Sigworth, Open channel noise I. Noise in acetylcholine receptor currents suggests conformational fluctuations, Biophys. J., **47**, 709 (1985).
14. F. Conti and E. Wanke, Channel noise in nerve membranes and lipid bilayers, Q. Rev. Biophys., **8**, 451 (1975).
15. F. Conti, The relationship between electrophysiological data and thermo-dynamics of ion channel conformations, Neurol. Neurobiol., **20**, 25 (1986).
16. T.L. Hill, An introduction to statistical thermodynamics, Addison-Wesley, Reading, Mass (1960).
17. A.L. Hodgkin and A.F. Huxley, The dual effect of membrane potential on sodium conductance in the giant axon of Loligo, J. Physiol. (Lond.), **116**, 497 (1952).
18. C.M. Armstrong and F. Bezanilla, Inactivation of the sodium channel. II. Gating current experiments, J. Gen. Physiol., **70**, 567 (1977).
19. J.E. Kimura and H. Meves, The effect of temperature on the asymmetrical charge movement in squid giant axons, J. Physiol. (Long.), **289**, 479 (1979).
20. C.M. Armstrong and W.F. Gilly, Fast and slow steps in the activation of Na channels, J. Gen. Physiol., **74**, 691 (1979).
21. R.D. Keynes, Voltage-gated ion channels in the nerve membrane, Proc. R. Soc. Lond., **220**, 1 (1983).
22. F. Conti, I. Inoue, F. Kukita and W. Stühmer, Pressure dependence of sodium gating currents in the squid giant axon, Eur. Biophys. J., **11**, 137 (1984).
23. K.T. Wann and A.G. Macdonald, The effects of pressure on excitable cells, Comp. Biochem. Physiol., **66A**, 1 (1980).
24. F. Conti, R. Fioravanti, J.R. Segal and W. Stühmer, Pressure dependence of the socium current of squid giant axon, J. Membr. Biol., **69**, 23 (1982).

25. F. Conti, R. Fioravanti, J.R. Segal and W. Stühmer, Pressure dependence of the potassium currents of squid giant axon, J. Membr. Biol., **69**, 35 (1982).
26. R. Benz, F. Conti and R. Fioravanti, Extrinsic charge movement in the squid axon membrane: effect of pressure and temperature, Eur. Biophys. J., **11**, 51 (1984).

Panel Discussions

September 4, 1986 — Riva del Garda, Italy

**Session 1. Understanding Brain Mechanisms and the Challenge
of Artificial Intelligence Machines: An Interdisciplinary Approach**

Session 2. Sixth Generation Computers: A Sketch

Panel:

B. Alder, G. Careri, E. Clementi, F. Conti, G.L. Hofacker, J. Ladik,
S. Lifson, E. Mingolla, C. Nicolini, M. Osborn, M.U. Palma,
F. Parak, G.N. Reeke, Jr., E. Sackmann, H.A. Scheraga, K. Schulten,
L. Stringa, D. Urry and G. Volpi

Editorial Comments: *In the following we are reporting excerpts from roundtable discussions on the topics listed above. The moderator for these panel discussions was Berni Alder (Lawrence Livermore Laboratory, Berkeley, CA, U.S.A.). Unfortunately, only parts of these discussions are reported herein due to the fact that the participants were not always within range of the microphone used to record these proceedings. The editors of this volume wish to beg your forgiveness for not having foreseen this problem.*

Session 1
Understanding Brain Mechanisms and the Challenge of Artificial Intelligence Machines: An Interdisciplinary Approach

ALDER: Well, I've agreed to be Chairman just because I can be objective since I know nothing about the subject that is going to be discussed. Let me just make a few opening remarks so that we put everything into perspective, the way I see it. Computers — that is serial computers — are at present basically at the limit of their design. That is you can no longer increase the speed of the basic clock (which is of the order of a nanosecond) and gain much more speed from serial computers because the speed of communication between the parts is limiting. Therefore, by necessity the computer people have to go into parallel computing in order to gain more power. There's no way you're going to get any orders of magnitude more power out of computers without going to parallel computers. Now there are a whole lot of different types of parallel computers being built, starting with the Cray people have now a 4-XMP, that is four parallel processors. In a few years we will see eight- and sixteen-processor Crays come out. In those cases the computer has an extremely powerful computer for each processor. There are also the kinds of computers that Enrico [Clementi] is doing where you have processors of medium power (compared to Cray), but the idea is to build many, many of them. The ultimate objective of these computers is the massively parallel computer. For these, presently, each microprocessor is — by necessity — extremely primitive. There already exists such computers which are 64,000 parallel elements, where each microcomputer is just basically a spinflip. So you have this whole range of computers being built.

On the other hand we have the brain, and the brain is a massively parallel processing computer. The question that I think is common to the brain and parallel computers, and what I would like to know from the biologists, is *what is the microprocessor in the brain?* The way I understand, it is not a single neuron but rather a collection of neurons of some sort. A second question is, what is its output? To put it briefly, what we want to know from the biologists is to learn the computer design: What is the microprocessor in the brain? How does it connect to other microprocessors? How do we get these unusual collective properties from billions and billions of these microprocessors in the brain? That, really, is the way I see the problem.

The third component which I think might be useful to start out this discussion with is how well can we model the brain and how have the physicists modeled the brain? We heard two talks about that yesterday — one by Klaus Schulten and the second by George Reeke — and those people told us what they can do, but they didn't tell us *how* they did it. I'm going to have two very brief introductory speakers from yesterday who are going to tell us how they modeled the brain and what the essential characteristics are of their model, namely its relation to a frustrated spin glass. So could I ask Prof. Klaus Schulten to start out and describe his Hamiltonian for a few minutes.

SCHULTEN: I have presented in my lecture yesterday two approaches to information processing, a formal algorithm suited for digital computers and a simulation of a network of neurons described by analog variables, modelled in close similarity to their physiological counterparts. The first approach dealt with the problem of recognition of stereodiagrams and employed a classical Hamiltonian spin system. This system had been invented by us exactly for the reason pointed out by the chairman, namely that one is seeking parallel algorithms which have the advantages you have heard about already.

The algorithm suggested is based on a Hamiltonian

$$H(\{S_{ij}\}) = E_{exchange}(\{S_{ij}\}) + E_{field}(\{S_{ij}\})$$

which has been designed such that the state of lowest energy corresponds to the best hypothesis about the 3-dim. world analogue to the stereodiagram. This analogue is presented by a square array of spin variables $S_{ij} \in \{ -N, -N + 1, \ldots, N \}$ where i,j denote the discrete x,y-coordinates of the array. The values of the spins describe the z-coordinate of the real world scene captured in the stereodiagram. The algorithm determines the ground state configuration of spins

$\{S_{ij}:i1.MonteCarloannealing\}$ by means of a stochastic algorithm, Monte Carlo annealing, which can be implemented in a straightforward way on massively parallel computers. Most algorithms employed in artificial intelligence, particularly for the purpose of pattern recognition, are not parallel.

The energy H has two contributions, an exchange interaction

$$E_{exchange} = -J \sum_{<(i,j),(k,l)>} F(S_{i,j}, S_{k,l}),$$

$$F(S_{i,j}, S_{k,l}) = \begin{cases} 1, & \text{if } S_{i,j} = S_{k,l} \\ q, & \text{if } S_{i,j} = S_{k,l} \pm 1; \ q < 1 \\ 0, & \text{else} \end{cases}$$

which induces the tendency that neighboring spins assume nearly identical values, i.e. that the stereodiagram tends to be interpreted in terms of continuous surfaces of the real world scene. The stereodiagram contains two arrays $\{H_{k,l}^{left}\}$ and $\{H_{k,l}^{right}\}$ of black ($H_{k,l} = +1$) or white ($H_{k,l} = -1$) pixels corresponding to the left and right stereogram. This information is communicated to the spin system by means of the contribution

$$E_{field} = \sum_{(i,j)} G(S_{i,j}),$$

$$G(S_{i,j}) = G_0 \sum_{k=i-w}^{k=i+w} \sum_{l=j-h}^{l=j+h} \left| H_{k,l+S_{i,j}}^{left} - H_{k,l}^{right} \right|$$

to the Hamiltonian H. This contribution compares pixel squares of size $w \times h$ in the left and right stereograms shifted horizontally by $S_{i,j}$ units. The better the squares compare the lower is the energy contribution for that particular $S_{i,j}$-value.

The Monte Carlo annealing method searches for the energy minimum by a long series of trials to reorient the spins $S_{i,j}$. Trials which induce a lowering of the energy are always accepted. Trials which correspond to an increase of the energy by ΔE are accepted only with probability $\exp(-\Delta E/T)$ where T represents temperature. The annealing method systematically varies this temperature, just as a metallurgist does when he anneals a metal. The temperature T can be correlated with a certain noise level inherent in the algorithm searching for the recognition of stereodiagrams and, as such it plays a role also in real neural systems.

ALDER: Well let me just interject for one minute here what I think is the essential characteristic of the Hamiltonians, that they're frustrated in the sense that you have a positive interaction with the nearest neighbor and a negative interaction with the second nearest neighbor so the system has lots of minima . . . that's one of the characteristics, right?

SCHULTEN: Ah . . . almost right. The difference is that in this case the exchange interactions are all homogeneous, but the fields are inhomogeneous. But you also get a frustrated system since the exchange interactions try to make the spins all parallel, but the fields try to force them non-parallel.

ALDER: You need, somehow, a positive and negative effect and they nearly cancel and then get *lots* of minima.   That's one of the essential characteristics in a Hamiltonian.

SCHULTEN: In our first approach to artificial intelligence we did not care very much about the neural system. Our second non-Hamiltonian approach, however, started from the question: What is the smallest non-trivial computational unit in the brain? Our answer to this in building our model was not the extreme caricature of a neuron assumed to be either in an active or inactive state, i.e. to be like a spin 1/2 particle, but rather something with a little bit more internal structure.   Such internal structure was taken into account by a time-dependent electrical potential inside the neuronal call which obeyed the kinetic equation

$$\frac{dU_i(t)}{dt} = -\frac{U_i(t)}{T_R} + \rho[\Delta t_i]\left(\omega\sigma[A_i(t)] + \frac{\eta}{\sqrt{T_R/2}}\zeta(t)\right).$$

The first term describes the relaxation of the potential U_i of neuron i to the stationary state. The second term involves a noise contribution and takes account of the effect of synaptic signals which converge on neuron i. This term contains as a factor a function $\rho(t)$ which models the refractory ($\rho = 0$) and the sensitive ($\rho = 1$) periods of neurons. The kinetic equation contains the important scaling parameter ω which sees to it that the network is neither suffering from epileptic hyperactivity nor from complete quiescence. One of the main objectives of our work had been to obtain a mean field expression for ω. This expression allows us to keep a neural network in a proper regime of electrical activity when we alter the number of neurons, receptors, etc.

The second independent variable in our model are the axonic currents (from the cell bodies to other neurons) which are described by a mean activity function

$$G_k(\Delta t_k/\tau) = \exp\left(-\frac{\Delta t_k}{\tau}\right); \quad \Delta t_k = t - t_{0k}.$$

Whenever the potential exceeds the threshold value at time t_{0k}, G_k is set to the value 1 and the potential is set to a refractory value. The activity functions G_k enter through the sum

$$A_i(t) = \sum_{neurons\ k} S_{ik}(t)G_k(\Delta t_k/T_U) + \sum_{receptors\ j} R_{ij}G_k^R(\Delta t_k^R/T_U)$$

the kinetic equation for the potential.

The third, and conceptually most important, independent variable of the model network is the synaptic strength S_{ik} which describes the effect the signals from neuron i have on the electrical potential of neuron k. The synaptic strength is governed by the kinetic equation

$$\frac{dS_{ik}}{dt} = \begin{cases} -\dfrac{S_{ik}(t)-S_{ik}(0)}{T_S} + \Omega G_k\left(\dfrac{\Delta t_k}{T_M}\right)\kappa(G_i, G_k), & \text{if } S_u \geq |S_{ik}| \geq S_l; \\ -\dfrac{S_{ik}(t)-S_{ik}(0)}{T_S}, & \text{else} \end{cases}$$

with

$$\kappa(G_i, G_k) = \begin{cases} 1, & \text{if } G_i > G_k > e^{-1} \wedge \bar{v}_i \gg v_s \wedge \bar{v}_k \gg v_s; \\ -1, & \text{if } G_k > e^{-1} > G_i \wedge \bar{v}_i \ll v_s \wedge \bar{v}_k \gg v_s; \\ 0, & \text{else.} \end{cases}$$

The synaptic dynamics involve a relaxation process and a growth or decay term which depends on the synchronicity of the pre- and postsynaptic activity at synapses. The evolution equations of the network are highly non-linear and as a result of electrical input through the receptors yield a self-organized synaptic connectivity structure. This structure is the basis for information storage. Information recall is also realized through the non-linear dynamics. Partial inputs of stored information can be discriminated by the network: the stored information closest to some input selects the activity of the network, suppressing those states of activity which do not fit the input. The noise inherent in the dynamics can help the network in its exploration of its state space and in its selection of the information which best fits the input.

The important aspect of information recall in networks has been studied intensively in the framework of the Hopfield model. This model represents a system of spin 1/2 particles with spins $S_j \in \{-1, 1\}$ which interact through a heterogeneous, non-local exchange interaction as described by the Hamiltonian

$$H = \sum_{i, j} J_{ij} S_i S_j.$$

The exchange interaction models a connectivity which stores information in the system. The information stored and to be recalled is represented in terms of configurations of network spin states denoted by $\xi_j^{(\alpha)}$, $\alpha = 1,2,...M$ for M information states. The proper connectivity for the associative recall of this information is given by the symmetric $(J_{ij} = J_{ji})$ exchange interaction

$$J_{ij} = - \sum_\alpha \xi_i^\alpha \xi_j^\alpha.$$

The Hopfield model admittedly presents an extreme abstraction of real brains, however, it can be analyzed by methods of the statistical mechanics of spin glasses. Such analysis has provided most important information on the storage capacity, the role of noise and the range of suitable network parameters which could not have been obtained by time-consuming and, therefore, extremely limited computer simulations of more realistic network models. Cognitive science and artificial intelligence, rather than crucifying the Hopfield model and its variants for lack of agreement with petty details, should cautiously embrace these efforts since they provide some understanding of collective computation in systems with a large number of simple elements and share some properties of biological networks, such as fault tolerance, analog information processing, no system-wide clock cycle and high connectivity.

ALDER: Thank you very much. The idea is to have a frustrated system and then to generate from it a periodic pattern. You need a second order differential equation to get a periodic pattern because first orders don't do that, and then you want that pattern to be generated from almost all initial conditions. That's the idea. Now the second guy is going to criticize this model from the biological point of view, hopefully. So, would Dr. George Reeke please try to see how realistic these mathematical models are from the biological point of view, as well as present his own models.

REEKE: To begin with I'm going to just say a few words about some of the details in our own models that may not have come out in what I said yesterday and then, at the end, I'll try to contrast this with what some of the physicists have been doing. The first comment I want to make is that obviously exactly what you do in detail depends on what you're trying to model and what level you're trying to work at. It's the same old problem of trying to explain chemistry with the Schroedinger equation. Here you have neurons, there's a great deal known about them, there's a tradition of work going back to the last century on their dynamic properties, their electrical conductivities, how they change, etc. You quickly find that if you want to use the basic equations that describe what's going on at a synapse to make a model of some higher function, like pattern recognition, that very quickly it becomes impractical to compute all of these things. So, judgment is involved, and one tries to pick a set of equations that form an abstraction of the actual situation that contains its essential elements, but not unnecessary detail that's just put there because of the underlying chemistry that's going on.

Having said that, the first decision you have to make is whether to model individual action potentials or not, and in the work I've done we've said, well, we don't need to do that. Instead we can look at a number which corresponds to the average activity of the neuron over some period of time and, in the language Dr. Schulten was using, that would be a continuous rather than a discrete neuronal response. Obviously, this decision may have to be changed if one works with more elaborate rules for synaptic modification that depend on correlations of activity coming into multiple synapses from different places. There are indications that the precise timing of the action potentials as they arrive at a common point may be very important. So we have to be prepared, then, to move to a different response function. By response function, I mean the function that describes the response of a unit, given its inputs and current state.

$$s_i(t) = [a_i(t) + n(t) + \omega s_i(t-1)] \quad \begin{array}{l} \text{if } a_i(t) \geq \theta_p > 0 \text{ or} \\ \text{if } a_i(t) \leq \theta_N < 0 \end{array}$$

$$= [\qquad n(t) + \omega s_i(t-1)] \quad \text{if } \theta_N < a_i(t) < \theta_p$$

$$a_i(t) = \sum_{j=1}^{M} c_{ij}(t-1)\left[s_{\ell_{ij}}(t-1) - \theta_E\right] - \sum_{k=1}^{M'} \beta\left[s_{\ell'_{ik}}(t-1) - \theta_I\right]$$

where

$[x] = 0$ if $x \leq 0$; $= x$ if $0 \leq x \leq 1$; $= 1$ if $x \geq 1$
$s_i(t) =$ state of group i at time t; $0 \leq s \leq 1$
$\quad a_i =$ combined afferent excitation and inhibition on group i
$\quad n =$ noise drawn from a normal distribution
$\quad \omega =$ persistence parameter $= e^{-1/\tau}$
$\quad \theta_p =$ positive response threshold; $\theta_p > 0$
$\quad \theta_N =$ negative response threshold; $\theta_N < 0$
$\quad c_{ij} =$ "strength" of synapse from group j onto group i; $-1 < c_{ij} < 1$
$\quad \ell_{ij} =$ location of j'th input group to group i
$\quad \theta_E =$ effectiveness threshold
$\quad \beta =$ inhibition coefficient
$\quad \ell'_{ik} =$ location of k'th geometrically defined group inhibitory to group i
$\quad \theta_I =$ inhibition threshold
$\quad M =$ number of specific connections onto a group
$\quad M' =$ number of geometrically defined inhibitory connections onto a group.

Well, what we do — the terms that we consider important — are the inputs that come in, and now I'm going to use this thing [*points to square brackets*] to indicate that there's a threshold condition being applied. I'll explain the square bracket later in detail. Now we have some inputs coming from other cells and there's a transmission delay so we have the state of cell j at time t−1 multiplied by some connection strength term, I think that was s in Schulten's equation, and those are added up. Each of these is tested against the threshold, which I call θ_E for excitatory threshold — that's just to make the system nonlinear. If it's linear then it's very boring, so you need thresholds — now what the bracket means is that if the number inside the brackets is less than 0, throw it away, don't keep it — so, now, there may be several terms like this for different classes of connections. In other words, a neuron in a certain part of the brain may have inputs from sensory elements from other places, from nearby neurons, etc. Now one compromise we make with reality here is we allow the c_{ij}'s to be a mixture of plus and minus terms, and that's OK in terms of inputs onto a cell, cells can have excitatory or inhibitory inputs, but typically real cells in their outputs only excite other cells or only inhibit them, so in the brain you have so-called excitatory and inhibitory cells, but that seems to me not an important distinction. That's one of those things that arises from chemistry and by imagining that you talk about a small group of neurons working together, one can ignore that. Well, then you also have inhibitory terms which are just like these [*pointing to sum with β*]. There's absolutely no difference in what goes on here except that where the connections come from is determined a different way. These come from a nearby locality and have to do with stabilizing the system and so they all have a common coefficient β rather than an individual c_{ij}, just suggesting that there's nothing specific going on there, whereas the other ones can come from wherever we want them to and, clearly, what's different in one instance of the network or another, or in one network built for one purpose or another, is where these j's come from. Then you have noise which is generated by some rule or another, and then we add an $s_i(t-1)$ times some decay constant, ω, so that you have exponential decay in the absence of input.

So this equation is not terribly different from what Prof. Schulten is doing, actually. This is a fairly common kind of model that a lot of people use because it does seem to encompass the essence of what we currently think neurons do. Now, the key thing is how these c_{ij}'s change — I write it as a difference equation rather than a differential, but

$$c_{ij}(t) = c_{ij}(t-1) + \delta \cdot \phi(c_{ij}) \cdot (s_i(t-1) - \theta_{M_I}) \cdot (s_j(t-1) - \theta_{M_J})$$

where

$$\begin{aligned}
\delta &= \text{amplification parameter} \\
\phi(c) &= \text{saturation factor} \\
&= 1 + c^4 - 2c^2 \text{ if change has same sign as } c_{ij} \\
&= 1 \qquad \text{if change has opposite sign as } c_{ij} \\
\theta_{M_I} &= \text{amplification threshold for current group} \\
\theta_{M_J} &= \text{amplification threshold for input groups}
\end{aligned}$$

This δ is just a parameter. Any time I use a Greek letter it's a parameter that's set for a particular run. δ determines how much c_{ij} is going to change. Then we have a saturation term where the purpose of this term is to say that if a coefficient is already very large, then it's hard to make it still larger. That's just a stabilizing thing. Now there are lots of different things you can do here and this is just the simplest one where you say if neuron i is firing above some threshold and neuron j is firing above some threshold, then you strengthen the connection; and I tried to explain that yesterday. Unlike Prof. Schulten, we do not include a general decay term here because — you know, he says we all forget, but I think the amazing thing is the other way around — we remember things for forty years. In any event, we've tried decay terms and usually what that results in is that it's not totally compatible with the idea of selection because, as the c_{ij}'s all tend to go back toward zero, that's giving you an instructional kind of system where the response is totally patterned by what comes in. After some length of time it no longer remembers how it started out, whereas we feel that it's important that the initial conditions remain important and have a lot to do with what kind of patterns the system responds to. So instead, we stabilize the system by an appropriate choice of amplification rule. These brackets here are very ambiguous and what I mean to say is that you can set up different things like maybe if the second term is negative and the first term is positive, you can make c_{ij} decrease — there are different ways to do that.

The only thing I want to add to that is we're working with a lot more complicated functions that I think are based on what people are seeing in the nervous system, as I mentioned yesterday, putting in heterosynaptic effects where what's going on at nearby synapses is important, and it's clear that that happens, that voltage is induced in the membrane by nearby events and is conducted to a given synapse where it can, for example, alter the state of voltage-dependent ion channels.

I'll stop this now and go on to the contrast since you say you don't see much contrast. Actually, my differences are more with the Hopfield type model than with what Dr. Schulten presented, and that has to do with the approach that a lot of physicists take. If you're a physicist, what you want to do is find the simplicities in a system, a few simple rules from which you can derive what's going on. And that's true here, too. One wants simple rules. You know: how does the neuron respond, how does the connection strength change? Those are the rules; but then if you go beyond that and try to analogize with things like spin glasses, where somebody has made some beautiful equation that explains what's going on, and you say, "Gee, that looks a little bit like a neuron; why don't we see how far this analogy can carry?" Then I think you start to get off the track because you start making assumptions because they have nice mathematical consequences rather than because they have anything to do with the brain. And, if you want to study networks like that, it's a perfectly fine thing to do; you may develop a better robot, or whatever, and Hopfield has shown some nice things in his recent "Science" paper on how to solve the travelling salesman problem. But to say it's like the nervous system just because it solves some of the same problems is where we differ. The two really fundamental things that I think are wrong about those kinds of models is that they assume that stationary states are important and you're trying to minimize some energy Hamiltonian and find out what stationary state you drop into and then you equate that with some memory or some recognition event. But, what's going on in the brain is continuous activity — input is coming in all the time and you never have a chance to stop and wait for temperature to drop. What is temperature in the brain? What is this simulated annealing? You know, physiological measurements show that activity — of course there's all kinds of variations — but you don't see a lot of noise declining to a minimum with simulated annealing and then a pulse going back up and declining to the next minimum. No, we see input coming in all the time and motor activity coming out. And it's the motor activity that determines the meaning of the responses, not somebody looking down and saying this and this neuron responded. So I think that's the key and critical difference. There are other minor things that come out of this, like assuming that connections are symmetric because that makes the map come out nice, but real neurons do not have symmetric connections. So, again, I tend to feel that an assumption like that is dangerous and useless and, if you're studying the brain rather than just a nice system, you should make assumptions that are like what we know goes on in the brain.

ALDER: We have one more formal talk and then I want to throw the discussion open to anybody, particularly the audience. And that is a contribution from IBM-Rome which has really had remarkable success; namely, speech recognition. And the connection with this is that an individual has to speak to a machine for awhile, the machine has to learn his particular way of speaking English, his accent, and so on. And that has in fact been done by the same Hamiltonian basically, by using this Hebb mechanism, for speech recognition. So, I would like to ask G. Volpi to give his talk on speech recognition.

VOLPI: I'll try to give you a general idea of a speech recognition problem as it is underway now in a lot of different types of research and development. I would just like to point out which are the main components of a speech recognition task, and try to give you an idea of the current methodologies that are used to solve each of these components, and try to give you an idea of the methodologies that I think are very promising for the near future.

When a speaker pronounces a sequence of words, a sentence, it produces an acoustic wave and then the reproduction of an acoustic signal which is captured by a computer after it has expression, and then we have to process the acoustic signal in order to identify the sequence of the words that have been pronounced. The main problem is this: we know something about the production process of human speech but, on top of the basic mathematical model that is possible today to develop, we must superimpose lots of different phenomena due to different dialects used by the different speakers, by the intonation, the different stress, the different speed that a speaker uses to pronounce a word. And then we add the superimposition of different phenomena that can produce elements of random nature. All this must be taken into account here. The second problem is that the signal that this production produces is very rich. It contains an amount of information which is greater than the linguistic information which is contained in the message that the speaker has given. And then it is necessary to compress the amount of data that are contained in the speech signal and extract, simultaneously, a set of acoustic parameters that are very relevant to do the later recognition tasks. And, after this, with these parameters it is necessary to try to identify the sequence of words using the acoustic information which is contained in the signal and using also the language properties that are characteristics of the language that the speaker used to pronounce the sentence.

The first part, the signal compression, is based on some kind of spectral analysis in order to produce a compression. But, to extract the acoustic features which are relevant for the later recognition tasks, the present technology is based on some human peripheral auditory system model. We have a lot of results on the behavior of the peripheral auditory system and then we try to model this result in this component of a recognition system. And we take especially into account the selectivity that the human ear has in distinguishing or extracting information on different frequencies and the capability of the system to adapt to the different level of the signal. This is what is done presently in the most advanced recognition systems, in different ways trying to develop different models for the human peripheral auditory system.

If we obtain, after this operation, a sequence of acoustic parameters that are just in agreement according to this model, now we have to extract the information which is contained in this sequence of acoustic parameters. The first part is done by acoustic models. In every recognition system you have to predefine the set of words that you want recognized and you have to prepare a sort of dictionary and to associate with the dictionary the proper acoustic description. When you have the input of the sequence of acoustic parameters coming from the earlier modeling, you have now to compare this acoustic sequence with the information you have associated to each word in the dictionary. And, probably the most accurate technique is based on the concept of memory, the same as you associate to each word of the dictionary a sequence of acoustic parameters. Then, when you add an input in your sequence you do just a comparison using more or less different sophisticated techniques just to obtain the best fitting. But this is also the technique that is used, for instance, in a commercial system. But I think that the most effective technique that is used now is the code technique based on the so-called hidden Markov model. It is a procedure which is of a stochastic nature and it takes into account random elements in the production of a word or a sentence by a human. For each word we prepare a mathematical model, which is a final state chain, and it is based (this model) on some probability laws. And these probability laws are not predefined and are not based on previous knowledge, but we have to define these laws on data and we have to do statistics on the pronunciation of the same word and of different words by different speakers, and try to extract such good statistics to define the probability laws that we have in this model. But, I would like to say this: these models are appropriate for speaker dependent systems in the sense that if a speaker wants to use a system, he must train the system in advance in the sense that he must pronounce something in front of the system in order to personalize the statistics that are, or the probability laws that are computed in these mathematical models in order to have a good recognition. But, in this approach it is possible to create a sort of median speaker with a median probability law and then use only a few minutes of reading of a speaker in front of a microphone, reading a predefined text containing some linguistic and phonetic information, in order to obtain a good first approximation.

ALDER: Thank you very much. That concludes our formal presentations. I would now like to open the floor to comments and discussion from either other members of the panel or from the audience. Fine, let us begin with Professor Lifson.

LIFSON: I submit that in order to understand the bridge between physics and biology one has to understand how selection preceded life, how selection is a general property of matter, and not only animate matter. I'll explain what this means. Obviously, there is no selection in rivers or how rivers evolved; there is no selection in mountains; it is poetic to talk about selection of this kind. However, very little attention was given in chemistry to autocatalytic systems, and particularly to self-replicating, more complex systems like oligomer chains, surfaces, etc. The point is that when such a simple elementary physical or chemical system is autocatalytic, namely self-replicating, it achieves something which is very unique from the point of view of statistical mechanics.

What does statistical mechanics tell us? From out of infinite numbers of microscopic states what we observe in the physical world, or macroscopic world, is only the averages. Some unique phenomenon will never be observed. There is a finite probability that they can happen, but they will never be observed. Now, if systems are self-replicating and make a whole assembly of self-replicating systems, it can be proven mathematically that such systems have the following property: a self-replicating system either grows exponentially or decays exponentially and there is a very sharp limit. If several systems compete for the same substrates, then they cannot all forever grow exponentially, because the growth is limited and some of them will start to decay exponentially. This selection property is a very sharp one and therefore what will happen is that out of an infinite number of different self-replicating systems, if one has a selective advantage, it will be singled out and lo and behold you have a system which has an *a priori* probability of zero to exist, but by self-replication it becomes ubiquitous. This is one solution to one of the biggest riddles for the physical understanding of life. How come life is made of molecules, proteins, DNA, etc., which as polypeptides are as common as anything, but as specific sequences they are so uncommon that, according to Wigner it cannot be explained except by a miraculous argument.

Now, the point is that selection acting in the physical world creates systems which have the property of being adaptive. Why are they selected? Because they have fit in their environment. Adaptivity is the consequence of selectivity. And the consequence of selectivity itself is that the system appears to be as if it has a purpose, as if it has a function, as if it has a very particular property which is typically inanimate. From this point of view one can see how the transition between inanimate to animate introduces the principle of apparent purposefulness which is synonymous with being a product of selective advantage. Now, I could enlarge on this, but I want to go right now to the brain.

What amazed me and got my attention as very revolutionary is the notion that the brain is not only a part of an organism, a product of a phylogenetic selection process, but it is itself a product of selection during the development and the growth of the organisms. In this respect it is not, in principle, different from the immune system which has also an element of selectivity (on which I don't want to enlarge). I think that in any research of analogs between mathematical systems and biological, real brains, one must remember that something will certainly be missing, and something fundamental may be missing if the element of selection is missing. I would sum it up by kind of an aphorismic statement. If spinglasses are similar to brains, why is it that brains can understand spinglasses, but spinglasses cannot understand brains? The answer is clear; spinglasses are not the product of natural selection (and don't undergo natural selection), and therefore they could never understand brains. Why is the brain the way it is; because it is not only the product of phylogenetic natural selection but, as we learned yesterday, it is itself a product of natural selection. Natural selection, here, brings me to the last comment and this is, when you want to model the brain it is extremely important to understand the latest results in the biological study of neural systems and neurons. I think the most important thing I learned in the last year is the discovery by Kandel and his group that when Aplysia learns the difference between habituation and sensitization (two opposite reactions), in both cases the result of input to the neurons observed, is a change of the rate of firing of the neurons. The rate of firing is the process of learning; either they fire faster as the result of the input, and then it is sensitization, or else they fire slower and this results in habituation.

ALDER: Let me point out to you that in fact the spinglasses do do that as well. In fact, they learn that way.

LIFSON: That's very interesting — so maybe one day they'll understand the brain.

This also relates, I think, to the question of whether one could conceive the kind of connection between software and hardware such that certain kinds of inputs to the computer would have some modifications of the elements of which the computer is built. This would be a very new kind of memory which is not necessarily science-fiction.

ALDER: Well, but this is also part of what Enrico mentioned in the adaptive Monte Carlo which can apply to random processes but you can also apply it to connections between computers.

SCHULTEN: I would like to rephrase a major part of Professor Lifson's eloquent comment on the ubiquitous nature of natural selection in a more prosaic way. If you want to describe any system which creates order you need nonlinear equations with competitive terms, and this has been known since Turing. I think this is what we all actually do and since we are witnessing here something like the birth of a new field, I think it's very sad to see that there is some kind of divergence which is unnatural and that people are maybe stressing more minute differences in their approach rather than stressing the overall similarity between the approaches we have actually seen here.

ALDER: I tend to agree. There seems to have been a gap developed between the physicists and the biophysicists which I don't share either. I mean, I think that Prof. Lifson's model can be put into a mathematical basis.

SCHULTEN: We all use nonlinear equations with very similar structures and maybe we should use that language and compare our equations.

ALDER: And we just said that the nonlinear equation that has the fastest reaction rate wins and you've got . . . method of evolution. Right. Oh, wait a minute, there's somebody who hasn't had a chance, Professor Hofacker.

HOFACKER: May I come to your introductory remarks. You spoke about the brain and how it thinks, or functions, and even Hopfield in his first papers speaks of a computational device. We haven't said anything about brain processes. We have seen, so far, imaging and the stability of memory, but nobody has made any remark about how the brain functions when it thinks. I wonder whether there is anybody who can make a qualified statement on that. I would like to question the obvious conclusion that the brain looks like a parallel computer. We just don't know. I mean, is that the only architecture we can imagine for a neural network? It is certainly not a sequential device, but thinking of a brain as a central processor with various agents is not borne out by any observations.

ALDER: Do you have a positive outgoing suggestion?

HOFACKER: Not really ... I wouldn't be willing to stick out my neck that far. But one can look at problems arising with very complex parallel computers and draw conclusions as to the human brain. The former ones, if AI gifted, are subject to well known logical fallacies (e.g. in assessing their own knowledge). Our brain avoids these easily. We would have greatest difficulties designing a computer of similar complexity but logically as fail-safe. That is perhaps worth a thought.

ALDER: Let me just say one thing. I think that what the expectation is is that if you have massive parallel computing with a hierarchy of connections, new phenomena, such as brain function, might appear.

CLEMENTI: We have to be careful not to be too "idealistic." The realistic relation between an extremely sophisticated computer and the brain (actually, the smallest element of it) is pretty much like cutting down a stick (to help somebody walk) when you really need a fully developed leg. So let's not push it more than that for now. It is very difficult to make a machine as complicated as the most simple single neuron. However, even with this realistic point of view, we still want to learn from the brain in order to make a machine.

HOFACKER: I would like to comment that relatively simple devices can exhibit problems which are not of a trivial nature.

NICOLINI: Lunch is approaching and I don't want to go overboard, but I sit here and listen to these talks; my impression is that we have not made too much progress since John Hopfield proposed his model several years ago. I really think that we are going too far in making analogies between the computer and the brain. My comment is that we did not yet really study the brain; we do not yet have the analytical tools to study it at the required resolution. We just learned now how complex is the cell; we have just now learned to do recombinant DNA in sito; we are just learning the complexity of the most simple system system such as a single cell. We do not have the slightest idea of the complexity of the brain. We may be right or we may be wrong, parallel may or may not be the answer. However, to me the real emphasis should be on the experimental side.

[*Prof. Nicolini turning to Alder.*] I think that we can move onto the next topic, right? [*Alder nods positively and Nicolini continues.*] If we want to build a sixth generation computer, I would start from where we know how to work: from small molecules like biopolymers, considering that really the future computer generation has to be able to integrate high numbers of elementary circuits in a smaller and smaller area. And I think that we have to face the reality that maybe microelectronics, silicon or any other material of this type will never do it. We have to think maybe biopolymers will be the way.

We can engineer proteins, we can make crystals with absolute fantastic properties, we can use recombinant DNA cloning to make the material we want with exceptional properties at the Å level, and just by doing that we will surely learn a lot. A concrete way to go in that direction is via the molecular level to the complex level. In the meantime we may acquire a large amount of information on the brain and this will tell us how to make a model. I know John well; John is not naive to assume that, like some people claimed before, he has the answer to the problem. Clearly he made a hypothesis hoping to explain facts and data, and he asked biologists to tell him what was wrong. Well, six years passed and nothing or very little came out.

ALDER: There's been some small changes, yes.

NICOLINI: Not too many if we compare it to the speed with which other fields move, like the capability to do protein engineering, the capability to communicate between small molecules, soliton waves, biosensors, like chemset based L and B monolayers, and so on. They are concrete advancements towards molecular electronics and this is where I think the emphasis should be.

Session 2
Sixth Generation Computers: A Sketch

ALDER: Thank you. The suggestion is that we talk briefly about *sixth generation computers* and what they might look like. And we just heard the seventh or eighth generation computer might consist of biopolymers. It's true, naturally, when one wants to go to smaller and smaller chips one wants to go to more and more parallelism; and, what other things do people see in the sixth generation?

CLEMENTI: I shall present a primitive, tentative, and rudimentary sketch of the sixth generation computers. Let me stress that we are still in the *fourth* generation, moving toward the *fifth*, and talking about the *sixth* is therefore bound to appear premature. However, it is certainly not too early to get started if you consider the complexity of this project and the fact that it will require much interdisciplinarity — even more than we have witnessed at this roundtable and the present symposium. Parenthetically, let me add that some might dislike categorizing the evolution of computers into "generations," and I am sympathetic to such semantic apprehensions. However, the computer industry and computer science communities have, in practice, accepted the use of the terms "fourth" and "fifth generation" computers.

Incidentally, as everyone knows, the latter term refers to the type of advanced architecture and concepts coming mainly from Japan.

I will stress, in this discussion, some ideas, designs and prototypes originated mainly in the United States, some of which are the result of many years of development and will likely become industrial products; others are public domain concepts which have, by now, passed some level of feasibility; finally, some are either most preliminary or are simply *untested suppositions*.

In any event, the past is the *boundary condition* for the present and for the future and, by pragmatically recalling the boundary conditions, I will have carved out most of the elements necessary for "sketching" these *sixth generation* computers. To start with, they should be conceived as an extension of current research and development, certainly stressing parallelism. You have just heard from Berni (Prof. B. Alder) why we need parallelism. His main point is totally correct. However, let me add immediately that we need *super scalar* and *vector* too. A code will *start* and *end* using "scalar," but contained within this *start* and *end* there will be many *forks* and *joins*, the former being sequential. Both forks and joins can benefit much from the most advanced vector architecture. I am rather puzzled when I hear about proposals to move quickly into "massive" parallelism — today a rather virgin land — proposing nodes so simple that most of what we have learned from fast scalars and vectors would be neglected. It would seem reasonable to consider "scratching" these architectural foundations only *if* and *when* we have *proven* we can do better with massive parallelism alone. Let me stress that I am not much interested in special purpose machines designed to solve a *specific* problem, however important that problem might be. Indeed, I am confident that the "global simulation" approach I have presented in the previous lecture will become a standard in large simulations; if my pre-vision is correct, this evolution necessarily calls for an architectural strategy where *scalar, vector and parallel architectures must coexist in a synergic fashion*. Of course, sixth-generation machines could have specialized "boxes" attached for special-purpose applications. This is certainly feasible, has already been done, and likely will become an accepted and standard practice.

The main feature of this sixth generation computer is to assemble and merge, in unique composition, many ideas and concepts which have been tested or proposed in the 1970s and 1980s — a rather enormous potential still untapped by the computer industry. This is why I am stressing the "boundary conditions," particularly in data flow architecture, in borrowing concepts from systolic and optical computers, and especially advocating heavy use of expert systems and artificial intelligence at the operating system level, and improved input-output where pattern recognition and voice recognition are fully exploited. These are some of the obvious opportunities which will grow and become more mature during the next decade, leading to that synthesis which we have named the "sixth generation" computer.

More concretely, let us contrast the rigidity of most of today's parallel computers with the data flow architecture proposed in the mid-1970s. The sequencing of forks and joins should not be artificially *forced* through parallel architectures, as we do today, but should naturally occur as soon as operands fill the capacity of a given node. Ideally, the codes should be written in a language which knows about parallelism rather than adapting sequential languages to parallel design via the use of precompilers. Particular attention is also needed for the network-connecting nodes, where the compiler will set up the most appropriate *type* of networking for the job to be executed and the scheduler will decide on the specific connectivity among those "recommended" by the compiler. For additional literature on data flow machines we refer to Refs. 1 – 12.

Systolic architecture might also contribute to the sixth-generation computers, especially if some of the present limitations are going to be reconsidered. In particular, I am referring to those concerning communication restrictions to nearest neighbors and the simplicity of present processors.[13-19]

Very optimistic proposals have been advanced under the heading of "optical computing,"[20] and these represent the outcome of research and development in processing radar data (since the 1960s) and, later, in optical pattern recognition. It is somewhat too early to assess whether or not these approaches will yield reliable technologies, but we should not leave any stone unturned.

A major area for a possible important breakthrough is in component *materials*. The sixth generation computer will certainly continue to use silicon and gallium arsenide, but this limited inventory of materials should be expanded. The search for new inorganic materials will be combined with more and more interest in *organic* materials with semiconducting properties.[21] The potential of organic materials has not even been scratched today, and the interdisciplinary barrier is much to be blamed. We would like to suggest that organized, systematic, in-depth simulations could bring about new organic materials for electronic components. Methods of quantum chemistry extended to the study of 1-D polymers and 3-D solids[22] constitute important tools for a first step in these simulations. The "thinking" process is equivalent to what is presently being done by the drug industry; we should remember that for drug design the main obstacles in the past were, indeed, a lack of interdisciplinary expertise and the limitations in computer power. An interdisciplinary approach and "supercomputers," the obvious tool, are required in our search for efficient new materials.

Let me recall that there is already much interest in these directions. New organic materials for molecular electronic devices are being actively sought for by many institutions. In the area of *conductors* there is much interest in conducting organic polymers to be used, for example, for interconnections of entire circuits with electronic device functions. Examples are polyacetylene, polyphenylene, polypyrrole, polythiophene, and polyazulene.[23] In the area of *rectifiers* there have been suggestions for *molecular rectifiers*.[24] In the area of *memory* elements, proposals have been made to look into bistable molecules such as hemiquinone molecules.[25]

We can even be more bold and look forward to when some of the above materials will be competitive with those presently used, and even dream about biotechnology and recombinant DNA techniques as a way to prepare the new materials, especially in the area of conducting polymers. Possibly these dreams represent one of the reasons for the seemingly heterogeneous mixture of disciplines at this Symposium. Let me digress for a second. When we consider molecular biology and biophysics in the last twenty to thirty years, *two very distinct directions* stand out. The first direction, the one that has captured the imagination of scientists and the public at large, recognizes the very fundamental role of DNA and proteins in molecular genetics and looks forward to DNA technology to obtain new organic materials — possibly new forms of life. We are already working on this and venture capital is becoming more available as industrial applications become more and more feasible. In the last 40 years, DNA has moved from the "discover and research only" mode to a vigorous, novel, "applied" field. *The second direction,* presently moving with much energy, aims at a deeper understanding of *neuron-neuron networks* or, if you prefer, *brain models,* and of the mechanisms of biological functions like vision and hearing which, for scientists with yet a different background, becomes voice and pattern recognition or, more broadly, *artificial intelligence.*

The physicists, the theoretical and physical chemists, the mathematicians, the engineers — those who turned their interest years ago to the *first direction,* DNA, will now likely look toward this new direction, "brain," because there are so many motivations common to both directions. All in all, this is the reason why we are here in Riva del Garda at this panel discussion.

There is much reason for optimism and at the same time for prudent realism. For example, I can look at our "home made" 1.2 gigaflop peak performance supercomputer, lCAP-3, and improve it. There are several cycle times in the various nodes of our machine, precisely $\sim$180 nsec, $\sim$56 nsec, down to $\sim$20 nsec; "most trivially" we could remap lCAP-3 with the best known existing components and operate at a cycle time of $\sim$4 nsec. In this way we could build a new lCAP-3 with about 40 gigaflop peak performance. This is not all. Presently, we have assembled only two clusters; the above remapping will sharply decrease the physical size and open the door for more clusters, say up to eight. Now the improved machine could reach about $150-200$ gigaflops of peak performance. Notice that we have proposed only "relatively trivial" improvements requiring no new discoveries, thus feasible with a two- to three-year effort, even with a numerically modest group of scientists and engineers. *But* much more could be done with VLSI, new materials, better connectivity and networking, etc. Thus, to think in terms of teraflops within the next decade is simply a rational hope, based on modest gains in present leading edge technology.

Most clearly, computers will become much more sophisticated. But please, let us not exaggerate. *Yes,* the sixth-generation computers will take concepts also from the brain and will have much artificial intelligence. *No,* we are not planning to construct artificial brains; at very best, the relation of a sixth generation computer to the brain is equal to the relation of a most rudimentary crutch (a stick) to a human leg. On the other hand, we look forward to "novel" realizations which will most certainly be much more sophisticated than today's "supercomputer" and which will be based on research and development carried out, especially in the United States, during the 1970s and 1980s.

References

1. J.B. Dennis, "Data Flow Supercomputers," Computer, Nov. 1980, p. 480.
2. J.B. Dennis and G.G. Rong, "Maximum Pipelining of Array Operations on Static Data Flow Machines," Proc. of the Int. Conf. on Parallel Processing, Aug. 1984, p. 331.
3. W.B. Ackerman, "Data Flow Languages," Proc. of Nat. Comp. Conf., Vol. 48, 1979, p. 1087.
4. T. Agerwala and Arvind, "Data Flow Systems," Computer, Vol. 15, No. 2, Feb. 1982, p. 10.
5. D.D. Gajski, D.A. Padua, D.J. Kuck and R.H. Kuhn, "A Second Opinion on Data Flow Architecture and Languages," Computer, Vol. 15, Feb. 1982, p. 58.
6. J.R. Gurd, C.C. Kirkham and I. Watson, "The Manchester Prototype Dataflow Computer," Comm. of the ACM, Vol. 28, Jan. 1985, p. 34.
7. A.P.W. Bohm, J.R. Gurd and J. Seargent, "Hardware and Software Enhancement of the Manchester Dataflow Machine," Proc. of the Int. Conf. on Parallel Processing, Aug. 1985, p. 420.
8. I.M. Patnaik, R. Govindarajan and N.S. Ramadoss, "Design and Performance Evaluation of EXMAN: An EXtended MANchester Data Flow Computer," Trans. Comp., Vol c-35, No. 3, March 1986, p. 229.
9. I. Hartimo, K. Kronlof, O. Simula and J. Skytta, "DFSP: A Data Flow Signal Processor," Trans. Comp., Vol. c-35, No. 1, Jan. 1986, p. 23.
10. V.P. Srini, "An Architectural Comparison of Dataflow Systems," Computer, Vol. 18, No. 3., Mar. 1986, p. 68.
11. W.W. Carlson and K. Hwang, "Algorithmic Performance of Dataflow Multiprocessors," Computer, No. 18, Vol. 12, Dec. 1985, p. 30.
12. J.L. Gaudiot, "Structure Handling in Data Flow Systems," Trans. Comp., Vol c-35, No. 6, June 1986, p. 489.
13. H.T. Kung, "Why Systolic Architectures," Computer, Vol. 15, No. 1, Jan. 1982.
14. H.T. Kung and C.E. Leiserson, "Systolic Arrays (for VLSI)," in Proc. Sparse Matrix Symp. (SIAM), 1978, p. 256.
15. S.Y. Kung, K.S. Arun, R.J. Gal-Ezer, and D.V. Bhaskar Rao, "Wavefront Array Processor: Language, Architecture and Applications," Trans. Comp., Vol. C31, No. 11, Nov. 1982, p. 1054.
16. S.Y. Kung, "On Supercomputing with Systolic/Wavefront Array Processors," Proc. of the IEEE, Vol. 72, No. 7, July 1984.
17. G.J. Li and B.W. Wah, "The Design of Optimal Systolic Arrays," Trans. Comp., Vol. c34, No. 1, Jan. 1985, p. 66.
18. T. Leighton and C.E. Leiserson, "Wafer-Scale Integration of Systolic Arrays," Trans. Comp., Vol. c34, No. 5, May 1985, p. 448.
19. D.I. Moldovan and J.A.B. Fortes, "Partitioning and Mapping Algorithms into Fixed Size Systolic Arrays," Trans. Comp., Vol. c35, No. 1, Jan. 1986, p. 1.
20. (a) J.W. Goodman, Proceedings of the 1985 International Conference on Parallel Processing, IEEE Catalog No. 85CH2140-2, pp. 282, 334.
    (b) A. Huang, same as above, p. 290.
21. (a) D.O. Cowan and F.M. Wiggin, "The organic solid state," in Chemical and Engineering News, July 21, 1986 issue, p. 27.

(b) P. Chandhari, "Electronic and Magnetic Materials," in Scientific American, October, 1986 issue, p. 136.

(c) J.M. Rowell, "Photonic Materials," in Scientific American, October, 1986 issue, p. 146.

22. (a) G. Del Re, J. Ladik, and G. Biczo, Phys. Rev., **155**, 997 (1967).

(b) J.M. Andre, L. Gouverneur, and G. Leroy, Int. J. Quant. Chem., **1**, 427 and 451 (1967).

23. J.L. Bredas and G.B. Street, Accts. of Chem. Res., **18**, 309 (1985).

24. A. Aviram and M.A. Ratner, Chem. Phys. Lett., **29**, 277 (1974).

25. A. Aviram, P.E. Seiden, and M.A. Ratner, in "Molecular Electronic Devices," F.L. Carter, ed., Marcel Dekker, p. 5 (1982).

Physicists Explore Human and Artificial Intelligence

J. Buhmann, R. Divko, H. Ritter and K. Schulten*
Physik-Department
Technische Universität München
D-8046 Garching

1. Historical Sketch of Brain Theory

The foundation of modern brain theory[1] is based on the epochal work of the physiologist Sherrington and the anatomist Cajal at the beginning of the twentieth century. Both established the modern view of neural networks as heterogeneous systems composed of single subunits, the neurons. They rejected the theory of Golgi and others that the brain is a continuous net of axons and neurons. Sherrington investigated the electrical firing of neurons and introduced the terminus "synapse" for the connection between the individual neurons. These ideas which drove away the animal ghosts of the continuum theory have been spectacularly confirmed half a century later by electron microscopy photographs of neurons and synapses.

In the forties of this century two mathematicians, McCulloch and Pitts, formulated a mathematical theory which allowed us to describe the behavior of neurons as boolean units. In this theory the complex dynamics of neurons is modeled by a logical element which switches to the "on"-state if enough afferent spikes excite the neuron, and, otherwise, rests in the "off"-state. The afferent excitation is summed up linearly and compared with a threshold value to determine the neural state at the next time step. The theory of McCulloch and Pitts is based on the conviction that information is transmitted by action potentials between the neurons. Only a firing neuron can communicate with other nerve cells. The most prominent principles of the hypothesis that neurons can be described by logical units hold also in the light of modern neurobiology. These fundamental ideas have survived until today and have entered in nearly all theories of neural networks.

Other important biological findings about nerve nets refined the picture of neural networks. Dale stated in the thirties that neurons can produce only one type of neurotransmitter and, therefore, can either inhibit or excite connected neurons, not both. At the same time the unidirectional natures of synapses as connections with a prominent direction was discovered. Twenty years later detailed electron micros-

---

*   Institute of Theoretical Physics, University of California, Santa Barbara, CA 93106.

copy photographs proved these findings and revealed the structure of synapses with the synaptic cleft, the presynaptic axon terminals and the postsynaptic dendritic membrane. The presynaptic membrane contains vesicles with neurotransmitters which are ejected into the synaptic cleft and change the electrical properties of the ionic channels in the postsynaptic membrane. Thereby, a postsynaptic electrical potential is induced.

The physiologist Hebb outlined a framework of how the various neurons can act together. He introduced the notion of "neural assembly" as a group of cooperating and densely connected neurons. The neurons of such an assembly should excite each other and should strengthen their mutual synapses under the influence of their activity states. Coincidences of pre- and postsynaptic spikes should be the condition for synaptic growth. Hebb was the first scientist who looked for serious concepts to understand the dynamics of neural networks and who tried to connect the neurons and their dynamics with the behavior of higher vertebrates or men. Hebb's coincidence hypothesis has entered in many modern theories of learning and in our days becomes confirmed by physiological discoveries on the molecular structure of synapses and on the mechanisms of their plasticity.

During the ten years from 1955 to 1965, often characterized as the "golden decade of cybernetics," all these ideas were introduced in various models of neural networks, a prominent one of these proposed by Caianello. He evolved his theory of "Thought Processes and Thinking Machines" on the basis of McCulloch's, Pitt's and Hebb's ideas. At the same time Rosenblatt constructed a layered network model with simple connectivity between the layers to recognize and associate patterns. His "Perceptron," however, showed remarkable limitations later proven by Minsky and Papert. The failure of the "Perceptron" and other problems disappointed the hope that thinking and intelligent behavior of humans could be explained with the help of fast computers and on the basis of cooperating neural assemblies in a few years. The area of "Artificial Intelligence" originally located within theoretical biology separated from the neuronal basis and pursued a more abstract, algorithmic approach.

New impulses, originating from thermodynamics far from equilibrium, from non-linear optics and from non-linear mechanical systems, have initiated a renaissance of brain theory in the late seventies. The experience with simple hydrodynamical systems which show complex formation of structure and internal patterns has suggested that the principles of self-organization, cooperation and competition between units with non-linear dynamics could also be essential features in the neural development and information processing of the brain. This more dynamical view has solved the problems of how the information of the neural wiring is genetically stored and how the local variability of the brain structure and fault tolerance of the brain function can be explained. The formation of neural projections between neural nets, i.e. the retinotopic projection[2] and the building of memories with associative and fault-tolerant properties[3,4] were simulated by non-linear dynamics which involves the electrical activity of neurons as well as changes of the synaptic connectivity.

Hopfield[5] found a concise description of an associative memory, in some respects the "harmonic oscillator" of brain theory. His system of spin-like neurons with dense connectivity obeys a Hamiltonian dynamics with dissipation, i.e. the network settles down in the next local minimum found in the phase space. The structure of the phase space reveals many minima and is equivalent to that of a spin glass, a current research topic of condensed matter physics. The results from spin glass physics have influenced brain theory and initiated new investigations. The emerging technology of massively parallel computers with thousands of processors also stimulates the research in this area.

In our own work we have followed both avenues in the history of brain theory, the avenue of modelling the brain in close agreement to physiological principles, and the avenue of experimenting liberally with digital algorithms which are crude caricatures of the way the brain processes information. In Section 2 of this paper we present results of computer simulations of neural assemblies made up of neuronal units modelled in close analogy to their physiological counterparents. The main new result of our work is that electrical noise apparent in 11 physiological recordings of neural tissue appears to play an important role in higher brain function. In Sections 3 and 4 we investigate algorithms which reproduce brain structure and function, albeit in a way which is hardly reminiscent of the biological system. The algorithms in Section 3 are concerned with the self-organization of information representation (mapping) and the control of motor tasks in the brain and in robots. The algorithms in Section 4 address the problem of optical pattern recognition. The results presented in Sections 2, 3, and 4 have been published previously in the proceedings of the conference "Neural Networks for Computing."[6]

2. Autoassociative Neural Network with "Physiological" Neurons

Recently, we simulated the activity and function of neural networks with neuronal units modelled after their physiological counterparents.[7] Neuronal potentials, single neural spikes and their effect on postsynaptic neurons were taken into account. The neural network studied was endowed with plastic synapses. The synaptic modifications were assumed to follow Hebbian rules, i.e. the synaptic strengths increase if the pre- and postsynaptic cells fire a spike synchronously and decrease if there exists no synchronicity between pre- and postsynaptic spikes. The time scale of the synaptic plasticity was that of mental processes, i.e. a tenth of a second, as proposed by von der Malsburg.[8] In this Section we present the model network with deterministic dynamics and we extend our previous study and include random fluctuations of the neural potentials. Such fluctuations can always be observed in electrophysiological recordings.[9] We will demonstrate that random fluctuations of the membrane potentials raise the sensitivity and performance of the neural network. The fluctuations enable the network to react to weak external stimuli which do not affect networks following deterministic dynamics. We argue that fluctuations and noise in the membrane potential are of functional importance in that they trigger the neural firing if a weak receptor input is presented. The noise regulates the level of arousal. It might be an essential feature of the information processing abilities of neuronal networks and not a mere source of disturbance to be suppressed. We will demonstrate that the neural network investigated here reproduces the computational abilities of formal associative networks.[2-4]

The neural system investigated is composed of a set of interconnected neurons, the membrane potentials of which evolve according to deterministic rules and according to stochastic fluctuations. The connections to sensory organs or to other neural networks are taken into account by a primary set of receptors which send input to the neurons.   The receptor-neuron connections form a local, static projection of the activity pattern presented by the receptors as modelled by a one-to-one or a center-surround connectivity. The system is schematically presented in Fig. 1.

2.1 Dynamics of the Membrane Potential

The dynamics of the membrane potentials involves two processes, the relaxation of the membrane potential and the neural interaction as determined by the somatic integration rule. Axonal spikes are generated whenever the membrane potential reaches a threshold value. The postsynaptic excitation by presynaptic spikes is described by an exponential activity function with decay time $T_U = 1ms$

$$G_k(\Delta t_k/\tau) = \exp\left(-\frac{\Delta t_k}{\tau}\right) \tag{2.1}$$

$\Delta t_k = t - t_{0k}$ measures the time that has elapsed since the last spike of neuron k at t_{0k}.

The kinetic equations of the membrane potentials $U_i(t)$ which also include the stochastic fluctuations are given by a system of non-linear coupled Langevin equations

$$\frac{dU_i(t)}{dt} = -\frac{U_i(t)}{T_R} + \rho[\Delta t_i]\left(\omega\sigma[A_i(t)] + \frac{\eta}{\sqrt{T_R/2}}\xi(t)\right). \tag{2.2}$$

The first term in Eq. 2.2 approximates the relaxation of the membrane potential $U_i(t)$ to its resting value $U_0 = 0mV$ within a time interval $T_R = 2.5ms$. The second term in Eq. 2.2 describes the communication of the postsynaptic cell i with the connected neurons and receptors, and adds a Gaussian white noise $\xi(t)$ with the strength $\eta/\sqrt{T_R/2}$. The noise produces a Gaussian distribution of the membrane potential $U_i(t)$ with mean value $U_0 = 0mV$ and variance $\eta = 10mV$. Afferent impinging activities in addition to the noise are integrated to the total postsynaptic excitation $A_i(t)$. The activity of the presynaptic neurons k or receptors j are weighted by the time-dependent synaptic strengths $S_{ik}(t)$ or the static receptor connection strengths R_{ik}, respectively,

$$A_i(t) = \sum_k S_{ik}(t)G_k(\Delta t_k/T_U) + \sum_k R_{ik}G_k^R(\Delta t_k^R/T_U). \tag{2.3}$$

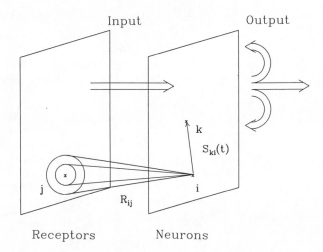

Figure 1. Schematic presentation of the neural model investigated: Receptors send spikes to a network of neurons. The resulting activity of the neural network is affected by an activity-dependent alteration of the synapses $S_{ik}(t)$, i.e. the network experiences a feedback as indicated.

The sigmoidal function $\sigma[A_i(t)]$ with a linear behavior for small $A_i(t)$ and a saturation value for strong activity prevents potential changes which are unphysiologically large. The total and relative refractory periods are taken into account by the function $\rho[\Delta t_i]$ which suppresses the sensitivity of neuron i to afferent excitation during a total refractory period $T_F = 5ms$. The function also lets the neuron gradually regain its sensitivity to incoming excitation or inhibition during a relative refractory period of $5ms$.

The continuous time evolution of the potential in our model is interrupted if the neuron reaches the threshold $U_T = 30mV$ and fires a spike. Instantaneously the membrane potential is set to a value normally distributed around the refractory potential $U_F = -15mV$. In this event the time of the last spike t_{0i} is updated and the memory function $G_i(\Delta t_i/T_U)$ is set to the value 1. This behavior is represented as follows:

$$\text{if } U_i(t) \geq U_T, \quad \text{then } \begin{cases} t_{0i} = t, \\ U_i(t) \approx U_F, \\ G_i(\Delta t_i/\tau) = 1. \end{cases} \tag{2.4}$$

The reaction of a neuron to a receptor input depends on the coupling constant ω and the connection strength R_{ik}. In the case of strong coupling the excited neuron will always reach the threshold whereas weak coupling causes only small postsynaptic potentials which never reach the threshold. Figure 2 shows the proba-

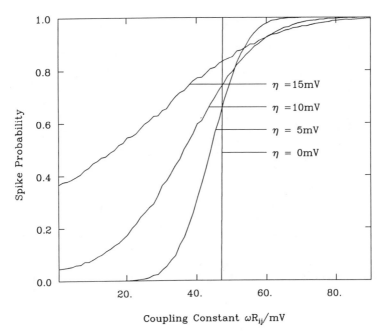

Figure 2. The probability to reach the threshold within 5*ms* after a receptor spike depends on the coupling between receptors and neurons. The gain of the curve strongly depends on the noise level η. In our computer simulations we have employed in most cases the value $\eta = 10mV$ and $\omega R_{ik} = 45mV$.

bility that a neuron which received a receptor spike at $t = 0ms$ will fire within 5*ms*. This probability is presented as a function of the coupling strength ωR_{ik} for three different noise levels ($\eta = 0,6,10mV$). Due to the synapse dynamics the mean spike probability of the neuron $\omega \overline{A_i}(t)$ is time-dependent and can be shifted by learning.

2.2 Synaptic Plasticity in the Stochastic Neural Network

In our neural network with stochastic firing we introduced a plasticity of the synapses on a time scale of 0.2–0.5s.[2] According to the Hebbian rules the synaptic dynamics were assumed to depend on the synchronicity or asynchronicity of the pre- and postsynaptic spikes. In addition to the Hebbian rules we require for synaptic modifications in the present study that the mean spike frequencies $\overline{\nu_i}$, $\overline{\nu_k}$ of both neurons exceed considerably the spontaneous spike rate $\nu_s \approx 5s^{-1}$. If both neurons satisfy this condition in the case of synchronous firing the synapse can be strengthened. If only the presynaptic neuron fires with a high spike rate the synapse $S_{ik}(t)$ is weakened after each presynaptic spike. Details are described in Ref. 2.

The plasticity of the synapse with the strength $S_{ik}(t)$ connecting neuron k to neuron i is governed by the equation

$$\frac{dS_{ik}}{dt} = \begin{cases} -\dfrac{S_{ik}(t)-S_{ik}(0)}{T_S} + \Omega G_k\!\left(\dfrac{\Delta t_k}{T_M}\right)\kappa(G_i, G_k), & \text{if } S_u \geq |S_{ik}| \geq S_l; \\[2ex] -\dfrac{S_{ik}(t)-S_{ik}(0)}{T_S}, & \text{else} \end{cases} \tag{2.5}$$

with

$$\kappa(G_i, G_k) = \begin{cases} 1, & \text{if } G_i > G_k > e^{-1} \ \wedge \ \bar{v}_i \gg v_s \ \wedge \ \bar{v}_k \gg v_s; \\ -1, & \text{if } G_k > e^{-1} > G_i \ \wedge \ \bar{v}_i \ll v_s \ \wedge \ \bar{v}_k \gg v_s; \\ 0, & \text{else.} \end{cases} \tag{2.6}$$

Equation 2.5(a) holds both for excitatory and inhibitory synapses. The first term describes a relaxation process which leads to the gradual loss of stored information. The second term effects a change of the synaptic strength. The influence of this term decays exponentially with the presynaptic activity $G_k(\Delta t_k/T_M)$. The short decay time $T_M = 2.5ms$ guarantees the Hebbian synchronicity condition for synaptic changes. The function $\kappa(G_i, G_k)$ switches between increase of the synaptic strength ($\kappa = 1$), decrease ($\kappa = -1$) and passive relaxation ($\kappa = 0$) of the synapses to the initial value $S_{ik}(0)$. The characteristic time Ω^{-1} determines the time scale for synaptic modifications. The values assumed for Ω^{-1} were in the range $0.2-0.5s$.

2.3 Learning and Association of a Pattern

The neural network presented showed remarkable associative properties in spite of the stochastic fluctuations of the membrane potentials. Starting from a homogeneous structure of synaptic connections with equal numbers of excitatory and inhibitory neurons the network learned a pattern presented by the receptors and associatively reconstructed the original pattern when only incomplete or disturbed patterns were presented.

The simulations of the network were carried out in three different stages. During a first stage which lasted $0.3-1.5s$ the neural network had to learn the pattern **brain**, synchronously presented by the receptors with a frequency of $50s^{-1}$. A homogeneous background noise with a spike rate of $10s^{-1}$ was superimposed on the pattern. The coupling constant ωR_{ii} was set to $45mV$ which effected the firing of about 75 percent of excited neurons. In a second stage lasting $50ms$ the receptors rested quiescent and the electrical activity of the network relaxed to the spontaneous spike rate. During a third stage the receptors presented the test pattern **bra n** which differed from the originally learned pattern by the letter i being left out.

Figure 3a shows the activity of the network at the beginning of the learning phase. At $t = 180ms$ the receptors corresponding to the pattern **brain** had just fired. Within $3ms$, 75 percent of the excited neurons reach the threshold and fire. The other neurons are only gradually excited and fail to fire. The network reaction to a

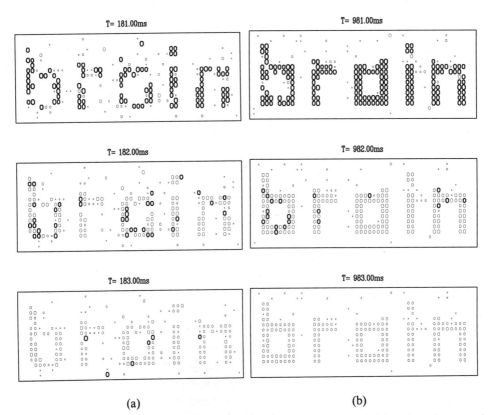

Figure 3. The activity function $G_i(\Delta t_i/T_M)$ is shown for an untrained (a) and an instructed (b) network. At the beginning of the learning session (t = 181ms, 182ms, 183ms) 75 percent of the excited neurons fire after a receptor spike. At the end of the learning stage (t = 981ms, 982ms, 983ms) nearly all excited neurons have synchronized their firing behavior and reach the threshold.

receptor input at the end of the learning stage is shown in Fig. 3b. Due to the acquired excitatory synaptic connections between neurons receiving input directly from the pattern **brain** (pattern neurons) the assembly reacts more synchronously and the fault level, given by the number of pattern neurons which fail to fire, nearly vanishes.

The success of the learning session is documented in Fig. 4. The incomplete test pattern **bra n** is associatively restored by the network. The neurons representing the missing letter **i** react with a delay time of $1-3ms$, i.e. they fire nearly synchronously with the neurons excited by the test pattern.

The synchronization of the neural activity and the associative abilities of the network can be understood on account of the synaptic structure acquired during the learning session. Figures 5a and 5b show the afferent synapses of neuron (37,4) [presented by a star] after the training. All the neurons representing the pattern **brain** have developed saturated excitatory or inhibitory synapses to the reference neuron. During the association task the excitatory synapses saturated at a strength

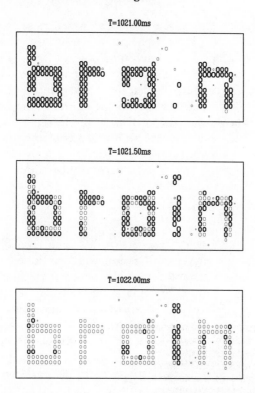

Figure 4. Network activity during the association task: The network associates the missing letter **i** by excitatory interaction within 2 milliseconds (t = 1021ms, 1021.5ms, 1022ms).

value S_u support the firing of the reference cell, whereas the inhibitory synapses saturated at $-S_l$ do not prevent the reference cell from firing. Afferent synapses of the reference cell (37,4) coming from a background neuron rest at the initial synaptic strength.

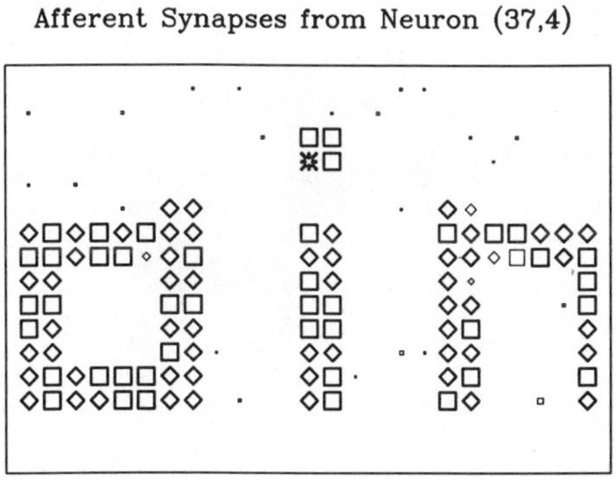

Afferent Synapses from Neuron (37,4)

Figure 5. The size of the squares and the diamonds encodes the changes $S_{ik}(t)-S_{ik}(0)$ which the excitatory and inhibitory synapses acquired during the learning session, respectively.

Due to fluctuations of the membrane potential which raise the sensitivity of the neurons the network can also learn a pattern which at any given time is only partially presented by the receptors. At each time interval the invisible fraction of the pattern (50 percent of the receptors) is chosen randomly. The uninstructed network has to learn the total pattern from the detected spike coincidences. The evolution of the synapses is demonstrated for the case of the afferent synapses of neuron (37,4) which represents the dot on the letter *i*. During the learning stage which lasts 3.7*s* the network has built up a synaptic structure which contains the information of the whole pattern (Fig. 6). This simulation demonstrates that the synchronization of all pattern receptors at any given time is not a necessary condition for learning.

2.4 Conclusion

We have presented a model neural network with a high level of endogenous noise acting on the cellular potentials. This noise, which is inherent in all biological neurons, does not destroy the abilities of the network to learn and associatively reconstruct patterns. On the contrary, the noise controls the level of arousal and makes the network capable to react to a weak receptor input otherwise neglected. We argue that noise has a functional importance in neural systems. The explicit simulation of single spikes allows us to test the influence of single neural events which are averaged over by mean spike rate models.[4] In addition, the nonspecific influence of large neural nets (neural activity bath) on small neural assemblies can also be studied by stochastic dynamics.

On the basis of the Hebbian rules which detect synchronicities between pre- and postsynaptic spikes, a second condition for synaptic changes is introduced to protect the synaptic structure against destruction by spontaneous activity. The mean spike rates $\bar{v}$ of the pre- and postsynaptic neurons have to exceed considerably the spontaneous spike rate v_s for an increase of the synaptic strengths. For a decrease of the synaptic strengths the postsynaptic spike rate must be considerably below v_s. With this modified rule the network can also learn highly noisy patterns and patterns which are presented by a partially asynchronous receptor activity.

Afferent Synapses from Neuron (37,4)   Afferent Synapses from Neuron (37,4)

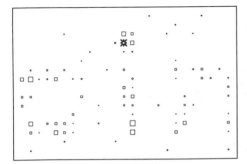

Figure 6. Evolution of the afferent synapses of neuron (37,4) for the time $t = 1s$ (a) and $t = 3.6s$ (b) during the learning stage. 50 percent of the pattern **brain** is invisible.

3. Topology Conserving Mappings in the Brain

It has been known for a long time that the brain has a modular structure. This is especially conspicuous in the neocortex, which is the brain structure most predominant in man, and which forms essentially a folded, large, two-dimensional neural sheet. Different cognitive abilities, such as touch, audition, vision, and motor capabilities reside in precisely circumscribed regions of this sheet, the so-called brain areas. These areas can be subdivided further, corresponding to different subtasks of each modality. Each area consists of a two-dimensional arrangement of groups of neuronal cells, called cortical columns, with each group devoted to the processing of a small part of the overall information impinging from input pathways upon its area. For the processing of information in the brain, neural activity is mapped between external sensory regions and cortical areas, or between two cortical areas. It is a very important organizational feature of the brain that these mappings are continuous, i.e. that neighboring neuronal groups are connected to neighboring cortical sites of the cortical or sensory regions to which they are mapped. As a consequence, neuronal wiring in the brain preserves the relation of neighborhood, i.e. the neuronal wiring establishes mappings which are topology conserving.

Due to the huge amount of neurons to be connected, the cortical "wiring diagram" cannot be prespecified in detail genetically but must self-organize during the ontogeny and maturation of the brain. In fact, it has been observed that the evolution of such mappings is shaped by sensory experiences. This has been revealed, for example, by neurophysiological experiments which show that the somatosensory map is plastic even through adulthood and can adapt to a changing sensory environment, e.g. brought about by changing "calibration" of the sensors or sensory injury and loss.[10-12]

To understand the principles inherent in the evolution of such mappings, we have studied the formation of the connections between the touch receptors distributed over the skin of a hand and the somatosensory cortex, using a mathematical model originally put forward by Kohonen.[10,11] In this model, the somatosensory cortex is represented as an array of vectors $u(x)$, each vector $u(x)$ corresponding to a neuronal group situated at location x in the cortex. The value of $u(x)$ specifies the position of the receptive field of the group, which is the region of the skin the group is connected to. Initially the connections are completely unordered, i.e. each group is connected to a sensor at a random position in the skin. Each touch stimulus, delivered to the hand at a location y, is assumed to create a local region of excitation in the neuronal sheet, which is centered around that particular neuronal group whose receptive field lies closest to the location of the stimulus, i.e. whose vector $u(x)$ matches y best. All neuronal groups in this activated region are now supposed to readjust their connectivity to get their receptive fields closer to the location y of the stimulus, which is mathematically realized by shifting their vectors $u(x)$ towards the stimulus location y by an amount decreasing with increasing distance from the center of the excited region. It turns out that the cumulative effect of such local adjustments, caused by a sufficiently long lasting train of touch stimuli, is capable of establishing a completely ordered mapping between the

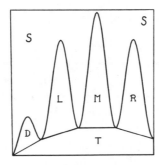

Figure 7. Model of the hand surface used in the simulation.

neuronal groups and the touch receptors. This is illustrated in Figs. 7 through 12 in a simulation for the case of the mapping between the surface of a (model-) hand (shown in Fig. 7) and an array of 30×30 vectors, representing a cortical region of 900 neuronal groups. In Fig. 8 we see the initial state of the connectivity, depicted in two different ways. In the upper diagram the array of neuronal groups is projected onto the hand surface via the vectors $u(x)$: each array element x is shown

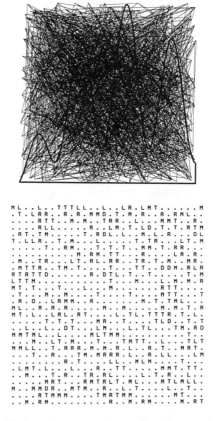

Figure 8. Initial state of the connectivity between hand surface and cortical array.

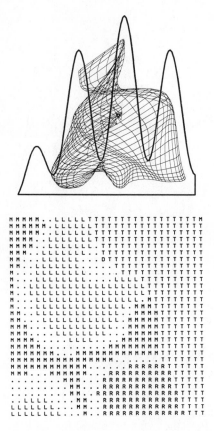

```
M M M M M . . L L L L L T T T T T T T T T T T T T T T T T T M
M M M M M M . L L L L L L T T T T T T T T T T T T T T T T T T
M M M M M . . L L L L L L L T T T T T T T T T T T T T T T T T
M M M M . . L L L L L L L T T T T T T T T T T T T T T T T T T
M M M M . . L L L L L L L . T T T T T T T T T T T T T T T T T
M M M . . . L L L L L L L . . T T T T T T T T T T T T T T T T
M M . . . . L L L L L L . . . . D T T T T T T T T T T T T T T
M M . . L L L L L L L . . . . . . T T T T T T T T T T T T T T
M . . . . L L L L L L L L . . . . . T T T T T T T T T T T T T
M . . . . L L L L L L L L L . . . . L L L L T T T T T T T T T
M . . L L L L L L L L L L L L L L L L L T T T T T T T T T T T
M . . L L L L L L L L L L L L L L L L L T T T T T T T T T T T
M . . . L L L L L L L L L L L L L L L . M T T T T T T T T T T
M . . . L L L L L L L L L L L L L L . M M M T T T T T T T T T
M M . . . L L L L L L L L L L L L L . . M M M M T T T T T T T
M M . . . L L L L L L L L L L L L L . M M M M M T T T T T T T
M M M . . . . L L L L L L L L L . . . M M M M M T T T T T T T
M M M M . . . . . L L L L . . . . M M M M M M T T T T T T T T
M M M M M M M . . . . . . . M M M M M M M M T T T T T T T T
M M M M M M M M M M M M M M . . . . . . . T T T T T T T T T
M M M M M M M M M M M . . . . . R R R R R R T T T T T T T
M M M . . . M M M M M M . . R R R R R R R R R T T T T T T
. . . . . . . . . M M M . . . R R R R R R R R R R T T T T T
. . . . . . . . M M . . R R R R R R R R R R R T T T T T
. . L L L L . . . . M M . . R R R R R R R R R R R R T T T T
L L L L L L L . . . M M . . R R R R R R R R R R R T T T T
L L L L L L L L . . . M . . R R R R R R R R R R R R T T T
```

Figure 9. Connectivity between hand surface and cortical array after 500 adaptation steps due to model stimuli.

as a point at location $u(x)$ in the hand surface and the projected positions of elements which are nearest neighbors in the array are connected by lines. The lower diagram shows the array itself. For each element a letter indicates the position of the skin receptor it is linked to, with the letters D,L,M,R,T referring to the regions given in Fig. 7. Dots indicate elements so far unconnected to receptors in the hand (for mathematical convenience these elements are given values of the vector u lying in the region S not belonging to the hand). Both diagrams show that the initial connectivity is completely unordered. The following pictures show the gradual evolution of an ordered mapping under a train of touch stimuli scattered randomly over the hand surface. Figure 9 shows partial order already after 500 touch stimuli, whereas in Fig. 10, after 20,000 touch stimuli, a very regular connectivity has evolved. If finger M is amputated, i.e. in the continuation of the algorithm, no further touch stimuli originate from region M (see Fig. 11), the connections between the cortex and the receptors are redistributed such that the neuronal groups which were connected to the removed region M finally get devoted to the adjacent areas L, R and T (Fig. 12). This behavior is in good qualitative agreement with physiological experiments, e.g. carried out in the brains of monkeys.[13-15]

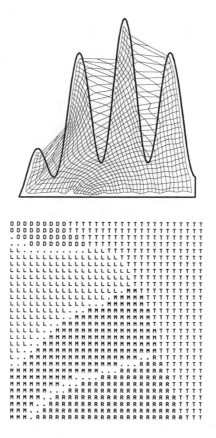

```
D D D D D D D D D T T T T T T T T T T T T T T T T T T T T T
D D D D D D D D D T T T T T T T T T T T T T T T T T T T T T
. D D D D D D D D D T T T T T T T T T T T T T T T T T T T T
. . . D D D D D D D D D T T T T T T T T T T T T T T T T T T T
L L . . . . . . . . . . L L L T T T T T T T T T T T T T T T T
L L L L L L L L L L L L L L L L T T T T T T T T T T T T T T T
L L L L L L L L L L L L L L L L L L T T T T T T T T T T T T T
L L L L L L L L L L L L L L L L L L L T T T T T T T T T T T T
L L L L L L L L L L L L L L L L L L L T T T T T T T T T T T T
L L L L L L L L L L L L L L L L L . M M T T T T T T T T T T T
L L L L L L L L L L L L L L L . M M M M T T T T T T T T T T T
L L L L L L L L L L L . . . M M M M M M T T T T T T T T T T T
L L L L L L L L L . . . M M M M M M M M T T T T T T T T T T T
L L L L L L L . . M M M M M M M M M M M T T T T T T T T T T T
L L L L L L . . M M M M M M M M M M M M M T T T T T T T T T T
L L L L L . . M M M M M M M M M M M M M M T T T T T T T T T T
L L L L . . M M M M M M M M M M M M M M M T T T T T T T T T T
L L L . . M M M M M M M M M M M M M M M M T T T T T T T T T T
L L . . M M M M M M M M M M M M M M M M . . R T T T T T T T T
L . . . M M M M M M M M M M M M M . . R R R R T T T T T T T T
. . M M M M M M M M M M M M M . . . R R R R T T T T T T T T
M M M M M M M M M M M M M . . . R R R R R T T T T T T T T
M M M M M M M M M . . . . R R R R R R R R R T T T T T
M M M M M M M . . . R R R R R R R R R R R R T T T T T
M M M M M M . . . R R R R R R R R R R R R R T T T T T
M M M M M . . R R R R R R R R R R R R R R R R T T T T
M M M M . . R R R R R R R R R R R R R R R R R T T T T
M M M . . R R R R R R R R R R R R R R R R R R T T T T
M M M . R R R R R R R R R R R R R R R R R R R T T T
```

Figure 10. Ordered connectivity between hand surface and cortical array after 20,000 adaptation steps.

There is neurophysiological evidence that the role of such mappings need not be restricted to the mediation between sensory input and a cortical target field, but that they can be involved as well in the generation of output. The presence of topology conserving mappings of variables concerned with motor responses, e.g. the mapping of the next movement vector for a saccade of the eyes to a location of increased activity in the superior colliculus,[15,16] suggests that topology conserving mappings may also subserve the acquisition and execution of motor tasks. In learning a new motor capability, initially the movement has to be generated fully consciously and step-by-step until it becomes increasingly automatic with further practice. This process of learning could be brought about by an internal teacher, who teaches, by a series of consciously generated movement instances, a topology conserving mapping the correct input-output relation for the task to be performed. As a consequence of the gradual formation of the map, the brain can take over the part of the teacher to an ever larger extent, rendering the execution of the task more and more automatic until, finally, the internal teacher becomes completely dispensable and thus free to dedicate itself to other tasks. We studied this process by the model task of learning how to balance a pole against gravity.[17] The motion of the pole was simulated in discrete time steps and monitored by an array of

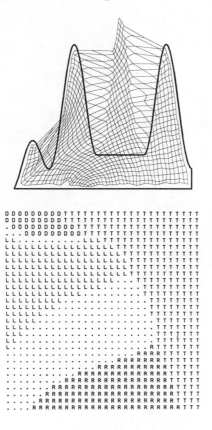

```
D D D D D D D D D T T T T T T T T T T T T T T T T T T T T
D D D D D D D D D T T T T T T T T T T T T T T T T T T T T
. D D D D D D D D D D T T T T T T T T T T T T T T T T T T
. . . D D D D D D D D D D T T T T T T T T T T T T T T T T
L L . . . . . . . . . . . L L L T T T T T T T T T T T T T T
L L L L L L L L L L L L L L L L L T T T T T T T T T T T T T
L L L L L L L L L L L L L L L L L L L T T T T T T T T T T T
L L L L L L L L L L L L L L L L L L L L T T T T T T T T T T
L L L L L L L L L L L L L L L L L L L L T T T T T T T T T T
L L L L L L L L L L L L L L L L L L L L T T T T T T T T T T
L L L L L L L L L L L L L L L L L L . . . T T T T T T T T T
L L L L L L L L L L L L L L L L . . . . . . T T T T T T T T
L L L L L L L L L L L L L L . . . . . . . . . T T T T T T T
L L L L L L L L L . . . . . . . . . . . . . . . T T T T T T
L L L L L L . . . . . . . . . . . . . . . . . . T T T T T T
L L L L L . . . . . . . . . . . . . . . . . . . T T T T T T
L L L L . . . . . . . . . . . . . . . . . . . . T T T T T T
L L L . . . . . . . . . . . . . . . . . . . . T T T T T T T
L L . . . . . . . . . . . . . . . . . . . . . T T T T T T T
L . . . . . . . . . . . . . . . . . . . . . R T T T T T T T
. . . . . . . . . . . . . . . . . . . . . R R R R T T T T T
. . . . . . . . . . . . . . . . . . . . R R R R R R T T T T
. . . . . . . . . . . . . . . . . . R R R R R R R R R T T T T
. . . . . . . . . . . . . . . . . R R R R R R R R R R R T T T T
. . . . . . . . . . . R R R R R R R R R R R R R R R T T T T
. . . . . . . . R R R R R R R R R R R R R R R R R T T T T
. . . . . . R R R R R R R R R R R R R R R R R R R T T T T
. . . . . R R R R R R R R R R R R R R R R R R R R T T T T
. . . . R R R R R R R R R R R R R R R R R R R R R T T T
```

Figure 11. State of Fig. 4 after removal of region M, i.e. the middle finger.

vectors $u(x)$. Initially a teacher generated a sequence of balancing forces capable of balancing the pole. As in the previous example, each vector codes for a certain state of the pole, here represented by two successive pole inclinations, separated by one time step of the simulation. In addition to this "sensory part," a third component codes for an "action": here the action specifies the value of a force, which is to be applied over one time step, whenever the state of the pole matches the "sensory part" of the vector. The "sensory input" to the array is now generated by the motion of the pole and consists of a sequence of pairs of successive pole inclinations. At each time step the unit x^*, whose "sensory part" matches the state of the pole most closely, is taken to be the center of the region of local adjustment of the values $u(x)$ in the array. Such an adjustment consists of refining the "sensory parts" of the vectors of all units lying within a small neighborhood of unit x^*. These units are matched more closely to the latest pair of successive pole inclinations and at the same time their "action part" is altered towards the force value supplied by the teacher. In the course of time the balancing force delivered by the teacher at each time step is replaced gradually by the output evoked by the "action part" of the array element residing in the center of the region of adjustment. Finally the contribution of the teacher vanishes and the control is exerted by the array alone. Figure 13 shows the gradual improvement of the balancing capability

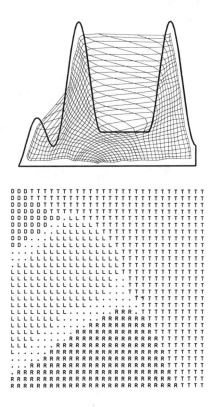

```
D D D T T T T T T T T T T T T T T T T T T T T T T T T T T T T T T T T
D D D T T T T T T T T T T T T T T T T T T T T T T T T T T T T T T T T
D D D D D T T T T T T T T T T T T T T T T T T T T T T T T T T T T T T
D D D D D T T T T T T T T T T T T T T T T T T T T T T T T T T T T T T
D D D D D D D . L L T T T T T T T T T T T T T T T T T T T T T T T T T
D D D D D D . . L L L L T T T T T T T T T T T T T T T T T T T T T T T
D D D D D . . L L L L L L T T T T T T T T T T T T T T T T T T T T T T
D D D . . . L L L L L L L T T T T T T T T T T T T T T T T T T T T T T
D D . . . L L L L L L L L L T T T T T T T T T T T T T T T T T T T T T
. . . . L L L L L L L L L L L T T T T T T T T T T T T T T T T T T T T
. . . L L L L L L L L L L L L L L T T T T T T T T T T T T T T T T T T
. L L L L L L L L L L L L L L L L L T T T T T T T T T T T T T T T T T
L L L L L L L L L L L L L L L L L L L T T T T T T T T T T T T T T T T
L L L L L L L L L L L L L L L L . . T T T T T T T T T T T T T T T T T
L L L L L L L L L L L L L L L . . . T T T T T T T T T T T T T T T T T
L L L L L L L L L L L L L L . . . . . . T T T T T T T T T T T T T T T
L L L L L L L L L L L L L L . . . . . . . . T T T T T T T T T T T T T
L L L L L L L L L L L . . . . . . . . R R R . T T T T T T T T T T T T
L L L L L L L L . . . . . . . R R R R R R R T T T T T T T T T T T T T
L L L L L L L . . . . . . R R R R R R R R R T T T T T T T T T T T T T
L L L L L . . . . . R R R R R R R R R R R T T T T T T T T T T T T T T
L L L . . . . R R R R R R R R R R R R R R T T T T T T T T T T T T T T
L . . . . . R R R R R R R R R R R R R R R R T T T T T T T T T T T T T
. . . . . R R R R R R R R R R R R R R R R R T T T T T T T T T T T T T
. . . R R R R R R R R R R R R R R R R R R R R T T T T T T T T T T T T
. R R R R R R R R R R R R R R R R R R R R R R R T T T T T T T T T T T
R R R R R R R R R R R R R R R R R R R R R R R R T T T T T T T T T T T
R R R R R R R R R R R R R R R R R R R R R R R R T T T T T T T T T T T
```

Figure 12. Readaptation of connectivity after 50,000 further iterations, subsequent to state in Fig. 11.

of the array at three different stages of learning. For both the "sensory" and the "action" parts of all array elements, random entries were taken as starting values. After learning for 100 seconds of simulated time, the array can balance the pole moderately well (bottom diagram of Fig. 13). The performance has improved after 500 seconds (middle diagram) until, after 1000 seconds (low diagram), only very little fluctuation around the unstable vertical pole position remains during a balancing trial.

In the course of the learning process the array has achieved two things simultaneously: First, it has established a mapping between inputs and appropriate responses by having learned the relation between the state of the pole and a necessary force to keep the pole balanced. Second, it has distributed the values of the "sensory parts" of its vectors u such that they populate only that region of the state space of the pole, which is actually visited by the motion.

However, this kind of learning still requires a teacher. Where does the teacher get its instructions from? Even motions such as that of a simple invertible pendulum may require preplanning if there are restrictions upon the available control force. For instance, if the initial state of the pendulum is the stable resting state, it may be desired to turn it into the inverted unstable equilibrium position by applying a

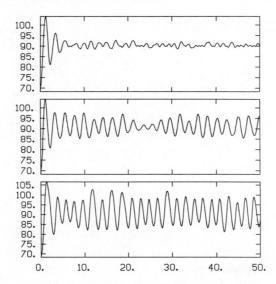

Figure 13. Balancing Capability of Array after different Periods of Learning. Each diagram shows the time evolution of pole inclination θ (vertical) during the first 50 seconds under the control of the array after releasing the pole from a resting position with $\theta = 70°$. *Lower diagram:* after 100 sec of training; *Middle diagram:* after 500 sec of training; *Top diagram:* after 1000 sec of training.

suitable torque at its pivot. In this situation the direct approach of simply turning it up may fail if the admissible torque for the task is too weak. Instead, the pendulum first has to be swung back and forth several times before the weak torque suffices to complete the motion. To plan a motion trajectory for tasks of this kind, we have considered a formulation of the task as a path search problem in the phase space of the system.[18] The solution of this problem can then be achieved using a physical analogy from the realm of diffusion. The computation can be performed in a fully parallel and "neuronal plausible" manner. In order to apply this method, the phase space of the system is discretized into a lattice of nodes, and possible transitions between nodes are represented as directed links. The presence of a link depends on the equation of motion and the available constraints on the control force. The given initial state and the desired final state of the motion can then be represented by a starting node A and an end node B in the lattice. The problem to find a trajectory between these two states is now transformed to the search for a lattice path, the path consisting of a series of oriented links connecting A and B. This latter search problem is solved by considering B as the source of a fictitious substance, which spreads over the lattice by diffusion. The steady state concentration of this substance can be calculated by a simple relaxation algorithm. The path sought is found by starting at A and following the steepest gradient of the concentration until B is reached.

The application of the method to the example mentioned above, i.e. moving a pendulum from the stable to the unstable position, is presented in Figs. 14 through 16. Figure 14 shows the discretized portion of the phase space of the pendulum, with the links indicating transitions possible with a maximum torque of 0.5 (a torque value of 1 corresponds to holding the pendulum fixed in horizontal position). If the

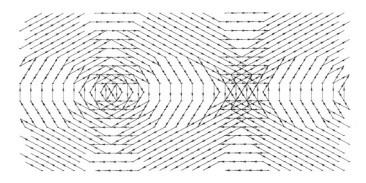

Figure 14. Discretized portion of the phase space for the pendulum described in the text. The angle θ is measured with respect to the vertical axis.

task is to bring the pendulum from the downward resting position into the upward resting state, a torque of 0.5 is not sufficient to achieve this via a direct path. Instead, the pendulum first has to be swung through a few oscillatory cycles to accumulate kinetic energy. The trajectory found by the above method, therefore, first spirals outward from its left starting point until it reaches its destination (Fig. 15). The resulting inclination angle versus time of the pendulum is shown in Fig. 16.

Although, in its present formulation, the method consists of a "neural-like" computation, it is unlikely that this particular algorithm may be realized in this manner in actual brains. Nevertheless it is important to explore ways to formulate tasks commonly solved by our nervous system in a manner amenable to highly parallel, neural-like computation, as only by experimenting within a broad spectrum of different algorithmic alternatives can we hope to finally come closer to the actual working principles of the brain.

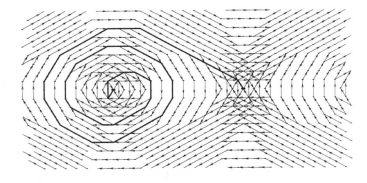

Figure 15. Trajectory found by the algorithm. Starting point A is the left focus of the spiral, corresponding to the pendulum resting downward ($\theta = 180°$, $\dot{\theta} = 0$). Target point B is the right end point of the trajectory, corresponding to the upward position of the pendulum ($\theta = 0°$, $\dot{\theta} = 0$).

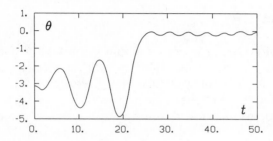

Figure 16. Time evolution of the angle θ of the pendulum corresponding to the simulation of the trajectory in Fig. 15.

4. Stochastic Spin Models for Pattern Recognition

4.1 Introduction

In this Section we exploit, for the recognition of patterns, the properties of physical spin systems to assume long range order and, thereby, to establish a global interpretation of patterns. For this purpose we chose spins which can take a discrete set of values to code for local features of the patterns to be processed (feature spins). A Hamiltonian is chosen for the system which entails a field contribution and interactions between the feature spins. The field incorporates the information on the input pattern. The spin-spin interaction represents *a priori* knowledge on relationships between features, e.g. continuity properties. The Hamiltonian is chosen such that the ground state of the feature spin system corresponds to the best global interpretation of the pattern. The ground state is reached in the course of local stochastic dynamics, this process being simulated by the method of Monte Carlo annealing.[19] Our study is related to work presented in in Refs. 20 and 21.

Spin systems are characterized by a set of values for the spin variable $S_{i,j}$, a lattice on which they are defined, and by an interaction energy. In the two-dimensional Ising model the spins take the values ± 1 and the interaction energy is defined by the Hamiltonian

$$E = -J \sum_{<(i,j),(k,l)>} S_{i,j} S_{k,l} - \sum_{(i,j)} H_{i,j} S_{i,j} \qquad (4.1)$$

where the brackets indicate summation over nearest neighbors. In the ferromagnetic case ($J > 0$) the first term, the exchange interaction, gives a negative contribution if neighboring spins point in the same direction. This term creates a tendency for an alignment of all the spins. The second term describes the interaction of the spins with a local magnetic field $H_{i,j}$ tending to align the spins locally with the field. The regularizing effect of the exchange interaction will be utilized in the following to solve pattern recognition tasks under the constraint that pattern features are expected to vary continuously.

4.2 Feature Spins

For the purpose of picture processing, the spins are chosen to code local features of a pattern. Examples for attributes coded by such feature spins are intensities, disparities between corresponding points in a stereogram or edges of different directions. Several different types of feature spins, interacting with each other and with external fields, may be needed to solve a specific pattern recognition problem.

At finite temperatures the feature spin system shows fluctuations like its physical counterpart. Certain values of a feature spin at a certain lattice point are more probable than others. One may consider the value of a feature spin as the hypothesis that the picture has a certain local attribute at this point.

At high temperatures all hypotheses are equally probable. After carefully cooling down to low temperatures (simulated annealing[19]) the fluctuations eventually disappear. At zero temperature the feature spins take definite values indicating the final global hypothesis about a pattern.

The final hypothesis, i.e. the ground state, achieved by the system after cooling down to low temperatures depends on the interaction among the feature spins as well as on the interaction with the external field. The interaction among the feature spins contains an *a priori* global knowledge on relationships to be expected to hold between the features of a pattern. Correct interpretations of patterns must meet certain constraints, e.g. the constraints of continuity, which have to be realized by the feature spin configurations in the final hypothesis. Such configurations can be achieved by a properly chosen interaction between the feature spins. For example, a Potts model type interaction between intensity spins yields a smooth change of brightness. The external field serves to communicate the pattern to be processed to the system of feature spins. Examples for pattern attributes coded by the external fields are local brightness or edges of various directions.

4.3 Stereo Vision

Whereas a certain degree of depth vision can be obtained from perspective distortion or from hidden parts of a scene, full stereo vision is a result of binocular perception. The projections of an object in both eyes differ slightly from each other. This difference (disparity) allows the reconstruction of the three-dimensional information. Figure 17 shows an image pair of dot patterns appearing completely random when viewed monocularly. But when viewed one through each eye the two pictures fuse showing a three-dimensional structure (square hovering over the ground). The absence of higher level structures in the patterns shows that disparity alone can be used to obtain three-dimensional information from a stereogram.

To obtain stereo information the disparity of corresponding points in the two retina projections of an image must be determined. The problem is to assign correspondences between points of the two pictures. This is a difficult task because of the so-called "false target problem"[22] occurring in its extreme in Julesz patterns.[23]

Figure 17. Julesz pattern[23] with 50% black dots. This random-dot stereogram of 50×50 pixels is generated by copying the right image from the left one, shifting a square-shaped region of 30×30 pixels slightly to the left and filling the gap caused by the shift with a new random pattern.

Every black pixel could correspond to every other black pixel. To restrict all possible combinations of points from both pictures to physically plausible correspondences the following matching conditions must hold:

• Compatibility: Black dots can only match black dots and vice versa.

• Continuity: The physical feature disparity varies smoothly almost everywhere over the image.

• Uniqueness: Except in rare cases each point from one image can match only one point from the other image.

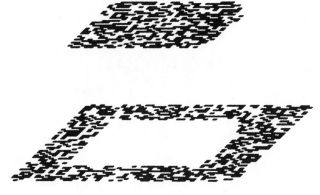

Figure 18. Ground state of the disparity spin system corresponding to the Julesz pattern in Fig. 17.

Figure 19. Julesz pattern of an eight level pyramid.

The following spin model is designed to find the correspondences between the pixels of both pictures of a random-dot stereogram and to measure the disparities. This information will be contained in the ground state of the spin system.

4.4 A Spin Model for Stereo Vision

The feature coded by a spin is the disparity of corresponding pixels. A disparity spin with a value $S_{i,j} \in \{0, \pm 1, \dots, \pm N\}$ at lattice site (i,j) stands for the hypothesis of a correspondence between the pixel (i,j) in the right picture of the stereogram and the pixel $(i, j + S_{i,j})$ in the left picture shifted $S_{i,j}$ units to the right. Both pixels are assumed to correspond to the same original point of an object and to have the disparity $S_{i,j}$.

The Hamiltonian of the disparity spin system is split into two contributions

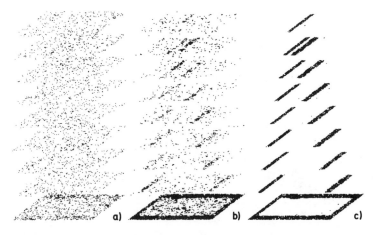

Figure 20. Behavior of the disparity spin system for the Julesz pattern in Fig. 19.

$$E_{total} = E_{exchange} + E_{field}. \tag{4.2}$$

These terms reflect the continuity and the compatibility conditions, respectively. The distance of an observer to a point on the surface of an object is a smoothly varying property. To achieve a corresponding property for the values of the disparity spin, a Potts model interaction is chosen

$$E_{exchange} = -J \sum_{<(i,j),(k,l)>} F(S_{i,j}, S_{k,l}),$$

$$F(S_{i,j}, S_{k,l}) = \begin{cases} 1, & \text{if } S_{i,j} = S_{k,l} \\ q, & \text{if } S_{i,j} = S_{k,l} \pm 1; \ q < 1. \\ 0, & \text{else} \end{cases} \tag{4.3}$$

If the value of a disparity spin is $S_{i,j}$ and if this hypothesis is correct, the pixel (i, j) in the right picture and the pixel $(i, j + S_{i,j})$ in the left picture have identical surroundings. If the disparity hypothesis $S_{i,j}$ is wrong the surroundings may be completely different. Therefore, the comparison of the neighborhoods of the two points assumed to correspond to each other indicates a possibly correct match.

Whereas many features can (and for real pictures must) be used for comparison, we restrict ourselves in the present application to the most simple choice and compare the pixel intensities in a square shaped region only. Comparison is established by the following energy contribution

$$E_{field} = \sum_{(i,j)} G(S_{i,j}),$$

$$G(S_{i,j}) = G_0 \sum_{k=i-w}^{k=i+w} \sum_{l=j-h}^{l=j+h} \left| H_{k,l+S_{i,j}}^{left} - H_{k,l}^{right} \right|. \tag{4.4}$$

Here $H_{i,j}^{left}$ and $H_{i,j}^{right}$ denote the intensity of the pixel (i, j) in the left and in the right picture of the stereogram and $G_0 = [(2h + 1)(2w + 1)]^{-1}$ is a normalization constant. Correct disparity spin configurations are characterized by low energy contributions.

For the input pattern shown in Fig. 17 the disparity field obtained is presented in Fig. 18. Starting from the temperature $T = 1.5$ the annealing process was stopped at a low temperature $T = 0.01$. For the interaction parameters in Eq. 4.3 we have assumed the values $J = 2$ and $q = 0$ and for the maximal disparity the value $N = 5$.

A more complicated stereogram containing the three-dimensional information of an eight level pyramid is shown in Fig. 19. The solutions of the disparity spin system

($N = 10$) with interaction parameters $J = 2$ and $q = 0.2$ are shown for three temperatures in Fig. 20.

At the high temperature $T = 0.8$ the fluctuations of the disparity spin values are large. This is demonstrated by a snapshot of the dynamics shown in Fig. 20a. Figure 20b illustrates that at the intermediate temperature $T = 0.3$ the system still fluctuates; however, the disparity field already indicates the presence of different disparity planes. Figure 20c shows that at the low temperature $T = 0.1$ the fluctuations almost disappeared and that the disparity spin system achieves the correct interpretation of the Julesz pattern, an eight level pyramid.

4.5 A Spin Model for Picture Restoration

Restoration of noisy pictures can be simplified if expected relations between picture attributes are known. As an example, we consider a chessboard-like pattern as input for a picture restoring system. There are several *a priori* qualities present in such a pattern: the intensity in a square is constant, at a square's border are straight edges in a vertical or a horizontal direction, and edges are continuous. A system of feature spins instructed with this knowledge can restore noisy chessboard patterns. Such system entails three kinds of feature spins

- Intensity spins which take the values ± 1 for black and white colors, respectively.

- Horizontal edge spins which take the values $+1$ for intensity changes from white to black, the value -1 from black to white and the value 0 in the case of an absence of any edges.

- Vertical edge spins which follow corresponding rules.

The intensity spins are defined on a square lattice. The edge spins are located between neighboring intensity spins.

The Hamiltonian of the picture restoring spin system can be written

$$E_{total} = \begin{cases} E_{i-ifield} + E_{h-hfield} + E_{v-vfield} \\ + E_{i-i} + E_{h-h} + E_{v-v} + E_{i-h} + E_{i-v} \end{cases} \tag{4.5}$$

where the indices i,h,v refer to intensity, horizontal and vertical edge spins, respectively, with the corresponding fields *ifield, hfield, vfield*. To implement the continuity condition for the intensity spins $I_{i,j} \in \{-1, +1\}$ an Ising-like interaction is assumed. To implement the continuity property of edges, an attractive interaction in the proper direction is employed for the horizontal edge spins $H_{i,j} \in \{-1, 0, +1\}$ as well as for the vertical edge spins $V_{i,j} \in \{-1, 0, +1\}$.

$$E_{i-i} = -J_i \sum_{<(i,j),(k,l)>} I_{i,j} I_{k,l} \tag{4.6}$$

Table I. Compatibility condition table $T(i,j,k)$ where i,j denote pairs of intensity spins and k denotes the edge spin in between.

i	j	k	T
1	1	0	1
−1	−1	0	1
1	−1	−1	1
−1	1	1	1
1	1	1	0
−1	−1	1	0
1	1	−1	0
−1	−1	−1	0
1	−1	0	0
−1	1	0	0
1	−1	1	0
−1	1	1	0
1	−1	−1	0
−1	1	−1	0

$$E_{h-h} = -J_h \sum_{(i,j)} \delta(H_{i,j+1}, H_{i,j}) \,;$$

$$E_{v-v} = -J_v \sum_{(i,j)} \delta(V_{i+1,j}, V_{i,j}) \,.$$

(4.7)

To obtain compatibility between the hypotheses of edge spins and intensity spins an interaction energy favoring consistent configurations is added

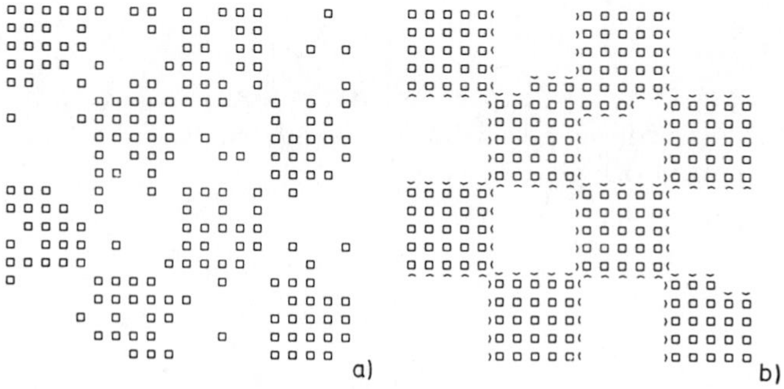

Figure 21. Input pattern and restored chessboard pattern for the feature spin system described by Eqs. 4.5 through 4.10.

$$E_{i-h} = -J_{i-h} \sum_{(i,j)} T(I_{i,j}, I_{i+1,j}, H_{i,j}) \, ;$$

$$E_{i-v} = -J_{i-v} \sum_{(i,j)} T(I_{i,j}, I_{i,j+1}, V_{i,j}) \, . \qquad (4.8)$$

Here T contains the compatibility condition listed in Table I.

The pattern to be processed is coded as a field $F_{i,j} \in \{ +1, -1 \}$ corresponding to black and white pixels at position (i, j). The interaction between intensity spins and the field is chosen like in the Ising model

$$E_{i-ifield} = -J_{ifield} \sum_{(i,j)} I_{i,j} \, F_{i,j}. \qquad (4.9)$$

The field for edge spins codes intensity changes. The interaction between edge spins and the corresponding fields is

$$E_{h-hfield} = -J_{hfield} \sum_{(i,j)} \delta(H_{i,j}, (F_{i+1,j} - F_{i,j})/2) \, ;$$

$$E_{v-vfield} = -J_{vfield} \sum_{(i,j)} \delta(V_{i,j}, (F_{i,j+1} - F_{i,j})/2) \, . \qquad (4.10)$$

As input for the picture restoration spin system we chose a distorted chessboard pattern, measuring 20×20 pixels. Twenty percent of the pixels were randomly reversed from black to white and vice versa. The resulting pattern is presented in Fig. 21a. The aim of the restoration process is to find the chessboard closest to this picture. The interaction parameters assumed for the restoration are $J_i = 1$, $J_h = J_v = 4.5$, $J_{i-h} = J_{i-v} = 4.5$, $J_{ifield} = 1$, $J_{hfield} = J_{vfield} = 3$. The temperature was lowered in 12 steps from an initial value of $T = 8$ to the final value of $T = 0.05$. Figure 21b shows the result: the chessboard pattern has been restored to a very large degree.

Acknowledgments

This research was supported in part by the National Science Foundation under Grant No. PHY82-17853, supplemented by funds from the National Aeronautics and Space Administration.

References

1. G. Palm, A. Aertsen, Brain Theory, Springer, Berlin-Heidelberg, New York (1986); H. Gardner, The Mind's New Science, Basic Books, New York (1985).

2. C. von der Malsburg, Self-Organization of Orientation Sensitive Cells in the Striate Cortex, Kybernetik, **14**, 85 (1973).

3. L.N. Cooper, Proceedings of the Nobel Symposium on Collective Properties of Physical Systems, ed. Lundqvist, Academic, New York, p. 252 (1973).

4. T. Kohonen, Self-Organization and Associative Memory, Springer, Berlin-Heidelberg, New York, p. 128 (1984).

5. J.J. Hopfield, Neural Networks and Physical Systems with Emergent Collective Computational Abilities, Proc. Natl. Acad. Sci. USA, **79**, 2554 (1982).

6. J.J. Hopfield, Neural Networks for Computing, American Institute of Physics Publication, in press.

7. J. Buhmann, and K. Schulten, Associative Recognition and Storage in a Model Network of Physiological Neurons, Biol. Cybern., **54**, 319 (1986).

8. C. von der Malsburg, The Correlation Theory of Brain Function, Internal Report 81/2, Dept. Neurobiologie, MPI f. Biophysikalische Chemie, Göttingen (1981).

9. M. Abeles, Local Cortical Circuits, Springer, Berlin-Heidelberg, New York (1982).

10. T. Kohonen, Self-Organized Formation of Topologically Correct Feature Maps, Biol. Cybern., **43**, 59 (1982).

11. T. Kohonen, Analysis of a Simple Self-Organizing Process, Biol. Cybern., **44**, 135 (1982).

12. H. Ritter and K. Schulten, On the Stationary State of Kohonen's Self-Organizing Sensory Mapping, Biol. Cybern., **54**, 99 (1986).

13. J. Fox, The Brain's Dynamic Way of Keeping in Touch, Science, **225**, 820 (1984).

14. J.H. Kaas, M.M. Merzenich and H.P. Killackey, The reorganization of somatosensory cortex following peripheral nerve damage in adult and developing mammals, Ann. Rev. Neuroscience, **6**, 325 (1983).

15. E.R. Kandel and J.H. Schwartz, Principles of Neural Science, Elsevier, New York (1985).

16. B.E. Stein, H.P. Clamann and S.J. Goldberg, Superior Colliculus: Control of Eye Movements in Neonatal Kittens, Nature, **??**, ?? (1980).

17. H. Ritter and K. Schulten, Topology Conserving Mappings for Learning Motor Tasks, Proceedings of the 1st Conference on Neural Networks and Computing, Utah, in press (1986).

18. H. Ritter and K. Schulten, Planning a Dynamic Trajectory as Path Finding in Phase Space, Wopplot Proceedings, Munich, submitted 1986.

19. S. Kirkpatrick, C.D. Gelatt and M.P. Vecchi, Optimization by Simulated Annealing, Science, **220**, 671 (1983).

20. S. Geman and D. Geman, IEEE Tran. Pattern Anal. Machine Intell., Vol. **PAMI-6**, 721 (1984).

21. P. Kienker, T.J. Sejnowski, G.E. Hinton and L.E. Schuhmacher, Separating Figure from Ground with a Parallel Network, in press (1986).

22. D. Marr, Vision, Freeman (1982).

23. B. Julesz, Foundations of Cyclopean Perception, Univ. Chicago Press (1971).

Recognition Automata Based on
Selective Neural Networks

George N. Reeke, Jr. and Gerald M. Edelman
The Rockefeller University
New York, New York 10021, USA

Abstract

The critical requirement for learning and other higher mental functions is the prior ability to categorize objects and events based on sensory signals reaching the brain. The theory of neuronal group selection provides an explanation for this ability based on a kind of Darwinian selection operating in somatic time on groups of interconnected neurons. These groups are formed during development with varied and overlapping abilities to respond to patterns of environmental stimulation. They are connected in networks to form repertoires of recognizing elements, and these repertoires, many of which are arranged in maps, are further arranged in parallel hierarchies which communicate with each other to carry out categorization with generalization according to attributes of adaptive significance to the organism. A key feature of such systems required to maintain spatiotemporal continuity is reentry of output signals, both within the system and globally, via changes in sensory input resulting from interactions of the system with the environment. These forms of reentry provide a basis for context-dependent figure/ground discrimination and for perceptual invariance to object transformations such as those resulting from motion.

Another key element of the selective paradigm is a form of differential amplification of groups which contribute to responses of adaptive value. In selective neural networks, such groups are selectively modified to enhance the organism's response to future instances of the same or similar stimuli. This modification consists of enduring changes in the efficacies of the synaptic connections between cells. Rules for these changes based on known biochemical and biophysical properties of neurons have been devised and their behavioral properties studied.

Working computer models of categorizing automata based on these principles have been constructed. Examples are presented to demonstrate their ability to carry out a variety of tasks involving recognition, categorization, generalization, and visual tracking. The computer models give insight into how biological pattern recognizing systems might operate and point the way toward construction of improved recognition automata.

Introduction

Attempts to understand human brain function, or at least some parts of it, have been made from every possible perspective, from the purely abstract and mathematical, through the experimental and physiological, to the highly personal and psychoanalytical. Yet, for all these approaches, there is a general consensus that our present state of knowledge remains incomplete and unsatisfactory. In particular, there appears to be a large and perhaps unbridgeable gap between the molecular level of reductionistic science and the presumably "higher" levels at which are carried out those unique functions that make us human.

While this gap undeniably exists, we believe its size is exaggerated as a direct result of our propensity to think in terms of fashionable models. In every age, at least as long as it has been realized that the brain is the organ of thought, men have looked to the most advanced technical artifacts of their times as a model for the brain. Thus, for the nervous system, we had the pneumatic model and later the telephone exchange model, which played a useful role at the time the elementary facts of neuronal signal conduction were being worked out. Today the computational model is popular.[1-3] It takes its strongest form among those in the artificial intelligence community who assume that mental algorithms exist,[4] that they can be deduced from a study of the information processing requirements of mental acts, and that these algorithms are the only important thing, being independent of any particular hardware on which they are implemented. According to this view, we should be able to reproduce natural intelligence, given a fast enough computer, once we deduce all the necessary algorithms.

The common feature of all these models that leads to an unbridgeable gap is essentially that they deal only with constructed mechanisms for acquiring and processing information, and not with the fundamental way that the categories of information or the mechanisms come to exist or the way they are coupled, through the behavior of the organism, back to the external world. Such models implicitly require the participation of intelligent agents of the very kind they are trying to explain, in the form of the little electrician who wires the system in just the way that is needed to make it function perfectly, and his cousin, the homunculus, who analyzes the results of the information processing on a little computer screen and directs the behavior of the organism according to what he sees. Of course, to explain the operation of the electrician and the homunculus requires additional information processing mechanisms, leading to an infinite regress and failure of the model to be convincing.

In order to develop a satisfactory theory, it is necessary to go beyond pure information processing to consider 1) how perceptual categories may be discovered in a world that does not come parsed into objects, much less into useful classes of objects, 2) how memory can function in the absence of computable representations and retrieval keys, and 3) how actions can be decided on the basis of imperfect information and myriads of competing associations. Regardless of whether we approach the problem from the physiological or the algorithmic point of view, it is clear that the explanatory gap cannot be closed unless we consider the entire biological system, its evolutionary origins and development, and its relationships to its environment.

In this paper, we summarize such an approach, one that takes into account principles of ontogeny and phylogeny that have served well to illuminate other areas of biology. This approach takes account of Darwin's principle of natural selection, and it suggests that another selection principle, applied to the nervous system in somatic time, can provide a unifying framework for understanding the higher cognitive functions of the brain. This approach does not involve algorithms, at least in the sense of preplanned sequences of computational events, because it recognizes that a major responsibility of the nervous system is to deal with unexpected events.

As far as we know, the nervous system is not programmed except by certain evolutionary patterns of ethological significance. We take the point of view here that there is no neural equivalent of the ASCII code for information storage and transmission—the only meaning of neuronal firing is in the effect it produces on other neurons or muscles. Logic itself is not given, but must be discovered, where it is needed at all, by the organism. (There are many organisms that get along perfectly well without logic—including some of us at times.) Memory is taken to be reconstructive and associative, not descriptive and addressable like that in a computer. To put the problem most bluntly, the system doesn't even know what the problem is—it just has to survive in the world as it is.

To begin our exploration of these problems, let us ask what is the first problem the nervous system must solve? Surely it is that the world does not come neatly parsed into labelled categories. Thus, the nervous system has to determine that the world contains separate objects (because certain groups of points move around together, or have the same color or texture), form categories from them, and then classify objects and events into those categories. By categories, we mean things like "tiger," "good to eat," and so on. Why are categories so important? Because without them, there would be no way to make associations; the entire collection of sensory signals received by the nervous system at any one time could never be related to that at another time. Without categories and associations, memory would be useless, and learning would be impossible.

How, then, do we categorize things? Most of us have rather a naive view of this problem. We tend to believe we have lists of features that characterize classes.[5] For example, if something is red, good to eat, and has a crunchy texture, it is probably an apple. However, this simple notion quickly runs into difficulties. For instance, to take an example from Wittgenstein,[6] what is the list of characteristics that defines the term "game"? What does a child throwing a ball have to do with Swedish naval exercises in the Baltic Sea? To answer this question, we must recognize that classes have flexible definitions that change according to the context in which we use them. For a simple example, consider Fig. 1. Here are several prototype objects in each of two classes, and, at the bottom, a single test object to be classified with the others. Is it a member of class I (being large) or of class II (being square)? Or is it neither (being shaded)? Patently, no decision can be made—the necessary information is not present in the figure. The large shaded square might

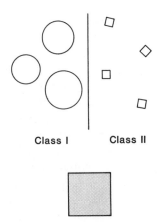

Class I   Class II

Figure 1. Two classes of objects, large circles and small squares. How is the large square to be classified?

well belong under different circumstances to class I or to class II or to some other class. The point is that in everyday life, we make such classifications by taking into account the relevant circumstantial data, and we are not surprised when the same object is classified differently at different times, for example, the penny as money, as electrical fuse, as screwdriver.

Another common misconception is that language is required to define categories. This notion is shown to be false by experiments that demonstrate categorization and generalization occurring in animals lacking language. For example, Herrnstein et al.[7] showed that pigeons could be taught to distinguish scenes (presented as projections) containing objects of a particular class, such as trees, from scenes not containing such objects. The classes did not need to be ones found in the everyday environment of pigeons; fish, for example, formed a well discriminated class.

In fact, one must add a caution to this tale. It would be a mistake to assume that the categories that are significant to pigeons are the same ones that are significant to us. If the theory we are going to present is correct, all categorizations must be ultimately species-dependent, because the ability of an organism to discriminate and categorize stimuli must depend on its phylogenetic endowment of sense organs and the associated feature-detecting and feature-correlating neuronal networks. These in turn are the result of natural selection, operating to make the species fit for its environment. The categories distinguished by a particular species will be just those whose recognition may be of specific adaptive value to members of that species; they can have no absolute validity for the reason that absolute categories do not exist in the world without the use of language. Of course, other categories may be incidentally distinguished by the available neuronal machinery, as in the case of the fish by the pigeons. Careful examination may reveal the incidental nature of these categories, as in the case of experiments in which drawings of the "Charlie Brown" cartoon characters were presented to pigeons.[8] The characters could be recognized only if background was removed, and even then the recognition depended only on the particular sets of features present in the drawings and not on their mode of combination. Thus, a drawing of Charlie Brown with his feet attached directly to his head was not distinguished from his more accustomed posture. These results suggest that the pigeons were not interpreting the drawings as representations of humans, but only as collections of abstract two-dimensional shapes.

In short, categorization may be done in different ways by members of different species, and is not even always done the same way by all members or any member of a single species. Humans, in particular, use multiple strategies or combinations of strategies depending on the class in question and the context in which classification occurs.[5] Three paradigmatic approaches will serve to illustrate the range of possibilities: The classical (Platonic) view is that of the ideal concept, characterized by a list of necessary and sufficient conditions to which new observations may be compared for classification. The probabilistic view sees the world in multiple dimensions, with classification accomplished by measuring the overlap of new observations with nearby Bayesian clouds of probability density. Finally, the exemplar view admits disjunctive categories by permitting new observations to be compared with one or more exemplar instances of each candidate category. The point is not that these approaches cannot be imitated by computer programs, for all of them have been. Rather, it is that we use more than one of them, not because we are careless or illogical, but because the need for them is built into the underlying structure of the world. For this reason, both the probabilistic and the exemplar approaches have been built into the automata we will describe.

It should be emphasized that learning need not be a factor in categorization, although it can of course lead to superior performance. What is required is that a response be associated with a pattern of sensory information corresponding to a class of objects or events in the environment, and that the response have adaptive value for the organism. Conventional responses, such as names, can be acquired only through learning, but before learning can occur, it is necessary to have categories or concepts to which the learned responses can be attached. The systems we will describe here have the ability to carry out categorization prior to learning. They thereby emulate the ability of biological systems to respond to stimuli never encountered before, yet they have no programs that embody specific information about particular stimuli.

In attempting to develop a theory to account for these capabilities, we would do well to begin by considering some basic facts of neurobiology, in the hope that they will point us in the right direction. First, nervous systems are organized as networks with distinct areas having different connectivities, apparently specialized for different functions. Often these networks form maps. This anatomy suggests a heterarchical arrangement in which subnetworks with different functions interact to give more complex functions which they do not possess alone. There is apparently a high degree of parallelism in the responses of the individual neurons. Extensive overlaps of dendritic and axonal arborizations suggest a functional degeneracy in which there are many alternative paths between any two points in the network, even in a map, and there are thus many units capable of responding to any given input pattern. No individual neuron appears to be indispensable for any function, suggesting that only patterns of response over many neurons can have functional significance.[9] Neurons have limited speed and dynamic range, suggesting that algorithms involving loops or high-precision computations cannot be carried out.

The final, and most important observation, is the enormous diversity of neuronal populations. It can easily be calculated that there is not sufficient information in the DNA of higher animals to specify uniquely all the connections of their neurons. Thus, indeterminate, epigenetic mechanisms must operate during development to determine the fine structure of the nervous system. The variation is seen at all phylogenetic levels but, if anything, is greater in higher forms. This is not what one would expect if the system were exquisitely optimized to carry out cognitive functions by explicit programming. In a computer, such variation in wiring would be a severe problem; in a brain, one would like to think that nature, with her usual parsimony, has made the system just "good enough" to work, suggesting that the unavoidable variation is part of the solution, and not part of the problem.

The features that we have pointed out, particularly the structural variation in the face of a need to deal with an unpredictable environment, are all characteristics of selective systems. They are the same ingredients found in the immune system[10] and indeed in evolution itself. These and other considerations have prompted the idea embodied in the neuronal group selection theory[11] that the brain is in fact a selective system operating in somatic time and this is necessary to provide the adaptability needed to survive in a hostile environment.

The Neuronal Group Selection Theory

The neuronal group selection theory postulates that two kinds of selection events play critical roles in shaping the development of the nervous system at different stages. During embryogenesis, selection among competing neuronal cells and their processes, elaborated through the basic developmental mechanisms of cell adhesion and movement, differential growth, cell division, and cell death, determines the anatomical form and patterns of synaptic connectivity of the nervous system. Later, during postnatal experience, selection among diverse pre-existing groups of cells, accomplished by differential modification of synaptic strengths or efficacies, without changes in the connectivity pattern, shapes the behavioral repertoire of the organism according to what is of adaptive value to it in its econiche. It is this latter form of selection that we shall discuss in the remainder of this chapter.

The selectional model is presented schematically in Fig. 2 and contrasted with an instructional scheme based on an information-processing model. There are three basic requirements for any selective system: (1) there must exist an *a-priori* collection of variant entities capable of responding to relevant environmental states (such a collection is called a "repertoire"); (2) there must be extensive opportunities for members of the repertoire to encounter the environment; and (3) there must exist a mechanism for differential amplification of the prevalence or strength of response of those members of the repertoire which are in some sense favored in their interactions with the environment. In Fig. 2, we have given a highly schematic synopsis of how these requirements might be met in the nervous system. The reper-

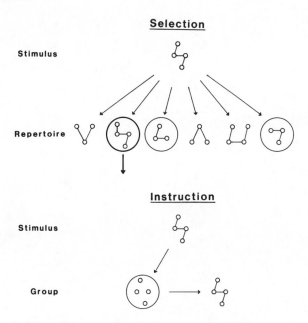

Figure 2. Schematic illustration contrasting selectional (top) and instructional (bottom) models for the establishment of neuronal group responses. Small circles connected with lines represent specific patterns of neuronal activity and the matching characteristics of groups that respond to them, and have no connection with neuroanatomy. In selection, groups whose specificity matches a given stimulus more or less well (groups surrounded with larger circles) respond to that stimulus; only those groups whose responses are of adaptive value for the organism (heavy large circle) are modified for stronger future response to similar stimuli (heavy arrow). In instruction, groups have no particular *a-priori* specificities (group without joining lines, bottom); specificity is dictated by interaction with stimuli. *(Reprinted from Ref. 21 with permission.)*

toires are taken to comprise groups of perhaps 50 to 10,000 neurons, capable, as a result of their interconnections, of responding to particular patterns of activity arriving at their dendritic synapses. These interconnections are formed during development, prior to experience. The inputs to which the groups respond originate ultimately at the sense organs (encounter with the environment), but frequently will have been relayed first through other neuronal groups. In higher forms, reinstitution of responses to similar previous situations will significantly influence present responses. Amplification takes the form of modification of the strengths of synaptic connections, leading to increased speed or strength of the selected responses when similar overall patterns of stimulation occur again later.

In contrast, in an instructional system (Fig. 2, bottom), there are no pre-established responses. The connection strengths between units are modified in such a way as to optimize the response of the system to each given combination of inputs. While at first glance this approach appears attractive, particularly for its economical use of units, it has several serious problems. In particular, it has proved difficult to derive general rules to give optimal synaptic modifications, especially if the rules are constrained to use only local information available at each synapse. The most promising rules found to date[12] involve "simulated annealing," a process in which the

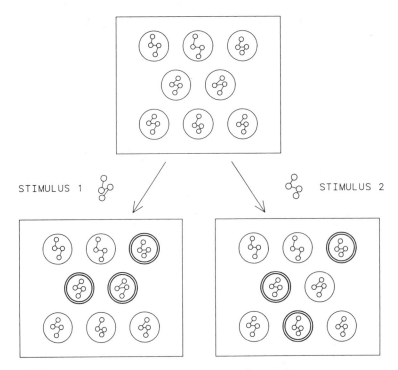

Figure 3. Degeneracy. The box at the top represents a repertoire containing groups with overlapping specificities for recognition (same conventions as in Fig. 2). In response to two similar stimuli (stimulus 1, left, and stimulus 2, right), the patterns of response include some unique groups and some groups in common (large circles outlined twice, bottom). The responses may similarly differ to presentation of the same stimulus on different occasions or in different contexts. These differences provide the basis for subtle context-dependent shadings in behavior.

system must undergo cyclical changes in operating conditions that appear hard to reconcile with the requirement in animal nervous systems to reach responses continuously and in real-time without pauses for "relaxation" of the networks. In addition, instructional systems are paralyzed in the face of unanticipated, novel situations, a flaw that could be fatal for an animal encountering a new kind of predator for the first time. Selective systems, in contrast, will always give some response, based on a combination of previously existing specificities.

This ability of selective systems to respond to any input implies the existence of a relationship between the sizes of the repertoires and the specificities of the individual groups. If recognition is too specific, there cannot be enough groups in a finite repertoire to recognize all possible stimuli and the system must fail; similarly, if specificity is too broad, stimuli with significant differences may be confused, and again the system must fail. The specificities must therefore be intermediate, implying that several groups may respond more or less well to any given stimulus. This phenomenon, which we call *degeneracy* (Fig. 3), is critical to an understanding of selective recognition systems. Degeneracy assures that any perceptual problem has multiple potential solutions. Context determines exactly which combination of

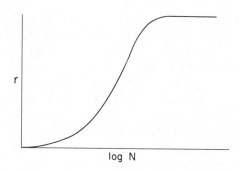

Figure 4. Probability of response as a function of repertoire size. Assuming the responses of all groups are independent, and the probability is p that any one group will respond to any one stimulus, then $r = 1 - (1 - p)^N$ is the probability that at least one group in a repertoire of N groups will respond to any one stimulus. By setting a permissible failure level, $1 - r$, and required selectivity level, p, for a particular ecological niche, one can arrive at an estimate of the necessary repertoire size, N. *(Reprinted from Ref. 21 with permission.)*

groups responds in a given situation, and therefore which solution is selected. Degeneracy also assures that the entire "space" of possible stimuli is covered, and that the system has the functional redundancy needed to make it "fail-safe" against the loss of individual groups.

Figure 4 illustrates how a degenerate recognizing system becomes capable of responding to any stimulus, even a novel one, once that system becomes sufficiently large. The critical assumption is that there is a small but nonvanishing probability that any single group will respond to any given stimulus that affects the relevant sensory modality. This assumption seems reasonable for a network connected in such a way that each group has access to at least some part of the spreading pattern of activation elicited by each stimulus. The algebraic laws for the combination of probabilities then assure that the overall likelihood of some degree of recognition approaches unity as the number of groups increases. In Fig. 4, this likelihood is calculated on the assumption that all the individual probabilities are equal, but the general shape of the curve will be the same for unequal probabilities. Of course, there is no guarantee that the responses obtained in this way are useful to the organism, but if the network is sufficiently large, there will always be *multiple* responses, from which the best may be selected and amplified.

One additional concept that is critical to the neuronal group selection theory is that of reentry, or the presentation of output signals, usually in a mapped arrangement, from one repertoire to another at the same or an earlier stage of neuronal processing. Reentry ensures consistency across the entire system with respect to the current state of the environment. It encompasses feedback, but is more general. Figure 5 illustrates two common forms of reentry in the nervous system. Near the bottom of the figure, two repertoires in different, parallel pathways that classify stimuli according to different criteria are cross-connected. Such interacting repertoires form "classification couples" which, by their mutual interaction, are able to perform classifications more complex than either could accomplish alone. Also suggested (long arrow at far right, bottom to top) is a mechanism by which the motor output of the organism influences sensory systems by changing the arrangement of objects or the organism's position in the environment. Both forms of reentry together assure that subrepertoires at all levels in the nervous system are constantly mapped to each other and to the outside world, obviating any need for the semaphores, time stamps, or other bookkeeping apparatus of multiprocessing computers.

The neuronal group selection theory is consistent with the biological facts we have summarized. It takes advantage of the unavoidable variance introduced by epigenetic events during the construction of neural networks to provide a plausible mechanism for categorization without programmed descriptions, homunculi, or reinforced learning. It is based on principles of selection similar to those that govern the evolution of species, using differential modification of synaptic weights in place of differential reproduction. Like any scientific theory, it must be tested by experiment; a discussion of the evidence already found in its favor[13] is beyond the scope of this chapter. Instead, we shall describe a series of automata that we have constructed in computer simulations in order to test the self-consistency of the theory as well as to demonstrate the ability of selective recognition systems to carry out interesting recognition and categorization tasks. Such models can be invaluable in helping to focus experimental questions for the biologist. Ultimately, we hope to learn from them how to construct machines capable of carrying out classification tasks far better than those that are now available.

Selective Recognition Automata

We have explored the properties of selective recognition systems in a series of automata. The first of these, called Darwin I, dealt with the process of recognition itself, using strings of binary digits as recognizands and recognizers.[14] We shall not describe Darwin I here, as the results have been incorporated in the later automata to which we now turn. The second automaton, Darwin II, is concerned with the recognition and classification of two-dimensional patterns presented on a retina-like array. A third machine, Darwin III, which is currently under test, will be described later.

The plan of construction for Darwin II is shown in Fig. 6. The recognizing elements of the automaton are abstractions of neuronal groups, connected in a series of sub-networks or repertoires. The arrangement of these repertoires, and the classes of possible connections between them, is fixed. For any particular instantiation of the automaton, however, the numbers, destinations, and initial strengths of the connections in each class can be established at will. Once established, the connectivity remains fixed, but the connection strengths vary in accord with rules for synaptic modification that serve as the mechanisms of group selection. Darwin II has no mechanism for generating motor output; performance is evaluated by examination of its internal states, directly or with appropriate statistics.

An important rule, one that distinguishes Darwin II from systems based on "frames,"[15] "schemata,"[16] or "scripts,"[17] is that no specific information about the

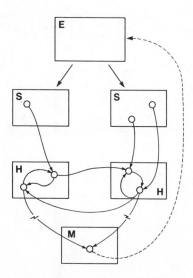

Figure 5. Highly simplified schematic drawing illustrating reentry in the nervous system. The box labelled E represents the environment. Boxes labelled S represent first-level sensory repertoires, which receive input (straight arrows) directly from sense organs. H and M represent, respectively, higher sensory repertoires and motor repertoires. Only a few cells (circles) and connections (curved arrows) are shown. As described in the text, reentry can involve direct connections between repertoires (arrows between two H repertoires), or can operate indirectly, through changes in the environment brought about by the behavior of the organism (dashed arrow).

stimulus objects to be presented is built into the system when it is constructed. Of course, general information about the kinds of stimuli that will be significant to the system (e.g. the fact that they will be line drawings) is implicit in the choice of feature-detecting elements that are used—this choice is akin to the specializations built into receptor organs of different species in the course of evolution.

At the top center in Fig. 6 is an input array or "retina" where the stimuli are presented. For convenience, these are usually letters of the alphabet presented as light or dark areas on a 16 × 16 grid, but the size of the array, the kind of patterns, and the number of gray levels (up to 256) can be varied at will. The automaton proper is shown below the input array. It consists of two parallel concatenations of repertoires making classifications according to quite different principles. These subsystems interact by reentry to give associative functions not possessed by either set alone. The subsystems are arbitrarily named "Darwin" and "Wallace" after the two main figures in the description of natural selection. The Darwin network (left) is designed to respond uniquely to each individual stimulus pattern, and loosely corresponds to the template matching or exemplar approach to categorization. The Wallace network (right), on the other hand, is designed to respond in the same way to different objects in a class, and loosely corresponds to the probabilistic matching approach to categorization. Of course, the particular response specificities of Darwin and Wallace are intended to be merely exemplary, and such stimulus characteristics as color, motion, or texture would also be represented in real nervous systems.

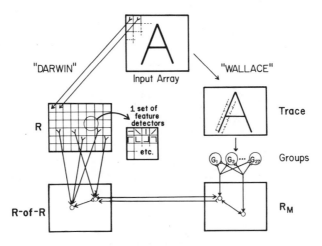

Figure 6. Simplified plan of construction for Darwin II. Boxes represent repertoires, circles represent neuronal groups, and arrows represent synaptic connections. The inset to the right of the R repertoire is an expanded view of one of the topographically mapped subrepertoires, indicated by the circle and curved arrow. The specific functions of the individual repertoires are described in the text. In a typical instantiation, R might contain 3840 groups (a 10 × 12 topographic map with 32 kinds of feature detectors at each point) with 16 inputs each, R-of-R, 4096 groups with 96 connections each, and R_M, 2048 groups with 15 inputs each, for a grand total of about 500,000 connections. The program is written in a combination of FORTRAN and IBM-370 Assembler language; simulation time on an IBM 3033 is about 20 microseconds per connection, or 10 seconds per cycle of a typical run. *(Reprinted from Ref. 21 with permission.)*

The two networks both have a hierarchical structure. Each has connected to the input array a level that deals directly with features of the stimulus, and below that an abstracting or combining level that receives its main input from the first level. The two abstracting repertoires are connected to form a classification couple (see Fig. 5). It is important that in such a couple, the earlier levels are not connected, for then the separate modes of classification would be confounded, with the loss of their distinct characteristics.

The first part of Darwin is the R or "recognizer" repertoire. It has groups that respond to local features on the input array, such as line segments oriented in certain directions or with certain bends (suggested by the inset in Fig. 6). Multiple sets of these groups are connected to the input array in a topographic map (suggested by the parallel arrows from the input array to R). As a result, patterns of response in R spatially resemble the stimulus patterns. The abstracting network connected to R is called R-of-R or "recognizer-of-recognizers." Groups in R-of-R are connected to multiple R groups distributed over the R sheet, so that each R-of-R group is capable of responding to an entire pattern of response in R. In the process, the topographic mapping of R is destroyed and R-of-R gives an abstract transformation of the original stimulus pattern. If the stimulus undergoes a change such as a translation to a new position on the input array, the pattern of response in R-of-R will be quite different. It is the responsibility of Wallace to deal with this

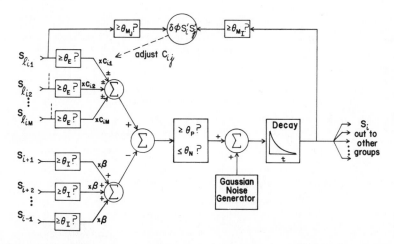

Figure 7. Logical structure of a group. For detailed mathematical description, see Edelman and Reeke,[18] from which this figure has been reproduced with permission.

translation problem; R-of-R is concerned with *individual* properties of a stimulus, and these include its relationship to the context in the background.

Wallace begins with a tracing mechanism designed to scan the input array, detecting object contours and tracing along them. It responds to correlations of features that distinguish objects as entities from the background by their spatial continuity. (This is something like what the eye does when it rapidly scans a scene to detect the objects present.) The result of the trace is the excitation of a subset of a set of groups (G_1, G_2, . . . , G_{27} in Fig. 6) that represent by their activity the particular correlations of features present in the input. These are connected in turn to an abstracting network, R_M, that responds to patterns of activity in $G_1 - G_{27}$ in much the same way that R-of-R responds to patterns of activity in R. Because the trace responds to the presence of lines or junctions of lines in the environment with little regard for their lengths and orientations, R_M is insensitive to both rigid and non-rigid transformations of stimulus objects and tends to respond to class characteristics of whole families of related objects.

The groups in Darwin II are constructed according to the same logical plan in all of the repertoires[18] (Fig. 7). The activity of each group, s_i, is a scalar function of the group's present inputs and past history. There are two different kinds of input connections. Specific connections (upper left) may come from the input array or from groups in the same or other repertoires. The sources of these connections are specified by lists, l_{ij}, in which j ranges over source groups and i over target groups. These lists are constructed differently for different repertoires according to their functional roles. Each input that exceeds a specified threshold value, θ_E, is multiplied by its individual connection strength, c_{ij}, (which may be positive or negative according to whether the connection is excitatory or inhibitory) and the product is added to the total input for the group. Connections of the second kind (lower left) are geometrically specified and nonspecific in the sense that they do not have individual weights. These connections have a function similar to lateral inhibition in neural networks. The inputs that exceed a specified threshold, θ_I, are multiplied by a common inhibitory coefficient, β, and then added into the total input for the group as for the specific connections.

Unless the combined inputs exceed an overall excitatory threshold, θ_P, or inhibitory threshold, θ_N, they have no effect on the group's activity. If above threshold, the input is added to the previous level of activity, which meanwhile has undergone simple exponential decay. In either case, a varying amount of noise is added to the response of the group by analogy with the noise found in real neuronal networks. The total obtained by combining all these terms is the activity of the group, s_i, which is made available to whatever other groups may be connected to this one (arrows at right, Fig. 7). The activities of all groups in a repertoire are calculated and then updated simultaneously in cycles.

The recognition specificity of each group depends on its connection list and connection strengths—the best response is obtained when the most active inputs are connected to synapses with high connection strengths. For other inputs, the group will respond more or less well, overlapping in specificity with other groups and conferring degeneracy on the system as a whole.

There are many ways that the connection strengths in a system of this kind could be modified to implement the amplification phase of selection. Figure 7 (top) shows one such scheme, known as the Hebb rule.[19] In this scheme, connection ij is strengthened if the activities of the pre- and post-synaptic groups, s_j and s_i, both exceed specified thresholds. In other words, a connection from an active input to an active group is strengthened, leading to a stronger response the next time a similar input is encountered. (In practice, cases must also be defined in which connection strengths are weakened, or all synapses are eventually driven to maximum strength and the system ceases to be selective.)

Darwin II provides a number of rules based on variations of the Hebb principle. All of these are strictly local in that the only variables allowed to affect the state of a particular synapse are the input to that synapse, the activity of the post-synaptic group, and the current strength of the synapse itself. These rules have the advantage of simplicity, but they deprive the automaton of the versatility conferred by heterosynaptic effects, in which activity at one synapse can affect the strengths of other, nearby, synapses. These, and other complications, are likely to occur in real neurons. Finkel and Edelman[20] have proposed a set of synaptic modification rules based on an extensive review of the available neurophysiological data. They suggest that presynaptic and postsynaptic changes can occur on different time scales and by different mechanisms. In their model, post-synaptic changes are heterosynaptic, being based on correlated activity at neighboring synapses, and relatively short-term in nature. Presynaptic changes, on the other hand, depend on average rates of transmitter release, and affect many or all of the synapses of a cell over long time scales. The net synaptic strength depends on a combination of both terms. Networks incorporating these rules display interesting dynamic effects, including competition among groups for the allegiance of cells lying near their borders.[20] The

Hebb-type rules, however, are adequate for the recognition and categorization tasks studied with Darwin II, and they were used in the experiments presented here.

Results Obtained with Darwin II

With appropriate stimulation sequences, Darwin II is capable of producing responses corresponding to behaviors such as categorization, recognition, generalization, and association. We shall present examples of only the first three of these here; a more detailed treatment is contained in Reeke and Edelman.[21] In examining these simulations, keep in mind the different design criteria of the Darwin and Wallace subnetworks: Darwin should give individual representations, that is, unique responses to each different stimulus and the same response to repeated presentations of the same stimulus, but stronger; Wallace should give class representations, that is, similar responses to different stimuli having common class characteristics. In the complete system, these individual and class representations interact to give associative recall of different stimuli in a common class.[21]

Figure 8 shows the responses obtained when two different A's and an X were presented on the input array under conditions in which the reentrant connections between Darwin and Wallace were inactive. The R responses (left column) resemble the shapes of the stimulus letters as a direct consequence of the topographic mapping built into the connections between the input array and R. At most positions on R, multiple groups may be seen responding with greater or lesser activity. These correspond to the different types of local feature detecting units which are replicated at each map position. The R-of-R responses (second column) are different for the three letters, corresponding to individual representations of the letters. In contrast, the R_M responses (third column) are very similar for the two stimuli that are in the same class (the two A's), and different for the other class (the X). In short, these patterns of response meet the criteria set out earlier for the Darwin and Wallace networks. The R_M responses, in their similarity for the two A's, demonstrate *categorization* without naming. (It should be noted that the particular categories arrived at by R_M for the different stimuli are dependent on the particular choice of trace correlations that R_M groups respond to, and might or might not agree with the categories we define for the letters. This corresponds to the evolutionarily dictated predispositions of organisms to attend to particular categories of stimuli in their environments.)

The effect of synaptic modification on these responses is shown in Fig. 9. Groups that respond above a certain level (the amplification threshold) are changed so as to give a stronger response (large circles at right); groups with weaker responses generally are changed so as not to respond at all after amplification. Groups that were not involved in the response to the stimulus in question remain unchanged, available for response to novel stimuli not yet encountered. These changes demonstrate *recognition*, or the enhanced response to a stimulus after it has been experienced before.

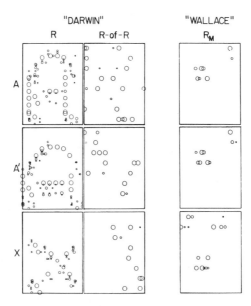

Figure 8. Responses of individual repertoires (R, R-of-R, and R_M, names at top) to three different stimuli. Circles represent groups; the radius of each circle is proportional to the response of the group. Groups responding at less than 0.5 of maximal response are omitted, leaving blank areas in the plots. The stimuli were a tall, narrow A (top row), a lower, wider A (middle row), and an X (bottom row). *(Reproduced from Ref. 21 with permission.)*

Amplification also improves the ability of the system to categorize, as shown in Table I. The numbers given in this table are ratios, before and after amplification, of the numbers of groups responding to two stimuli in the same class to the numbers responding to two stimuli in different classes. (The class membership information used to derive these statistics is not known to the automaton during the test.) The ratios, which are corrected for the numbers of common responders that would be obtained by chance alone under the various conditions, are a measure of the degree of commonality in the responses to stimuli that are in the same class. As such, they indicate, in agreement with the results shown in Fig. 8, that Wallace is far more effective at categorizing under these conditions than is Darwin, but that the Darwin responses to each stimulus are not entirely idiosyncratic. They also show that after selective amplification, the classifying ability of Wallace is improved nearly threefold, with a smaller improvement apparent in the results for Darwin. This occurs even though there is no feedback from the environment that would enable the system to "learn" which responses are "correct."

One can look for *generalization* by making similar tests with letters related to but not included in the set used during selective amplification. Generalization occurs if the responses to the novel letters are more like those to the training letters than would have been the case without amplification. In Wallace, generalization is already present without amplification as a trivial consequence of the class-responding property of that network, but in Darwin, generalization is not built in, and in fact cannot occur without the help of Wallace. Reentrant connections from

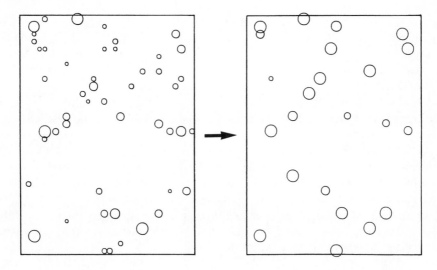

Figure 9. Effect of amplification on R-of-R responses. The response to a particular stimulus is shown on the first trial (left box) and after several presentations (right box).

R_M to R-of-R permit R_M to give this help in the following way: Among the groups in R-of-R responding to different stimuli of the same class will be some that happen to receive connections from R_M groups responding to these same stimuli. The extra input to these groups will increase their likelihood of exceeding the modification threshold, thus amplifying not only their connections from R_M, but also those from R. Because of the class-response property of R_M, these groups will tend to be the same ones for the various stimuli. Therefore, groups in R-of-R that respond to multiple members of the class will be favored for amplification. After a number of repetitions of this process, with letters from several different classes, the R-of-R responses will become more alike within each class. To the extent that novel stimuli in the same class share features in R with the stimuli used during amplification, responses to them will also become more similar, consistent with generalization.

Table I. Classification in Darwin II[a]

| | Repertoire | |
Time Tested	Darwin [R-of-R]	Wallace [R_M]
Initially	1.21	90.93
After selection[b]	1.41	241.30

[a]Repertoire sizes: R, 3,840 groups; others, 4,096 groups. Total connections: 368,840; no Darwin-Wallace connections. Stimuli used: 16 letters, 4 each of 4 kinds. Quantity shown is ratio of number of groups responding to two stimuli in same class to number of groups responding to two stimuli in different classes, corrected for numbers that would respond in each case by chance alone.

[b]Each of four stimuli was presented for eight cycles, then the entire set was repeated three times.

Results for a typical experiment of this type are shown in Table II. Sixteen different stimuli from four different classes (different letters of the alphabet) were used. Selective amplification was allowed to occur while each stimulus was presented for four cycles, repeating the entire set of 16 letters four times (256 cycles in all). A separate set of test letters, different but of the same four kinds, was presented before and after this process. (During these tests, amplification and the reentrant connections from R_M to R-of-R were disabled.) For these particular test letters, the degree of commonality of response within classes, as compared to that between classes, increased 3.4 times, from a ratio of 1.77 to 6.10, as a result of selective amplification. With a set of unrelated control letters, no such increase was seen, showing that the effect is specific, and not due to a general increase in similarity of response to all stimuli. R-of-R thus displays generalization in that, after it has had some experience with a number of letters of a particular kind, its response to novel letters of the same kind is more like that to the letters already "seen" than would have been the case without that experience.

In further experiments not described here,[21] reentrant connections in both directions between R-of-R and R_M give rise to *association* by linking individual responses in R-of-R to two different stimuli in the same class via R_M in such a way that presentation of one of the stimuli evokes elements of the response proper to the other. These demonstrations of categorization, recognition, generalization, and association all illustrate the point made in the introduction that these critical aspects of perception can, and indeed must, occur prior to conventional learning. They also show how systems based on more than one principle of categorization can be joined in "classification couples" to give modes of classification not available to any single system.

Taken together, the experiments with Darwin II demonstrate the self-consistency of neuronal group selection as a model for critical components of our perceptual faculties. In more recent work, we have begun to explore a new class of selective recognition automaton incorporating the ability to interact with the environment through motor output. Such interactions allow the automaton to signal its

Table II. Generalization in R-of-R.[a]

Stimuli	Intraclass Chance	Interclass Chance	Intraclass Interclass
Initially			
Training set	2.09	0.72	0.90
Test set	2.89	1.63	1.77
Control set	—	1.96	—
After selection[b]			
Test set	6.10	1.00	6.10
Control set	—	1.00	—

[a]Repertoire sizes: R, 3,840 groups; others, 1,024 groups. Connections to each R-of-R group: 96 from R, 64 from R-of-R, 128 from R_M Stimuli used: 16 letters, 4 each from 4 classes.

[b]Each of 16 stimuli was presented for 4 cycles, then entire set was repeated 4 times.

responses without need for the experimenter to interpret its internal states, but are mainly important for a more fundamental reason. Interaction with the environment completes a global reentrant loop by which the responses of the automaton affect its own sensory input, giving it access to much more powerful selective paradigms. Its responses can now be interpreted as having varying degrees of adaptive value for it, leading to the generation of more complicated behavioral sequences and the possibility of learning. These automata are being used to study problems involving motion, perceptual invariance, figure-ground discrimination, and memory, among other phenomena. One such experiment will be discussed briefly in the next section.

Preliminary Experiments with a New Class of Selective Automaton

To carry out investigations of the kind outlined in the last paragraph, an entirely new neural network simulation code, Darwin III, was written. This program, which will be described in detail elsewhere, is a flexible tool for generating neural networks with a wide variety of architectures and modelling them under a wide variety of stimulation protocols. It incorporatres a number of features not found in Darwin II, of which the following are most relevant to the experiments to be discussed: (1) Networks may be constructed from any number of repertoires, each of which may contain one or more layers of cells of different kinds. Each cell layer may have its own rules for connectivity and synaptic modification. (2) An environmental module permits stimuli to be generated by combining shapes from a library of letters and other objects. Stimulus objects can be independently moved about the input array or "flashed" on or off at will. (3) Two devices are provided for motor output from the system: a multijointed arm, and a movable head. Motions of the arm can affect the locations of objects in the environment, while motions of the head affect only their perceived positions on the input array. (4) "Value schemes" may be defined to specify the adaptive value to the automaton of its motor actions. Selective amplification may be made dependent on adaptive value via changes in synaptic modification parameters.

The neuronal group selection theory suggests that object motion is an important factor in the selective process, particularly in early visual learning, where it provides a major clue to the perceptual system that the world in fact can be parsed into separate objects. This view is consistent with experiments suggesting that human infants have a conception of objects as spatially connected, coherent, continuously movable entities.[22] For this reason, the first experiments with Darwin III have been aimed at gaining a better understanding of this parsing process, beginning with the simple ability to track a single moving object on a plain background.

The network architecture set up for this task is shown schematically in Fig. 10. Familiar from Darwin II are the input array and Darwin and Wallace networks with R and R-of-R at the upper left. R_M is replaced by a pair of simple scanning repertoires that respond to objects on the input array at larger or smaller distances from a central fovea. A rather elaborate R-out repertoire controls the motor system. There are no prearranged motor skills; performance can develop and improve only by selection from spontaneous movements generated by pairs of

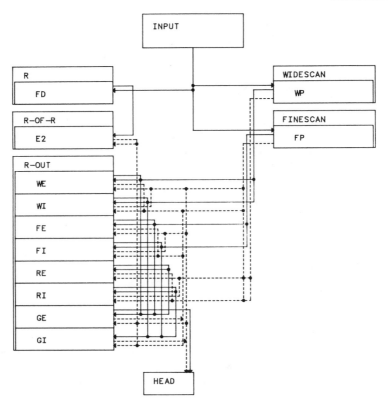

Figure 10. Network architecture for visual tracking. The input array is at the top, motor output to the movable head at the bottom. At either side are the neuronal repertoires (large boxes) with their various cell layers (smaller included boxes). Predominantly excitatory connection tracts are shown as solid lines, inhibitory tracts as dashed lines.

mutually inhibitory pattern generating layers. Separate layers are provided for large jumps (saccades) and fine tracking movements of the head. Connections from the scanning cells in Wallace are made indiscriminately to motor cells in these layers corresponding to all directions of head movement. Amplification of these connections is modulated by value schemes based respectively on the appearance of activity in a circumfoveal region and in the fovea itself. Thus, connections from a particular point on a scanning repertoire to a particular motor area will tend to be strengthened when the appearance of an object at that point is correlated with activity in that motor area that leads to foveation of the stimulus, and weakened when it does not. After a suitable period of experience with various moving stimuli, the system should "learn" to make the appropriate saccades and fine tracking movements with no further specification of its task than that implicit in the value scheme.

How this all works out in practice is summarized in Fig. 11. The key to this rather complex figure is given in the legend; the important thing to note is the modification of the connection strengths from scanning repertoires to motor areas represented in the crosses in each frame. Early in the experiment (Fig. 11a), these connections are weak, and motions are being generated mainly as a result of activity in the pattern generating layers. As time goes on (Figs. 11b and 11c), the strengths of these connections are selectively modified, and activity in the scanning repertoires increasingly comes to dominate the motion of the head. The automaton finally displays a system of behavior in which the head scans at random when no stimulus is visible, makes a rapid saccade to any stimulus that appears within the outer limits of its widest visual field, and finely tracks any stimulus that has successfully been foveated. During fine tracking, R and R-of-R are able to respond to the now centered object, permitting position-independent categorization to occur. Habituation eventually sets in, permitting occasional saccades to other parts of the visual field. After such a saccade, a new stimulus object, if present, may take over as tracking target.

Of course, this system makes mistakes. For example, amplification errors may occur when the stimulus crosses the edge of a visual field diagonally, and very large stimuli may confuse the tracking mechanism. By the use of standard engineering techniques, a better tracking system could surely be designed, one in which the correct motions for a spot of light at any location on the visual field would all be calculated in advance and incorporated in the logic of the design. The selective system takes longer to get right. But the advantage is that the machinery "doesn't know what it is for"—the same networks can accomplish various tasks depending on what we, acting as external agents of "evolution," decide they should find "adaptive." Equally simple training regimes should work for a wide variety of harder tasks, such as finding a desirable object in a background of distracting objects and picking it out with the arm, and so forth. The modular combination of subsystems capable of carrying out recognition tasks provides an attractive approach to the construction of automata with increasingly complex perceptual capabilities and should have eventual significance for efforts at machine vision. On the output side, the corollary for motor control is the ability to generate complex behavioral sequences by selection from simple and innate motion patterns, with obvious significance for developments in robotics and the control of complex systems.

Summary and Conclusions

The automata presented in this chapter are intended to illustrate certain aspects of the theory of selective recognition systems without attempting to model real nervous systems in a detailed way. The basic experimental data and the computer resources to do that are not yet at hand. Thus, the models do not provide evidence for the applicability of the theory to real nervous systems, evidence which can only come from experiment. Nonetheless, the theory provides satisfying explanations for the whole panoply of perceptual processes involving categorization, and, as inanimate examples of selective systems the models can help to demonstrate the self-consistency of the theory as an abstract description of these processes. Such demonstrations are important for understanding complicated biological systems and at the same time can provide real insight into the computer science problem of designing artificial systems with brain-like capabilities.

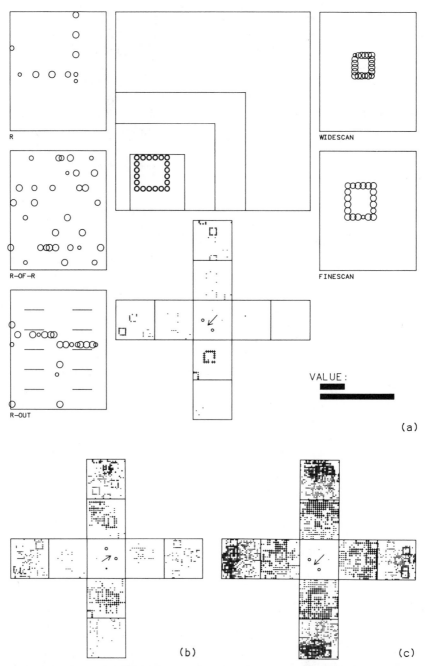

Figure 11. State of Darwin III at three stages in a visual tracking task: after 42 cycles (a), 182 cycles (b), and 796 cycles (c). At top center in (a) is the input array, with the stimulus (a square) delineated by small circles and the visual fields of the two scanning (R_M) and one foveal (R) repertoires outlined by successively smaller squares. Down the left and right sides of (a) are shown the activities in the various repertoires, keyed to their relative positions in Fig. 10 and represented as in Fig. 8. The behavioral values for saccade and fine tracking motions are shown at the bottom right in (a). The measures of value and the activity in R and R-of-R are large only when the stimulus has been foveated, as is the case here. The large cross at bottom center in (a) displays the strengths of the connections from

To get a better insight into what has and has not been accomplished with these models, it may be useful to compare them, purely as pieces of hardware, with existing or contemplated conventional engineering solutions to the same problems. The selective systems certainly have the advantage of simplicity, in that only a few kinds of basic units are required, and the connections among them can be largely random, at least in detail. There are similar advantages with respect to software development, which need barely be concerned with the final purposes of the system—it is only necessary to provide rules for the operation and modification of the repertoire elements, and to provide suitable value schemes. The evaluation of behavior implicit in these value schemes can in many cases substitute for a detailed analysis of the tasks to be carried out by the system. This is particularly important in view of the high cost of software development relative to hardware for today's computers. Finally, selective systems are inherently highly parallel. The most natural distribution of the computational load is to have a separate processor for each node in the network. Such nodes require only minimal coordination and synchronization, eliminating many of the difficult computer science problems relating to the programming of parallel processors.

There is, of course, a price to be paid for these advantages. In selective systems, *a priori* programming is to a large extent replaced by training through experience, a process which may require a large number of trials to reach satisfactory levels of performance. Selective systems also make inefficient use of the hardware, in the sense that adequately degenerate repertoires must inevitably contain many units that will never be used. In evolution, natural selection eventually brings the number of units into balance with the needs of the system. In artificial systems, this process could be speeded up by reviving an idea from the 1950s[23] to reconfigure unproductive units with new specificities for further selection. In any event, the continually declining cost of hardware may eventually make this problem unimportant.

In the studies presented here, we have shown that simple systems based on a selective principle can carry out interesting recognition and categorization tasks with potential applications in such areas as robotics and image analysis. In designing these systems, we have avoided the temptation to make choices based on mathematical elegance or facile analogies to solved problems in physics; we instead find elegance and simplicity in the beautiful solutions to these problems that nature has

---

the wide-range scanning repertoire (WIDESCAN) to the motor areas (outer four squares), from the narrow-range scanning repertoire (FINESCAN) to the motor areas (inner four squares), and finally the activities of the up, right, down, and left motor neurons and the resultant motion of the head (circles and arrow in central square). The four arms of the cross correspond (clockwise from top) to the up, right, down, and left motor areas. Connection strengths are shown for excitatory and inhibitory connections as '+' or '−' signs, respectively, with sizes proportional to the connection strengths. (b) and (c) show only the connection strengths at cycles 182 and 796 as in the bottom center part of (a). In a perfectly engineered system, the outer halves of the eight squares forming the arms of the cross would contain excitatory connections ('+' signs) and the inner halves would contain inhibitory connections ('−' signs). The actual development of the connections under selective amplification closely approaches this ideal arrangement by the end of the experiment.

worked out through natural selection. We believe that this approach provides a solid foundation for the further study of both natural and artificial intelligence.

Acknowledgments

We thank the International Business Machines Corporation for their support of this work.

References

1. A.R. Anderson, "Minds and Machines," Prentice-Hall, Englewood Cliffs, N.J. (1964).
2. D.E. Rumelhart and J.L. McClelland, "Parallel Distributed Processing. Vol. I: Foundations," MIT Press, Cambridge, Mass. (1986).
3. J.L. McClelland and D.E. Rumelhart, "Parallel Distributed Processing. Vol. II: Psychological and Biological Models," MIT Press, Cambridge, Mass. (1986).
4. D. Marr, "Vision," W.H. Freeman & Co., San Francisco, Calif. (1982).
5. E.E. Smith and D.L. Medin, "Categories and Concepts," Harvard University Press, Cambridge, Mass. (1981).
6. L. Wittgenstein, "Philosophical Investigations," the English text of the third edition, Macmillan, New York (1958).
7. R. Herrnstein, D. Loveland and C. Cable, Natural Concepts in Pigeons, J. Exp. Psychol. Animal Behav. Processes, 2, 285 (1976).
8. J. Cerella, Mechanisms of Concept Formation in the Pigeon, in "Analysis of Visual Behavior," D.J. Ingle, M.A. Goodale and R.J.W. Mansfield, eds., MIT Press, Cambridge, Mass. (1982).
9. K. Lashley, In Search of the Engram, in "Physiological Mechanisms in Animal Behavior" (Society of Experimental Biology Symposium No. 4), Academic Press, New York, p. 454 (1950).
10. G.M. Edelman, Antibody Structure and Molecular Immunology, Science, 180, 830 (1973).
11. G.M. Edelman, Group Selection and Phasic Reentrant Signalling: a Theory of Higher Brain Function, in "The Mindful Brain," G.M. Edelman and V.B. Mountcastle, eds., The MIT Press, Cambridge, Mass. p. 51 (1978).
12. S. Kirkpatrick, C.D. Gelatt, Jr., and M.P. Vecchi, Optimization by Simulated Annealing, Science, 220, 671 (1983).
13. G.M. Edelman and L.H. Finkel, Neuronal Group Selection in the Cerebral Cortex, in "Dynamical Aspects of Neocortical Function," G.M. Edelman, W.E. Gall and W.M. Cowan, eds., Wiley, New York, p. 653 (1984).
14. G.M. Edelman, Group Selection as the Basis for Higher Brain Function, in "Organization of the Cerebral Cortex," F.O. Schmitt, F.G. Worden, G. Adelman and S.G. Dennis, eds., MIT Press, Cambridge, Mass., p. 51 (1981).
15. M. Minsky, A Framework for Representing Knowledge, in "The Psychology of Computer Vision," P.H. Winston, ed., McGraw-Hill, New York, p. 211 (1975).

16. D.A. Norman and D.G. Bobrow, On the Role of Active Memory Processes in Perception and Cognition, in "The Structure of Human Memory," C.N. Cofer, ed., Freeman, San Francisco, Calif., p. 114 (1976).

17. R.C. Schank, The Role of Memory in Language Processing, in "The Structure of Human Memory," C.N. Cofer, ed., Freeman, San Francisco, Calif., p. 162 (1976).

18. G.M. Edelman and G.N. Reeke, Jr., Selective Networks Capable of Representative Transformations, Limited Generalizations, and Associative Memory, Proc. Natl. Acad. Sci. USA, 79, 2091 (1982).

19. D.O. Hebb, "The Organization of Behavior," Wiley, New York (1949).

20. L.H. Finkel and G.M. Edelman, Interaction of Synaptic Modification Rules within Populations of Neurons, Proc. Natl. Acad. Sci. USA, 82, 1291 (1985).

21. G.N. Reeke, Jr. and G.M. Edelman, Selective Networks and Recognition Automata, Ann. N.Y. Acad. Sci., 426, 181 (1984).

22. T.J. Kellman and E.S. Spelke, Perception of Partly Occluded Objects in Infancy, Cognitive Psych., 15, 483 (1983).

23. O. Selfridge, Pandemonium, a Paradigm for Learning, in "Mechanisation of Thought Processes, I, Natl. Phys. Lab. Symp. No. 10," H.M. Stationery Office, London (1959).

A Neural Theory of Preattentive Visual Information Processing: Emergent Segmentation, Cooperative-Competitive Computation, and Parallel Memory Storage

S. Grossberg* and E. Mingolla**
Center for Adaptive Systems
Boston University
111 Cummington Street
Boston, MA 02215

Abstract

A real-time neural theory of preattentive visual information processing is described. The theory employs specialized neural networks expressed by systems of nonlinear ordinary differential equations. These dynamical systems describe hierarchies of nonlinear filters and cooperative-competitive feedback processes which generate coherent emergent segmentations that are sensitive to a scene's global statistical properties. The emergent segmentations are controlled automatically by internal feedback loops rather than by external parameters. The theory shows how these circuits contextually resolve several basic uncertainties of local visual information processing. Historical antecedents of this theory within the recent decades of work on neural modelling are described, including the role of Liapunov functions in the analysis of memory storage.

1. Introduction: Neural Models of Vision

The foundations of science as a whole, and of physics in particular, await their next great elucidations from the side of biology, and especially from the analysis of sensations . . . Psychological observation on the one side and physical observation on the other may make such progress that they will ultimately come into contact, and that in this way new facts may be brought to light. The result of this investigation will not be a dualism but rather a science which, embracing both the

---

* Supported in part by the Air Force Office of Scientific Research (AFOSR 85-0149 and AFOSR F49620-86-C-0037), the Army Research Office (ARO DAAG-29-85-K-0095), and the National Science Foundation (NSF IST-84-17756).

** Supported in part by the Air Force Office of Scientific Research (AFOSR 85-0149).

organic and inorganic, shall interpret the facts that are common to the two departments.

$-$ E. Mach (1914)[1]

The problems surmounted by mammalian preattentive visual systems and the designs through which solutions are achieved have implications beyond the concerns of the vision specialist. The need to rapidly segment and group functional wholes from background using noisy and varying inputs, often called the "figure-ground" problem, has resulted over the course of evolution in a parallel computational architecture of extraordinary flexibility, robustness, and power. It may surprise the nonspecialist in vision, however, to learn of the *generality* of the principles embodied in visual neural circuit designs. These principles involve the context-sensitive separation of signal and noise from spatially patterned signal sources, the hierarchical resolution of local measurement uncertainties in massively parallel systems designed to cope with such input sources, and the emergence of quantized and coherent representations of the external world within such massively parallel systems.

Preattentive vision is presently an active and exciting research area in several disciplines for a number of reasons. On the one hand, the potential economic payoffs for automating even a portion of the early visual processes are immense, and accordingly many specialized artificial intelligence or machine vision algorithms have been designed in recent years. While these algorithms have had some successes, the simply-stated problems of segmenting and grouping units within input patterns have proven surprisingly intractable. Machine vision algorithms are characteristically *brittle,* performing with marginal adequacy provided stringent constraints on input environments are observed, and failing badly due to deviations from required input conditions. Additionally, many machine algorithms rely on internally generated predictive signals from stored templates of the inputs the algorithm is expected to encounter. Their processes are thereby not *preattentive* in the sense of unfolding automatically without regard to stored expectations or symbolically-mediated scrutiny.

Preattentive vision is an attractive area for scientific investigation apart from commercial incentives, however. Many decades of excellent experimental research have provided a large quantitative data base. Most of that research has been performed in two main disciplines: psychophysics and neurophysiology. From the former, we have precise and often paradoxical measures of the performance of animals and humans in segmenting and grouping visual stimuli which need not be familiar to the subject. From the latter, we have a rather detailed mapping of the connections among visual neurons through the first several synaptic layers.

This large experimental data base on preattentive vision has for decades resisted theoretical unification. To see why this is so, consider one of the most simply stated problems in early vision, that of localizing edges or contours in the input. At a first pass, we would expect to have visual experiences of contours whenever there are physical contours in the visual input; that is, whenever there are spatial discontinuities in a pattern of luminances. Unfortunately for theoretical approaches based solely on simple filtering, however, we also characteristically perceive vivid contours in images in which no corresponding luminance discontinuities exist. Regions separated by visual contours also occur in the presence of: statistical differences in textural qualities such as orientation, shape, density, or color;[2,3] binocular matching of elements of differing disparities in random dot stereograms;[4] accretion and deletion of texture elements in moving displays;[5] and in classical "illusory contours."[6]

The contour percepts just named may strike the nonspecialist in vision as arcane laboratory curiosities not characteristic of "normal" or everyday vision. Nonetheless, just as many of the most fundamental discoveries in physics occurred by considering phenomena of marginal relevance to "everyday" spatial or temporal scales, our work has clarified that anomalous or illusory percepts in many cases hold the key to understanding the adaptive brain designs used in vision. Moreover, as we show later in this chapter, there is nothing obvious about the visual system's mechanisms for detecting even the ostensibly simple and straightforward *actual* luminance discontinuities from the retinal uptake process that constitutes the earliest visual measurement stages.

The major points of interest that we believe our work holds for scientists not working in vision is captured by the words "emergent" and "cooperative-competitive" in this chapter's title. Indeed, the emergence of complex order in the self-organizing dynamics of biological systems is characteristic of many of the most fundamental phenomena under investigation in a variety of disciplines today. Because of the large and structured data base of vision, because so many of us can so readily experience paradoxical visual phenomena for ourselves, and because the resource constraints for models of visual processes are so stringent, preattentive visual segmentation and grouping phenomena offer a unique proving ground for new theoretical ideas about biological computation.

2. Biological versus Metaphorical Approaches to Neural Modelling

In his closing remarks to the Symposium, Professor Hofacker pointed out that two broad approaches to neural network modelling are currently in practice. In one approach, already established models of physical systems are modified to a greater or lesser degree in order to attempt explanations of some brain process. This approach is often characteristic of investigators whose training and research experience has been in scientific areas other than brain theory until their attention was turned to brain processes. It has the advantage of beginning with formalisms whose own dynamics are relatively well understood because of prior successful applications to non-brain systems.

The second approach involves the discovery and development of new neural models in order to more adequately explain and predict data concerning the dynamics of actual neural systems, while also analyzing the formal functional and computational capabilities of these models. Since our own approach is of the latter type, we briefly emphasize why we proceed in such a manner.

A. Brain and Environmentally Adaptive Self-Organization

The brain is *not* hopelessly complicated. Because brains are among the most complex structures ever studied by human scientists, many non-experts feel that almost *any* metaphor of their function is as likely to be correct — or incorrect — as any other. For example, some modellers suggest that essential aspects of brain processes are like simulated annealing, notwithstanding the patent nonexistence of a brain process capable of interpreting an externally controlled temperature that regulates the system towards a phase transition. Natural proponents of such models have tended to come from the ranks of those working with the actual physical systems for which such models are appropriate.

Likewise, as pointed out by Professor Reeke in his talk, the brain has a long history of being modelled by whatever technological system is most advanced at the time, be it telephone switchboard or von Neumann computer. These metaphors have in common an alarming tolerance for gross and fundamental discrepancies between known characteristics of actual brains and the characteristics of the model systems. In no other area of modern science have such patent violations of explanatory fit between model and data been tolerated.

B. The State of Theoretical Knowledge

We are *not* totally ignorant. The claim that we are is often a corollary of the premise that the brain is hopelessly complicated. This belief in the inefficacy of brain theory is held even by many experimentalists working on brain. Due to the importance of understanding why this is so for enabling a theorist to do productive work in brain theory, Grossberg began his 1982 book with a discussion of the scientific issues that have led to this confusing sociological situation. The book[7] begins:

> How is psychology different from physics? What new philosophical and scientific ideas will explicate this difference? Why were the inspiring interdisciplinary successes of Helmholtz, Maxwell, and Mach a century ago followed by a divergence of psychological and physical theory rather than a synthesis? Why has physics rapidly deepened and broadened its theoretical understanding of the world during this century, while psychology has spawned controversy after controversy, as well as dark antitheoretical prejudices?

> The difference between psychology and physics centers in the words evolution and self-oganization. Classical physical theory focusses on a stationary world and the transitions between known physical states. Studies of mind and brain focus on a nonstationary world in which new organismic states are continually being synthesized to form a better adaptive relationship with the environment. These new states can thereupon be maintained in a stable fashion to form a substrate for the synthesis of yet more complex states in a continuing evolutionary progression. Perhaps no better example of the evolutionary process exists than language learning, which is one of the defining characteristics of human civilization.

Whereas physics has gradually fashioned a measurement theory for a stationary world, psychology needs to discover an evolutionary measurement theory, or universal developmental code. Wherease physics has been well served by linear mathematics, the evolutionary psychological processes (development, learning, perception, cognition) depend on nonlinear mathematics. Since the time of Helmholtz, Maxwell and Mach, nineteenth century linear mathematics has stood ready to express and analyze the intuitive insights of physicists interested in electromagnetic theory, relativity, and quantum theory. Students of mind cannot turn to a well-developed body of appropriate mathematics with which to express their deepest intuitions. New nonlinear mathematics must be found that is tailored to these ideas. Scientific revolutions wherein both physical intuitions and mathematical concepts need to be developed side-by-side are especially complex and confusing, but they also offer special intellectual rewards. In the present instance, understanding self-organizing systems is a necessary step towards understanding life itself, both in its individual and collective forms.

Brain studies play a central role in this pursuit for more than the egocentric reasons that brains are the crucibles of all human experience. The brain is a universal measurement device acting on the quantum level. Data from all of our senses — even a few light quanta! — are synthesized by our minds into a common dynamical coin that supports a unitary experience, rather than a series of dislocated experiential fragments. This universality property is the scientific reason, I believe, that brain studies are starting to play a role as central to evolutionary studies as black body radiation played in the development of quantum theory. This universality property clarifies the usefulness of brain theory laws towards explaining a growing body of data about living systems other than brains.

We find ourselves today in a paradoxical and disturbing situation. After physicists abandoned the study of mind, psychological experimentalists were left with an inappropriate world view for understanding each other's data. Personal experimental replication became a major source of security in an atmosphere of conceptual solipsism. Experimentalists dug into paradigms that were sufficiently narrow to maintain the replication criterion. Experimental approaches to mind hereby shattered into a heap of mutually suspicious fiefdoms, and mind theorists became persona non grata. The nature of the crisis and the opportunity facing the brain sciences suggests that a long-range dialog between data and theory should be fostered instead.

In fact, an independent discipline of brain theory has been undergoing rapid development during the past two decades to frontally analyze how brains can adaptively self-organize the behaviors which they control in response to complex and nonstationary fluctuations in their environments. These developments are illustrated by a number of recent books and review articles.[7-19] Many explanations and predictions derived from such neural models have withstood numerous experimental tests, a number of basic theorems about neural designs have been proven, and good working hypotheses are undergoing a period of rapid and intense development. It is no longer scientifically acceptable for people interested in brain modelling to propose a model *ex nihilo,* as if it were sufficient unto itself and immune from the fundamental scientific requirements of comparing its total explanatory power with that of other brain models or of justifying its compatibility with the full range of known data. The indigenous fruits of previous neural modelling efforts provide biologically appropriate and computationally powerful foundations for extending the effort of modelling brain processes.

C. Scientific Communication

The themes of overestimated complexity and underestimated achievement in studying brain have conspired in recent years to bring about an atmosphere among certain recent students of brain theory that, if unchecked, is sure to retard further progress. The most disturbing recent example of this trend is that previously discovered results from brain theory have lately been presented as if new to the physics community, notably in the articles of Hopfield[20,21] and Hopfield and Tank.[22] How these results are embedded within the greater neural modelling literature is summarized below.

3. The Problem of Absolute Stability of Associative Learning and Parallel Memory Storage: The Role of Symmetry and Energy Dissipation

One of the core mathematical problems whose understanding subserves the successful design of neural models of distributed information processing, associative learning, and parallel memory storage is a problem of stability. What classes of neural models are capable of *absolutely stable* computations; that is, computation whose stability is not destroyed by changing the inputs, numerical parameters, or initial states of the system within the broad constraints that define the model class? Absolute stability theorems are of special importance in systems capable of self-organizing in response to nonstationary environments. In such systems, the effects of the environment on system inputs, parameters, and states often cannot be predicted in advance. An absolute stability theorem guarantees that key computational properties of the system will not be subverted by the very process of self-organization that is a primary function of the system's design. By its very nature, an absolute stability theorem requires a *global* analysis of the system's dynamical evolution. By a global analysis we mean one which characterizes how a system evolves starting from *any* physically realizable initial state.

The mathematical theory of global absolute stability for associative learning and long term memory storage by nonlinear neural networks was introduced in the late 1960s.[23-30] The mathematical theory of global absolute stability for pattern transformation and short term memory storage by cooperative-competitive nonlinear neural networks was introduced in the early 1970s.[11,31-38] The latter collection of theorems provided, in fact, the mathematical foundation for designing the neural networks for preattentive vision that are outlined herein. From the perspective of this extensive mathematical literature, and its links to the works of many other authors, notably Amari,[39] Geman,[40] and Kohonen,[18,41] the recent work of Hopfield has a retrospective ring. For example, Hopfield and Tank[22] claimed "a new concept for understanding the dynamics of neural circuitry" using equation

$$C_i \frac{dx_i}{dt} = - \frac{1}{R_i} x_i + \sum_{j=1}^{n} T_{ij} f_j(x_j) + I_i \ (i = 1 \dots n) \tag{1}$$

for the neuron state variables x_i. The concept is that the variables $x_i(t)$ approach equilibrium as $t \rightarrow \infty$ if the connections T_{ij} are symmetric ($T_{ij} = T_{ji}$). Hopfield and Tank also stated that "a nonsymmetric circuit . . . has trajectories corresponding to complicated oscillatory behaviors . . . but as yet we lack the mathematical tools to manipulate and understand them at a computational level," and that "the symmetry of the networks is natural because, in simple associations, if A is associated with B, B is symmetrically associated with A."

Each assertion is contradicted by known results about neural networks. In fact, associations are often asymmetric, as in the asymmetric error distributions arising during list learning.[42-46] Neural network models have explained these distributions[30,47] using Eq. 1 supplemented by an associative learning equation for the connections T_{ij}:

$$\frac{dT_{ij}}{dt} = - A T_{ij} + B x_i f_j(x_j) \tag{2}$$

Thus Eq. 1 is a classical neural network equation — called the *additive model* — which formed the foundation for Grossberg's theory of associative learning and long term memory storage. Moreover, due to the nonlinear term $x_i f_j(x_j)$ in Eq. 2, $T_{ij} \neq T_{ji}$, so that symmetry of associations is, in general, strongly violated.

In addition, the global stability theorems about associative learning and long term memory storage which were proved during the late 1960s and early 1970s included and generalized system $(1) - (2)$. Thus symmetry is not necessary to prove associative learning and long term memory storage by neural networks. Nor is symmetry needed to design stable neural networks for adaptive pattern recognition.[14,18,48,49] Methods have also been developed for analyzing the oscillatory behavior of neural circuits.[17,32,50-55] The relationship between symmetry and stability in neural networks is much more subtle and better understood than Hopfield and Tank[22] have acknowledged.

This conclusion does not deny that symmetry helps to analyze system (1). In fact, Cohen and Grossberg[31] discovered an energy function for a large class of neural networks that includes, as special case, the energy function which Hopfield[21] published to analyze the additive mode (1). Cohen and Grossberg noted that this class of neural networks was "designed to transform and store a large variety of patterns. Our analysis includes systems which possess infinitely many equilibrium points,"[31] examples of which were constructed by Grossberg.[7,33]

The Cohen-Grossberg[31] neural networks are defined by the equations

$$\frac{dx_i}{dt} = a_i(x_i)\left[b_i(x_i) - \sum_{j=1}^{n} c_{ij}d_j(x_j)\right] \quad (i = 1 \ldots n). \tag{3}$$

Given symmetric connections ($c_{ij} = c_{ji}$), these networks admit an energy function

$$V = -\sum_{i=1}^{n} \int^{x_i} b_i(\xi_i)d'_i(\xi_i)d\xi_i + \frac{1}{2}\sum_{j,k=1}^{n} c_{jk}d_j(x_j)d_k(x_k). \tag{4}$$

Along system trajectories,

$$\frac{d}{dt}V = -\sum_{i=1}^{n} a_i(x_i)d'_i(x_i)\left[b_i(x_i) - \sum_{k=1}^{n} c_{ik}d_k(x_k)\right]^2. \tag{5}$$

If $a_i(x_i) \geq 0$ and $d'_i(x_i) \geq 0$, then $\dfrac{d}{dt}V \leq 0$, which is the key property of an energy function. Cohen and Grossberg noted that "the simpler additive neural networks . . . are also included in our analysis."[31] System (3) reduces to the additive network (1) when $a_i(x_i) = C_i^{-1}$, $b_i(x_i) = -\dfrac{1}{R_i}x_i + I_i$, $c_{ij} = -T_{ij}$ and $d_j(x_j) = f_j(x_j)$. Then

$$V = \sum_{i=1}^{n} \frac{1}{R_i}\int^{x_i}\xi_i f'_i(\xi_i)d\xi_i - \sum_{i=1}^{n} I_i f_i(x_i) - \frac{1}{2}\sum_{j,k=1}^{n} T_{jk}f_j(x_j)f_k(x_k), \tag{6}$$

which includes the energy functions claimed by Hopfield[21] and Hopfield and Tank.[22] Cohen and Grossberg[31] also analyzed the much more difficult and physiologically important cases where the cells obey membrane, or shunting, equations and the signal functions $d_j(x_j)$ may have output thresholds. Thus the "new concept" in Hopfield and Tank[22] is a recent special case of an established neural network theory.[31,33,35-37]

Hopfield and Tank also erroneously asserted that: "Unexpectedly, new computational properties resulted . . . from the use of nonlinear graded-response neurons instead of the two-state neurons of the earlier models."[31] It has long been understood that two-state neuronal models differ computationally from graded-response models with sigmoid signal functions.[17,32,33,38,50-57] In fact, Grossberg[33] emphasized the importance of sigmoid signal functions in neural networks after mathematically classifying how different signal functions influence input transformation and memory storage.

One of the two-state neuron models discussed by Hopfield[20] is the classical McCulloch-Pitts[58] model. In this model,

$$x_i(t + 1) = sgn\left[\sum_{j=1}^{n} A_{ij}x_j(t) - B_i\right], \tag{7}$$

where $sgn(w) = 1$ if $w = 0$ and -1 if $w \leq 0$. The McCulloch-Pitts[58] model is one of a family of neural models that has been intensively studied over a 40-year period. For example, Caianiello[59] used a state-equation of the form

$$x_i(t + \tau) = 1\left[\sum_{j=1}^{n}\sum_{k=0}^{l(m)} A_{ij}^{(k)}x_j(t - k\tau) - B_i\right] \tag{8}$$

where $1(w) = 1$ if $w > 0$ and 0 if $w \leq 0$. Rosenblatt[60] used a state-equation of the form

$$\frac{d}{dt}x_i = -Ax_i + \sum_{j=1}^{n}\phi(B_j + x_j)C_{ij}, \tag{9}$$

where $\phi(w) = 1$ if $w \geq \theta$ and 0 if $w < \theta$.

It is easily seen that a generalization of the McCulloch-Pitts model is a discrete approximation of a special case of the additive model, and is thus also subsumed under the Cohen-Grossberg[31] analysis. To see this, generalize Eq. 7 to the state-equation

$$x_i(t + 1) = g\left(\sum_{j=1}^{n} A_{ij}x_j(t)\right). \tag{10}$$

In Eq. 7,

$$g(w) = sgn(w - B_i). \tag{11}$$

Equation 10 is a discrete approximation to a continuous process operating in real-time. Rewriting Eq. 10 as

$$x_i(t + 1) - x_i(t) = -x_i(t) + g\left(\sum_{j=1}^{n} A_{ij}x_j(t)\right) \tag{12}$$

shows that this continuous process is

$$\frac{d}{dt}x_i = -x_i + g\left(\sum_{j=1}^{n}A_{ij}x_j\right).$$ (13)

Introduce the new variable

$$y_i = \sum_{j=1}^{n}A_{ij}x_j.$$ (14)

Then

$$\frac{d}{dt}y_i = -y_i + \sum_{j=1}^{n}A_{ij}g(y_i).$$ (15)

Equation 15 is an additive model with zero external inputs I_i. The Cohen-Grossberg[31] analysis includes continuous signal functions $g(y_i)$ with piece-wise derivatives, such as in Eq. 11. Thus the energy function for system (15) with signal function (11) and symmetric coefficients $A_{ij} = A_{ji}$ is a special case of Eq. 6, namely

$$V = -\frac{1}{2}\sum_{j,k=1}^{n}A_{jk}g(y_j)g(y_k).$$ (16)

The energy function (Eq. 16) generalizes in the continuous case the energy function described by Hopfield[20] for the discrete McCulloch-Pitts equation.

Members of the established neural modelling community have been shocked to see the classical additive Eq. 1 and the classical McCulloch-Pitts Eq. 7 called the Hopfield model by some scientists who are not familiar with the neural modelling literature. This would be analogous to a scientist unfamiliar with physics trying to rename the Schrodinger equation after someone who applied that venerable equation to analyze some atom or molecule in 1986. Neither of these classical equations of the neural modelling field can be renamed to honor work which represents a special case of more powerful results in our literature.

The fact that such a possibility even exists should alert new students of neural networks to the necessity of studying the neural network literature in a scholarly fashion to avoid wasting precious creative energies on recreating known results while the field as a whole surges ahead.

4. General Neural Design Principles versus Specialized Neural Architectures: The Hierarchical Resolution of Informational Uncertainty

The theorems which have been outlined or cited above — particularly those in Cohen and Grossberg,[31] Ellias and Grossberg,[32] Grossberg,[33,35] and Grossberg and Levine[38] — have been necessary, but far from sufficient, to develop an effective analysis of the preattentive visual system. This is because many brain systems are specialized architectures which have evolved over the millenia to adaptively cope with different classes of environments. Correspondingly, the functional anatomies of different brain regions — cerebellum, hypothalamus, visual cortex, hippocampus, peripontine reticular formation, superior colliculus, frontal eye fields, to name a few — exhibit characteristic differences that are preserved across many species. General theorems disclose some of the key design principles from which these specialized neural architectures have been fashioned. Finer considerations are needed to understand the evolutionary parameter selection processes whereby these general designs are specialized to form adaptive feedback exchanges with particular environments.

The specialized architecture of the preattentive visual system rapidly generates emergent segmentations, or groupings, of static or moving visual patterns even though these patterns may involve only inhomogeneities in the spatiotemporal statistics of their energy distributions. The preattentive system does this with about equal ease in response to many different kinds of inhomogeneity, whether of luminance, wavelength, motion, texture, shading, spatial scale, or stereo disparity. Thus a model of preattentive edge or boundary detection that can only deal with one type of visual information does not provide an adequate account of preattentive vision. It is also insufficient to argue that another model or "module" will take care of other image properties if only because the visual system, in the few tens of milliseconds it takes to perform preattentive grouping, cannot know *a priori* whether and for what part of the image to switch on the appropriate module. Instead, the visual system cooperatively uses partial information of many different types to synthesize its emergent segmentations.

Our work has begun to show how the preattentive visual system uses several types of signals to cooperatively reduce ambiguity. In particular, it identifies to several new uncertainty principles with which the visual system is designed to effectively deal, and shows how a carefully chosen real-time neural architecture overcomes these uncertainties using specialized combinations of hierarchically organized nonlinear filters and nonlinear cooperative-competitive feedback networks.

5. Emergent Segmentations from Cooperative-Competitive Interactions

One of the core problems of preattentive visual grouping concerns what the eminent psychologist Jacob Beck first called *emergent features*.[2,3,61-66] Examples of the visual system's segmentation of scenic elements into emergent features may be experienced during inspection of Figs. 1 and 2. From Fig. 1, one may experience long and continuous horizontal groupings even though there are no long and continuous horizontal luminance discontinuities in the image. Likewise one may experience global circular groupings from Fig. 2 even though no large circular luminance contrasts exist in that image. Such figures strikingly illustrate the context-sensitivity of visual measurement: What we experience from any part of an image depends on what surrounds that part. Thus preattentive vision mechanisms are *pattern* processors rather than *pixel* processors.

The cooperative-competitive feedback processes which generate these emergent segmentations cope with several of the types of informational uncertainty with which the visual system is confronted. One type of uncertainty is introduced by the relatively coarse and finite sampling of optical information by the visual system's earliest detectors. These detectors filter weighted averages of incident light energy in both the spatial and temporal domains in order to prevent the proliferation of a large number of highly specialized detectors which are incapable of reacting to several different types of visual information. Thus a certain amount of informational certainty is sacrificed at early processing stages to ensure system versatility, and recovered by cooperative-competitive feedback processes acting at later processing stages to restore system precision. In contrast, reliance on infinitesimal operators, notably differential operators, in visual filters creates a pressure towards the development of specialized models which are incapable of incorporating the design constraints that make biological vision systems so versatile.

The model's cooperative-competitive feedback interactions internally regulate the real-time grouping and convergence of the system to one of a very large number of possible stable equilibria. These nonparametric, internally regulated model mechanisms are quite unlike stochastic relaxation techniques, which rely on the independent, external manipulation of a noise parameter and predetermined probability distributions to regulate convergence to equilibrium. Consequently, stochastic relaxation techniques can, at best, sharpen expected properties of an image. They are unable to simulate the key property of preattentive vision: The automatic discovery of emergent image groupings which may never have been experienced before.

6. The Boundary Contour System and the Feature Contour System

The theory specifies both the functional meaning and the mechanistic interactions of the model microcircuits which comprise the macrocircuit schematized in Fig. 3. This macrocircuit is built up from two systems, the Boundary Contour System (BC System) and the Feature Contour System (FC System).

The BC System controls the emergence of a 3-D segmentation of a scene. This segmentation process is capable of detecting, sharpening, and completing boundaries; of grouping textures; of generating a boundary web of form-sensitive compartments in response to smoothly shaded regions; and of carrying out a disparity-sensitive and scale-sensitive binocular matching process. The outcome of this 3-D segmentation process is perceptually invisible within the BC System. Visible percepts are a property of the FC System.

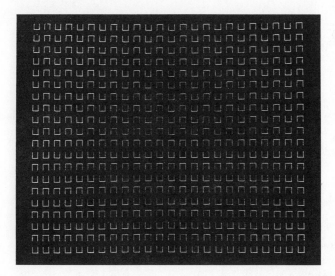

Figure 1. Textural grouping supported by subjective contours: Cooperation among end cuts generates horizontal subjective contours in the bottom half of this figure. *(Reprinted with permission from Ref. 3.)*

A completed segmentation within the BC System elicits topographically organized output signals to the FC System. These completed BC Signals regulate the hierarchical processing of color and brightness signals by the FC System (Fig. 3). Notable among FC System processes are the extraction of color and brightness signals that are relatively uncontaminated by changes in illumination conditions. These Feature Contour signals interact within the FC System with the output signals from the BC System to control featural filling-in processes. These filling-in

Figure 2. A Glass pattern: The emergent circular pattern is "recognized," although it is not "seen," as a pattern of differing contrasts. The text suggests how this can happen. *(Reprinted with permission from Ref. 112.)*

processes lead to visible percepts of color-and-form-in-depth at the final stage of the FC System, which is called the Binocular Syncytium (Fig. 3).

7. Preattentive versus Postattentive Color-Form Interactions

The processes summarized in Fig. 3 are preattentive and automatic. These preattentive processes may, however, influence and be influenced by attentive, learned object recognition processes. The macrocircuit depicted in Fig. 4 suggests, for example, that a preattentively completed segmentation within the BC System can directly activate an Object Recognition System (ORS), whether or not this segmentation supports visible contrast differences within the FC System. The ORS can, in turn, read-out attentive learned priming, or expectation, signals to the BC System. In response to familiar objects in a scene, the final 3-D segmentation within the BC System may thus be *doubly* completed, first by automatic preattentive segmentation processes and then by attentive learned expectation processes. This doubly completed segmentation regulates the filling-in processes within the FC System that lead to a percept of visible form. The FC System also interacts with the ORS. The rules whereby such parallel inputs from the BC System and the FC System are combined within the ORS have been the subject of active experimental investigation.[67-75]

The present theory hereby clarifies two distinct types of interactions that may occur among processes governing segmentation and color perception: preattentive interactions from the BC System to the FC System (Fig. 3) and attentive interactions between the BC System and the ORS and the FC System and the ORS (Fig. 4). The remainder of the chapter summarizes the monocular model mechanisms whereby the BC System and the FC System preattentively interact. This foundation is used to derive the theory's binocular mechanisms.[12,13]

8. Discounting the Illuminant: Extracting Feature Contours

One form of uncertainty with which the nervous system deals is due to the fact that the visual world is viewed under variable lighting conditions. When an object reflects light to an observer's eyes, the amount of light energy within a given wavelength that reaches the eye from each object location is determined by a product of two factors. One factor is a fixed ratio, or reflectance, which determines the fraction of incident light that is reflected by that object location to the eye. The other factor is the variable intensity of the light which illuminates the object location. Two object locations with equal reflectances can reflect different amounts of light to the eye if they are illuminated by different light intensities. Spatial gradients of light across a scene are the rule, rather than the exception, during perception, and wavelengths of light that illuminate a scene can vary widely during a single day. If the nervous system directly coded into percepts the light energies which it received, it would compute false measures of object colors and brightnesses, as well as false measures of object shapes. This problem was already clear to Helmholtz.[76] It

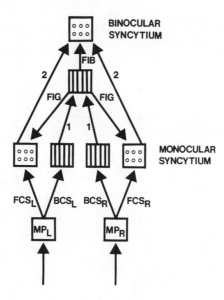

Figure 3. Macrocircuit of monocular and binocular interactions within the Boundary Contour System (BCS) and the Feature Contour System (FCS): Left and right monocular preprocessing stages (MP_L and MP_R) send parallel monocular inputs to the BCS (boxes with vertical lines) and the FCS (boxes with three pairs of circles). The monocular BCS_L and BCS_R interact via bottom-up pathways labelled 1 to generate a coherent binocular boundary segmentation. This segmentation generates output signals called filling-in generators (FIGs) and filling-in barriers (FIBs). The FIGs input to the monocular syncytia of the FCS. The FIBs input to the binocular syncytia of the FCS. References 115 and 116 describe how inputs from the MP stages interact with FIGs at the monocular syncytia to selectively generate binocularly consistent Feature Contour signals along the pathways labelled 2 to the binocular syncytia.

demands an approach to visual perception that points away from a simple Newtonian analysis of colors and white light.

Land[77] and his colleagues have sharpened contemporary understanding of this issue by carrying out a series of remarkable experiments. In these experiments, a picture constructed from overlapping patches of colored paper, called a McCann Mondrian, is viewed under different lighting conditions. If red, green, and blue lights simultaneously illuminate the picture, then an observer perceives surprisingly little color change as the intensities of illumination are chosen to vary within wide limits. The stability of perceived colors obtains despite the fact that the intensity of light at each wavelength that is reflected to the eye varies linearly with the incident illumination intensity at that wavelength. This property of color stability indicates that the nervous system "discounts the illuminant," or suppresses the "extra" amount of light in each wavelength, in order to extract a color percept that is invariant under many lighting conditions.

In an even more striking experimental demonstration of this property, inhomogeneous lighting conditions were devised such that spectrophotometric readings from positions within the interiors of two color patches were the same, yet the two patches appeared to have different colors. The perceived colors were, moreover,

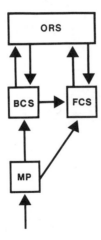

Figure 4. A macrocircuit of processing stages: Monocular preprocessed signals (MP) are sent independently to both the Boundary Contour System (BCS) and the Feature Contour System (FCS). The BCS preattentively generates coherent boundary structures from these MP signals. These structures send outputs to both the FCS and the Object Recognition System (ORS). The ORS, in turn, rapidly sends top-down learned template signals to the BCS. These template signals can modify the preattentively completed boundary structures using learned information. The BCS passes these modifications along to the FCS. The signals from the BCS organize the FCS into perceptual regions wherein filling-in of visible brightnesses and colors can occur. This filling-in process is activated by signals from the MP stage. The completed FCS representation, in turn, also interacts with the ORS.

close to the colors that would be perceived when viewed in a homogeneous source of white light.

These results show that the signals from within the interiors of the colored patches are significantly attenuated in order to discount the illuminant. This property makes ecological sense, since even a gradual change in illumination level could cause a large cumulative distortion in perceived color or brightness if it were allowed to influence the percept of a large scenic region. In contrast, illuminant intensities typically do not vary much across a scenic edge. Thus the ratio of light signals reflected from the two sides of a scenic edge can provide an accurate local estimate of the relative reflectances of the scene at the corresponding positions. We have called the color and brightness signals which remain unattenuated near scenic edges FC signals.

The neural mechanisms which "discount the illuminant" overcome a fundamental uncertainty in the retinal pickup of visual information. In so doing, however, they create a new problem of uncertain measurement, which illustrates one of the classical uncertainty principles of visual perception. If color and brightness signals are suppressed except near scenic edges, then why do we not see just a world of colored edges? How are these local FC signals used by later processing stages to synthesize global percepts of continuous forms, notably of color fields and of smoothly varying surfaces?

9. Featural Filling-In and Stabilized Images

Our monocular theory has developed mechanisms whereby contour-sensitive FC signals activate a process of lateral spreading, or filling-in, of color and brightness signals within the FC System. This filling-in process is contained by topographically organized output signals from the BC System to the FC System (Fig. 3). Where no BC signals obstruct the filling-in process, its strength is attenuated with distance. Our monocular model for this filling-in process was developed and tested using quantitative computer simulations of paradoxical brightness data.[78]

Many examples of featural filling-in and its containment by BC signals can be cited. A classical example of this phenomenon is described in Fig. 5. The image in Fig. 5 was used by Yarbus[79] in a stabilized image experiment. Normally the eye jitters rapidly in its orbit, and thereby is in continual relative motion with respect to a scene. In a stabilized image experiment, prescribed regions in an image are kept stabilized, or do not move with respect to the retina. Stabilization is accomplished by the use of a contact lens or an electronic feedback circuit. Stabilizing an image with respect to the retina can cause the perception of the image to fade.[79-83] The adaptive utility of this property can be partially understood by noting that, in humans, light passes through retinal veins before it reaches the photosensitive retina. The veins form stabilized images with respect to the retina, hence are fortunately not visible under ordinary viewing conditions.

In the Yarbus display shown in Fig. 5, the large circular edge and the vertical edge are stabilized with respect to the retina. As these edge percepts fade, the red color outside the large circle is perceived to flow over and envelop the black and white hemi-discs until it reaches the small red circles whose edges are not stabilized. This percept illustrates how FC signals can spread across, or fill-in, a scenic percept until they hit perceptually significant boundaries.

In summary, the uncertainty of variable lighting conditions is resolved by discounting the illuminant and extracting contour-sensitive FC signals. The uncertainty created within the discounted regions is resolved at a later processing stage via a featural filling-in process that is activated by the FC signals.

10. The Boundary Contour System and the Feature Contour System Obey Different Rules

Figure 6 provides another type of evidence that Feature Contour and Boundary Contour information is extracted by separate, but parallel, neural subsystems before being integrated at a later stage into a unitary percept. The total body of evidence for this new insight takes several forms: the two subsystems obey different rules; they can be used to explain a large body of perceptual data that has received no other unified explanation; they can be perceptually dissociated; when they are interpreted in terms of different neural substrates (the cytochrome-oxidase staining blob system and the hypercolumn system of the striate cortex and their prestriate

Figure 5. A classical example of featural filling-in: When the edges of the large circle and the vertical line are stabilized on the retina, the red color (dots) outside the large circle envelops the black and white hemidisks except within the small red circles whose edges are not stabilized.[79] The red inside the left circle looks brighter and the red inside the right circle looks darker than the enveloping red.

cortical projections), their rules are consistent with known cortical data and have successfully predicted new cortical data.[84-85]

Figure 6 illustrates several more rule differences between the BC System and the FC System. The reproduction process may have weakened the percept of an "illusory" square. The critical percept is that of the square's vertical boundaries. The black-grey vertical edge of the top-left pac-man figure is, relatively speaking, a dark-light vertical edge. The white-grey vertical edge of the bottom-left pac-man figure is, relatively speaking, a light-dark vertical edge. These two vertical edges possess the same orientation but opposite directions-of-contrast. The percept of the vertical boundary that spans these opposite direction-of-contrast edges shows that the BC System is sensitive to boundary orientation but is indifferent to direction-of-contrast. This observation is strengthened by the fact that the horizontal boundaries of the square, which connect edges of like direction-of-contrast, group together with the vertical boundaries to generate a unitary percept of a square. Opposite direction-of-contrast and same direction-of-contrast boundaries both input to the same BC System.

The FC System must, on the other hand, be exquisitely sensitive to direction-of-contrast. If FC signals were insensitive to direction-of-contrast, then it would be impossible to detect which side of a scenic edge possesses a larger reflectance, as in dark-light and red-green discriminations. Thus the rules obeyed by the two contour-extracting systems are not the same.

The BC System and the FC System differ in their spatial interaction rules in addition to their rules of contrast. For example, in Fig. 6 a vertical illusory boundary

Figure 6. A reverse-contrast Kanizsa square: An illusory square is induced by two black and two white pac-man figures on a grey background. Illusory contours can thus join edges with opposite directions of contrast. (This effect may be weakened by the photographic reproduction process.)

forms between the Boundary Contours generated by a pair of vertically-oriented and spatially aligned pac-man edges. Thus the process of boundary completion is due to an *inwardly* directed and *oriented* interaction whereby *pairs* of inducing BC signals can trigger the formation of an intervening boundary of similar orientation. In contrast, in the filling-in reactions of Figs. 4 and 5, featural quality can flow from each FC signal in all directions until it hits a Boundary Contour or is attenuated by its own spatial spread. Thus featural filling-in is an *outwardly* directed and *unoriented* interaction that is triggered by *individual* FC signals.

11. Illusory Percepts as Probes of Adaptive Processes

The adaptive value of a featural filling-in process is clarified by considering how the nervous system discounts the illuminant. The adaptive value of a boundary completion process with properties capable of generating the percept of a Kanizsa square (Fig. 6) can be understood by considering other imperfections of the retinal uptake process. For example, as noted in Section 9, light passes through retinal veins before it reaches retinal photoreceptors. Human observers do not perceive their retinal veins in part due to the action of mechanisms that attenuate the perception of images that are stabilized with respect to the retina. Mechanisms capable of generating this adaptive property of visual percepts can also generate paradoxical percepts, as during the perception of stabilized images or ganzfelds,[79,81-83] including the percept of Fig. 5.

Suppressing the perception of stabilized veins is insufficient to generate an adequate percept. The images that reach the retina can be occluded and segmented by the veins in several places. Broken retinal contours need to be completed, and occluded retinal color and brightness signals need to be filled-in. Holes in the retina, such as the blind spot or certain scotomas, are also not visually perceived[86-88] due to a combination of boundary completion and filling-in processes.[89] These completed boundaries and filled-in colors are illusory percepts, albeit illusory percepts with an important adaptive value. Observers are not aware which parts of such a completed figure are "real" (derived directly from retinal signals) or "illusory" (derived by boundary completion and featural filling-in). Thus in a perceptual theory capable of understanding such completion phenomena, "real" and "illusory" percepts exist on an equal ontological footing. Consequently, we have been able to use the large literature on illusory figures, such as Fig. 6, and filling-in reactions, such as in response to Fig. 5, to help us discover the distinct rules of BC System segmentation and FC System filling-in.[79,90-102]

12. Boundary Contour Detection and Grouping Begins with Oriented Receptive Fields

Having distinguished the BC System from the FC System, the rules whereby boundaries are synthesized are now stated with increasing precision.

In order to effectively build up boundaries, the BC System must be able to determine the orientation of a boundary at every position. To accomplish this, the cells at the first stage of the BC System possess orientationally tuned receptive fields, or oriented masks. Such a cell, or cell population, is selectively responsive to oriented contrasts that activate a prescribed small region of the retina, and whose orientations lie within a prescribed band of orientations with respect to the retina. A collection of such orientationally tuned cells is assumed to exist at every network position, such that each cell type is sensitive to a different band of oriented contrasts within its prescribed small region of the scene, as in the hypercolumn model of Hubel and Wiesel.[103]

These oriented receptive fields illustrate that, from the very earliest stages of BC System processing, image contrasts are grouped and regrouped in order to generate configurations of ever greater global coherence and structural invariance. For example, even the oriented masks at the earliest stage of BC System processing regroup image contrasts (Fig. 7). Such masks are oriented *local contrast* detectors, rather than edge detectors. This property enables them to fire in response to a wide variety of spatially nonuniform image contrasts that do not contain edges, as well as in response to edges. In particular, such oriented masks can respond to spatially nonuniform densities of unoriented textural elements, such as dots. They can also respond to spatially nonuniform densities of surface gradients. Thus by sacrificing a certain amount of spatial resolution in order to detect oriented local contrasts, these masks achieve a general detection characteristic which can respond to boundaries, textures, and surfaces.

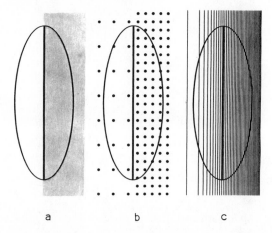

Figure 7. Oriented masks respond to amount of luminance contrast over their elongated axis of symmetry, regardless of whether image contrasts are generated by (a) luminance step functions, (b) differences in textural distribution, or (c) smooth luminance gradients (indicated by the spacings of the lines).

The fact that these receptive fields are *oriented* greatly reduces the number of possible groupings into which their target cells can enter. On the other hand, in order to detect oriented local contrasts, the receptive fields must be elongated along their preferred axis of symmetry. Then the cells can preferentially detect differences of average contrast across this axis of symmetry, yet can remain silent in response to differences of average contrast that are perpendicular to the axis of symmetry. Such receptive field elongation creates even greater positional uncertainty about the exact locations within the receptive field of the image contrasts which fire the cell. This positional uncertainty becomes acute during the processing of image line ends and corners.

13. An Uncertainty Principle: Orientational Certainty Implies Positional Uncertainty at Line Ends and Corners

Oriented receptive fields cannot easily detect the ends of thin scenic lines or scenic corners. This positional uncertainty is illustrated by the computer simulation in Fig. 8. The scenic image is a black vertical line (colored grey for illustrative purposes) against a white background. The line is drawn large to represent its scale relative to the receptive fields that it activates. The activation level of each oriented receptive field at a given position is proportional to the length of the line segment at that position which possesses the same orientation as the corresponding receptive field. The relative lengths of line segments across all positions encode the relative levels of receptive field activation due to different parts of the input pattern. We call such a spatial array of oriented responses an *orientation field*. An orientation field provides a concise statistical description in real-time of an image as seen by the receptive fields that it can activate.

In Fig. 8, a strong vertical reaction occurs at positions along the vertical sides of the input pattern that are sufficiently far from the bottom of the pattern. The con-

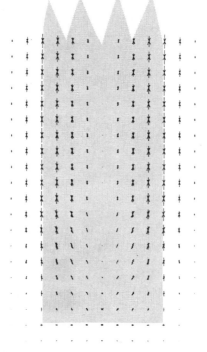

Figure 8. An orientation field: Lengths and orientations of lines encode the relative sizes of the activations and orientations of the input masks at the corresponding positions. The input pattern, which is a vertical line end as seen by the receptive fields, corresponds to the shaded area. Each mask has total exterior dimension of 16×8 units, with a unit length being the distance between two adjacent lattice positions. *(Reprinted with permission from Ref. 104.)*

trast needed to activate these receptive fields was chosen low enough to allow cells with close-to-vertical orientations to be significantly activated at these positions. Despite the fact that cells were tuned to respond to relative low contrasts, the cell responses at positions near the end of the line are very small. Figure 8 thus illustrates a basic uncertainty principle which says: Orientational "certainty" implies positional "uncertainty" at the ends of scenic lines. The next section shows that a perceptual disaster would ensue in the absence of hierarchical compensation for this type of informational uncertainty.

14. Boundary-Feature Trade-Off: A New Organizational Principle

The perceptual disaster in question becomes clear when Fig. 8 is considered from the viewpoint of the featural filling-in process that compensates for discounting the illuminant. If no BC signals are elicited at the ends of lines and at object corners, then in the absence of further processing within the BC System, Boundary Contours will not be synthesized to prevent featural quality from flowing out of all line ends and object corners within the FC System. Many percepts would hereby become badly degraded by featural flow. In fact, as Section 9 indicated, such featural flows occasionally do occur despite compensatory processing.

Thus basic constraints upon visual processing seem to be seriously at odds with each other. The need to discount the illuminant leads to the need for featural filling-in. The need for featural filling-in leads to the need to synthesize boundaries capable of restricting featural filling-in to appropriate perceptual domains. The need to synthesize boundaries leads to the need for orientation-sensitive receptive fields. Such receptive fields are, however, unable to restrict featural filling-in at scenic line ends or sharp corners. Thus, orientational certainty implies a type of positional uncertainty, which is unacceptable from the perspective of featural filling-in requirements. Indeed, an adequate understanding of how to resolve this uncertainty principle is not possible without considering featural filling-in requirements. That is why perceptual theories which have not clearly distinguished the BC System from the FC System have not adequately characterized how perceptual boundaries are formed. We call the design balance that exists between BC System and FC System design requirements the *Boundary-Feature Trade-Off*.

We now summarize how later stages of BC System processing compensate for the positional uncertainty that is created by the orientational tuning of receptive fields.

15. All Line Ends Are Illusory

Figure 9 depicts the reaction of the BC System's next processing stages to the input pattern depicted in Fig. 6. Strong horizontal activations are generated at the end of the scenic line by these processing stages. These horizontal activations are capable of generating a horizontal boundary within the BC System whose output signals prevent flow of featural quality from the end of the line within the FC System. These horizontal activations form an "illusory" boundary, in the sense that this boundary is not directly extracted from luminance differences in the scenic image. The theory suggests that the perceived ends of *all* thin lines are generated by such "illusory" line end inductions, which we call *end cuts*. This conclusion is sufficiently remarkable to summarize it with a maxim: *All line ends are illusory*. This maxim suggests how fundamentally different are the rules which generate geometrical percepts, such as lines and surfaces, from the axioms of geometry that one finds in the great classics of Euclid, Gauss and Riemann.

16. The OC Filter and the Short-Range Competitive Stages

The processing stages that are hypothesized to generate end cuts are summarized in Fig. 10. First, oriented receptive fields of like position and orientation, but opposite direction-of-contrast, cooperate at the next processing stage to activate cells whose receptive fields are sensitive to the same position and orientation as themselves, but are insensitive to direction-of-contrast. These target cells maintain their sensitivity to *amount* of oriented contrast, but not to the *direction* of this oriented contrast, as in our explanation of Fig. 6. Such model cells, which play the role of complex cells

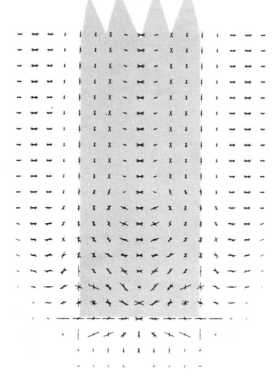

Figure 9. Response of the second competitive stage to the orientation field of Fig. 8: End cutting generates horizontal activations at line end locations that receive small and orientationally ambiguous input activations. *(Reprinted with permission from Ref. 104.)*

in Area 17 of the visual cortex, pools inputs from receptive fields with opposite directions-of-contrast in order to generate boundary detectors which can detect the broadest possible range of luminance of chromatic contrasts. These two successive stages of oriented contrast-sensitive cells are called the OC Filter.[104]

The output from the OC Filter successively activates two types of short-range competitive interaction whose net effect is to generate end cuts. First, a cell of prescribed orientation excites like-oriented cells corresponding to its location and inhibits like-oriented cells corresponding to nearby locations at the next processing stage. In other words, an on-center off-surround organization of like-oriented cell interactions exists around each perceptual location. The outputs from this competitive mechanism interact with the second competitive mechanism. Here, cells compete that represent different orientations, notably perpendicular orientations, at the same perceptual location. This competition defines a push-pull opponent process. If a given orientation is excited, then its perpendicular orientation is inhibited. If a given orientation is inhibited, then its perpendicular orientation is excited via disinhibition.

These competitive rules generate end cuts as follows: The strong vertical activations along the edges of a scenic line, as in Fig. 8, inhibit the weak vertical activations near the line end. These inhibited vertical activations, in turn, disinhibit horizontal activations near the line end, as in Fig. 9. Thus the positional uncertainty generated by orientational certainty is eliminated by the interaction of two short-range competitive mechanisms.

The properties of these competitive mechanisms help to explain many types of perceptual data. For example, they contribute to an explanation of neon color flanks and spreading[104] by showing how some BC signals are inhibited by boundary completion processes. They also clarify many properties of perceptual grouping, notably of the "emergent features" that group textures into figure and ground.[104] Such percepts can be explained by the end cutting mechanism when it interacts with the next processing stage of the BC System.

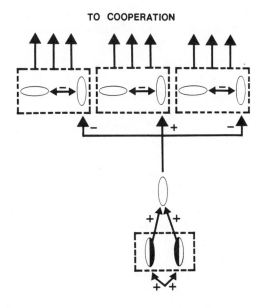

Figure 10. Early stages of Boundary Contour processing: At each position exist cells with elongated receptive fields of various sizes which are sensitive to orientation, amount-of-contrast, and direction-of-contrast. Pairs of such cells sensitive to like orientation but opposite directions-of-contrast (lower dashed box) input to cells that are sensitive to orientation and amount-of-contrast but not to direction-of-contrast (white ellipses). Collectively, these two stages consist of the OC Filter. These cells, in turn, excite like-oriented cells corresponding to the same position and inhibit like-oriented cells corresponding to nearby positions at the first competitive stage. At the second competitive stage, cells corresponding to the same position but different orientations inhibit each other via a push-pull competitive interaction.

17. Long-Range Cooperation: Boundary
Completion and Emergent Features

The outputs from the competition input to a spatially long-range cooperative process, called the *boundary completion* process. This cooperative process helps to build up sharp coherent global boundaries and emergent segmentations from noisy local boundary fragments. In the first stage of this boundary completion process, outputs from the second competitive stage from (approximately) like-oriented cells that are (approximately) aligned across perceptual space cooperate to begin the synthesis of an intervening boundary. For example, such a boundary completion process can span the blind spot and the faded stabilized images of retinal veins. The same boundary completion process is used to complete the sides of the Kanizsa square in Fig. 6. Further details about this boundary completion process can be derived once it is understood that the boundary completion process overcomes a different type of informational uncertainty than is depicted in Fig. 8.

This type of uncertainty is clarified by considering Figs. 11 and 12. In Fig. 11a, a series of radially directed black lines induce an illusory circular contour. This illusion can be understood as a byproduct of four processes: Within the BC System, perpendicular end cuts at the line ends (Fig. 9) cooperate to complete a circular boundary which separates the visual field into two domains. This completed boundary structure sends topographically organized boundary signals into the FC System (Fig. 3), thereby dividing the FC System into two domains. If different filled-in contrasts are induced within these domains due to the FC signals generated by the black scenic lines, then the illusory circle can become visible. No circle is perceived in Fig. 11b because the perpendicular end cuts cannot cooperate to form a closed boundary contour. Hence the FC System is not separated into two domains capable of supporting different filled-in contrasts.

Figure 12a shows that the tendency to form boundaries that are perpendicular to line ends is a strong one; the completed boundary forms sharp corners to keep the boundary perpendicular to the inducing scenic line ends. Figure 12b shows, however, that the boundary completion process can generate a boundary that is not perpendicular to the inducing line ends under certain circumstances.

18. Orientational Uncertainty and the
Initiation of Boundary Completion

A comparison of Figs. 12a and 12b indicates the nature of the other problem of uncertain measurement. Figures 12a and 12b show that boundary completion can occur within *bands* of orientations, which describe a type of real-time local probability distribution for the orientations in which grouping can be initiated at each position. These orientations include the orientations that are perpendicular to their inducing line ends (Fig. 12a), as well as nearby orientations that are not perpendicular to their inducing line ends (Fig. 12b). Figure 9 illustrates how such a band of end cuts can be induced at the end of a scenic line. The existence of such bands

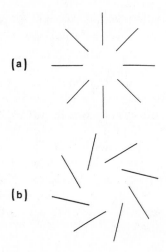

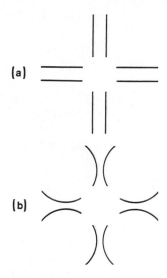

Figure 11. (a) Bright illusory circle induced perpendicular to the ends of the radial lines. (b) Illusory circle becomes less vivid as line orientations are chosen more parallel to the illusory contour. Thus illusory induction is strongest in an orientation perpendicular to the ends of the lines, and its strength depends on the global configuration of the lines relative to one another. *(Adapted with permission from Ref. 113.)*

of possible orientations increases the probability that spatially separated boundary segments can begin to group cooperatively into a global boundary. If only a single orientation at each spatial location were activated, then the probability that these orientations could precisely line up across perceptual space to initiate boundary completion would be vanishingly small. The (partial) orientational uncertainty that

Figure 12. (a) Illusory square generated by changing the orientations, but not the end-points, of the lines in Fig. 11a. In (b), an illusory square is generated by lines with orientations that are not exactly perpendicular to the illusory contour. *(Adapted with permission from Ref. 113.)*

is caused by bands of orientations is thus a useful property for the initiation of the perceptual grouping process that controls boundary completion and textural segmentation.

Such orientational uncertainty can, however, cause a serious loss of acuity in the absence of compensatory processes. If *all* orientations in each band could cooperate with *all* approximately aligned orientations in nearby bands, then a fuzzy band of completed boundaries, rather than a single sharp boundary, could be generated. The existence of such fuzzy boundaries would severely impair visual clarity. Figure 12 illustrates that only a single sharp boundary usually becomes visible despite the existence of oriented bands of boundary inducers. How does the nervous system resolve the uncertainty produced by the existence of orientational bands? How is a single global boundary chosen from among the many possible boundaries that fall within the local oriented bandwidths?

Our answer to these questions suggests a basic reason why later stages of Boundary Contour processing must send nonlinear feedback signals to earlier stages of Boundary Contour processing. This cooperative feedback provides a particular grouping of orientations with a competitive advantage over other possible groupings.

19. Boundary Completion by Cooperative-Competitive Feedback Networks: The CC Loop

We assume, as is illustrated by Fig. 6, that pairs of similarly oriented and spatially aligned cells of the second competitive stage are needed to activate the cooperative cells that subserve boundary completion (Fig. 13). These cells, in turn, feed back excitatory signals to like-oriented cells at the first competitive stage, which feeds into the competition between orientations at each position of the second competitive stage. Thus, in Fig. 13, positive feedback signals are triggered in pathway 2 by a cooperative cell if sufficient activation simultaneously occurs in both of the feedforward pathways labelled 1 from similarly oriented cells of the second competitive stage. Then both pathways labelled 3 can trigger feedback in pathway 4. This feedback exchange can rapidly complete an oriented boundary between pairs of inducing scenic contrasts via a spatially discontinuous bisection process.

Such a boundary completion process realizes a new type of real-time statistical decision theory. Each cooperative cell is sensitive to the position, orientation, density, and size of the inputs that it receives from the second competitive stage. Each cooperative cell performs like a type of statistical "and" gate, since it can only fire feedback signals to the first competitive stage if both of its branches are sufficiently activated. We call such cooperative cells *bipole* cells. The entire cooperative-competitive feedback network is called the CC Loop. The CC Loop can generate a sharp emergent boundary from a fuzzy band of possible boundaries for the following reason.[85,104]

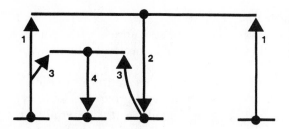

Figure 13. A cooperative-competitive feedback exchange leading to boundary completion: Cells at the bottom row represent like-oriented cells at the second competitive stage whose orientational preferences are approximately aligned across perceptual space. The cells in the top two rows are dipole cells in the cooperative layer whose receptive field pairs are oriented along the axis of the competitive cells. Suppose that simultaneous activation of the pair of pathways 1 activates positive boundary completion feedback along pathway 2. Then pairs of pathways such as 3 activate positive feedback along pathways such as 4. Rapid completion of a sharp boundary between the locations of pathways 1 can hereby be generated by a spatially discontinuous bisection process.

As in Fig. 9, certain orientations at given position are more strongly activated than other orientations. Suppose that the cells which encode a particular orientation at two or more approximately aligned positions can more strongly activate their target bipole cells than can the cells which encode other orientations. Then competitive cells of similar orientation at intervening positions will receive more intense excitatory feedback from these bipole cells. This excitatory feedback enhances the activation of these competitive cells relative to the activation of cells which encode other orientations. This advantage enables the favored orientation to suppress alternative orientations due to the orientational competition that occurs at the second competitive stage (Fig. 10). Cooperative feedback hereby provides the network with autocatalytic, or contrast-enhancing, properties that enable it to choose a single sharp boundary from among a band of possible boundaries by using the short-range competitive interactions. In particular, if in response to a particular image region there are many small-scale oriented contrasts but no *preferred* orientations in which long-range cooperative feedback can act, then the orientational competition can annihilate an emergent long-range cooperative grouping between these contrasts before it can fully form. Thus the CC Loop is designed to sense and amplify the preferred orientations for grouping, and to actively suppress less preferred orientations of potential groupings in which no orientations are preferred. This property is designed into the CC Loop using theorems which characterize the factors that enable cooperative-competitive feedback networks to contrast-enhance their input patterns, and in extreme cases to make choices.[32,33,38]

20. A Nonparametric Real-Time Statistical Decision Process

A preattentive BC System representation emerges when CC Loop dynamics approach a non-zero equilibrium activity pattern. The nonlinear feedback process whereby an emergent line or curve is synthesized need not even define a connected set of activated cells until equilibrium is approached. This property can be seen in Fig. 14, which illustrates how a sharp boundary is rapidly completed between a pair of noisy inducing elements by the spatially discontinuous bisection process in Fig. 13. This process sequentially interpolates boundary components within progressively finer spatial intervals until a connected configuration is attained.

Thus, the CC Loop behaves like an on-line statistical decision machine in response to its input patterns. It senses only those groupings of perceptual elements that possess enough "statistical inertia" to drive its cooperative-competitive feedback exchanges towards a non-zero stable equilibrium configuration. After a boundary structure does emerge from the cooperative-competitive feedback exchange, it is stored in short term memory by the feedback exchange until it is actively reset by the next perceptual cycle. While the boundary is active, it possesses hysteretic and coherent properties due to the persistent suppression of alternative groupings by the competition, the persistent enhancement of the winning grouping by the cooperation, and the self-sustaining activation by the feedback. In addition, the conjoint action of the OC Filter and the CC Loop reconcile two ostensibly conflicting types of perceptual computation. Inputs from the OC Filter to the CC Loop retain their "analog" sensitivity to amount-of-contrast in order to properly bias its operation to favor statistically important image groupings. Once the CC Loop responds to these inputs, it uses its nonlinear feedback loops and long-range cooperative bandwidths to generate a more structural and "digital" representation of the form within the image. Such a coherent boundary structure is qualitatively different from classical definitions of lines and curves in terms of connected sets of points or tangents to these points.

21. Spatial Impenetrability and Textural Grouping: Gated Dipole Field

Figure 15 depicts the results of computer simulations which illustrate how these properties of the CC Loop can generate a perceptual grouping or emergent segmentation of figural elements.[104] Figure 15a depicts an array of nine vertically oriented input clusters. Each cluster is called a Line because it represents a caricature of how a field of OC Filter output cells respond to a vertical line. Figure 15b displays the equilibrium activities of the cells at the second competitive state of the CC Loop in response to these Lines. The length of an oriented line at each position is proportional to the equilibrium activity of a cell whose receptive field is centered at that position with that orientation. The input pattern in Fig. 15a possesses a vertical symmetry: Triples of vertical Lines are colinear in the vertical direction, whereas they are spatially out-of-phase in the horizontal direction. The BC System senses this vertical symmetry, and generates emergent vertical boundaries in Fig. 15b. The BC System also generates horizontal end cuts at the ends of each Line, which can trap the featural contrasts of each Line within the FC System. Thus the emergent segmentation simultaneously supports a vertical macrostructure and a horizontal microstructure among the Lines.

In Fig. 15c, the input Lines are moved so that triples of Lines are colinear in the vertical direction and their Line ends are lined up in the horizontal direction. Both vertical and horizontal boundary groupings are generated in Fig. 15d. The segmentation distinguishes between Line ends and the small horizontal inductions that

REAL TIME BOUNDARY COMPLETION

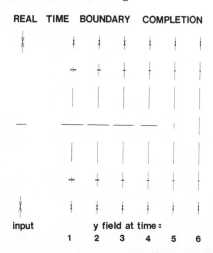

input   y field at time =
                1   2   3   4   5   6

Figure 14. Each column depicts the same band of positions at the second competitive stage (y field) at a different time during the boundary completion process. The input (leftmost column) consists of two noisy but vertically biased inducing line elements and an intervening horizontal line element. Line lengths are proportional to the activities of cells with the represented positions and orientational preferences. The competitive-cooperative feedback exchange triggers transient almost horizontal end cuts before attenuating all nonvertical elements as it completes a sharp emergent vertical boundary.

bound the sides of each Line. Only Line ends have enough statistical inertia to activate horizontal boundary completion via the CC Loop.

In Fig. 15e, the input Lines are shifted so that they become non-colinear in a vertical direction, but triples of their Line ends remain aligned. The vertical symmetry of Fig. 15c is hereby broken. Consequently, in Fig. 15f the BC System groups the horizontal Line ends, but not the vertical Lines.

Figure 15h depicts the emergence of diagonal groupings where no diagonals exist in the input pattern. Figure 15g is generated by bringing the three horizontal rows of vertical Lines close together until their ends lie within the spatial bandwidth of the cooperative interaction. In Fig. 15h, the BC System senses diagonal groupings of the Lines. Diagonally oriented receptive fields are activated in the emergent boundaries, and these activations, as a whole, group into diagonal bands. Thus these diagonal groupings emerge on both microscopic and macroscopic scales.

The computer simulations illustrated in Fig. 15 show that the CC Loop can generate large-scale segmentations without a loss of positional or orientational acuity. In order to achieve this type of acuity, the CC Loop is designed to realize the *postulate of spatial impenetrability*.[104] This postulate was imposed to prevent the long-range cooperative process from leaping over all intervening images and grouping together inappropriate combinations of inputs. The mechanism which realizes the postulate must not prevent like-oriented responses from cooperating across spatially aligned positions, since such grouping is a primary function of the cooperation. The mechanism does, however, need to prevent like-oriented responses from cooper-

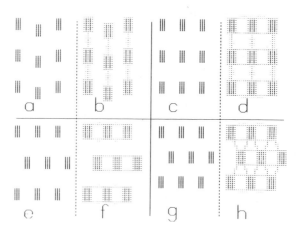

Figure 15. Computer simulations of processes underlying textural grouping: The length of each line segment is proportional to the activation of a network node responsive to one of twelve possible orientations. Parts (a), (c), (e), and (g) display the activities of oriented cells which input to the CC Loop. Parts (b), (d), (f), and (h) display equilibrium activities of oriented cells at the second competitive stage of the CC Loop. A pairwise comparison of (a) with (b), (c) with (d), and so on indicates the major groupings sensed by the network. *(Reprinted with permission from Ref. 112.)*

ating across a region of (approximately) *perpendicularly* oriented responses. In particular, it prevents the horizontal end cuts in Fig. 15 which are separated by the vertically oriented responses to each Line from activating a receptive field of a dipole cell. As a result, only end cuts at the Line *ends* can cooperate to form horizontal boundaries which span two or more Lines.

The postulate of spatial impenetrability can be realized by modelling the second competitive stage as a gated dipole field.[37,105] Figure 16 joins together the OC Filter with a CC Loop whose second competitive stage is a gated dipole field. Such a circuit was used to generate the computer output illustrated by Fig. 15.

In the gated dipole field of Fig. 16, the first competitive stage delivers inputs to the on-cells of the field. As previously described, such an input excites like-oriented on-cells at its own position and inhibits like-oriented on-cells at nearby positions. As previously described, on-cells at a given position compete among orientations at the second competitive stage. In addition to on-cells, a gated dipole field also possesses an off-cell population corresponding to each on-cell population. In the network in Fig. 16, on-cells inhibit off-cells which represent the same position and orientation. Off-cells at each position, in turn, compete among orientations. Both on-cells and off-cells are driven by a source of tonic activity, which is kept under control by their inhibitory interactions. Thus an input which excites vertically oriented on-cells at a given position can also inhibit vertically-oriented off-cells and horizontally-oriented on-cells at that position. In addition, due to the inhibition of like-oriented on-cells at nearby positions, vertically-oriented off-cells and horizontally-oriented on-cells can be excited due to disinhibition at these nearby positions.

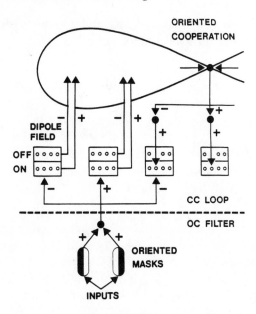

Figure 16. Circuit diagram of the Boundary Contour System: Inputs activate oriented masks of opposite direction-of-contrast which cooperate at each position and orientation before feeding into an on-center off-surround interaction. This interaction excites like-orientations at the same position and inhibits like-orientations at nearby positions. The affected cells are on-cells within a dipole field. On-cells at a fixed position compete among orientations. On-cells also inhibit off-cells which represent the same position and orientation. Off-cells at each position, in turn, compete among orientations. Both on-cells and off-cells are tonically active. Net excitation of an on-cell excites á similarly oriented cooperative receptive field at a location corresponding to that of the on-cell. Net excitation of an off-cell inhibits a similarly oriented cooperative receptive field of a bipole cell at a location corresponding to that of the off-cell. Thus, bottom-up excitation of a vertical on-cell, by inhibiting the horizontal on-cell at that position, disinhibits the horizontal off-cell at that position, which in turn inhibits (almost) horizontally oriented cooperative receptive fields that include its position. Sufficiently strong net positive activation of both receptive fields of a cooperative cell enables it to generate feedback via an on-center off-surround interaction among like-oriented cells. On-cells which receive the most favorable combination of bottom-up signals and top-down signals generate the emergent perceptual grouping.

Spatial impenetrability is achieved by assuming that active on-cells send excitatory signals, whereas active off-cells send inhibitory signals, to the similarly oriented receptive fields of bipole cells (Fig. 16). Consequently, if horizontally oriented on-cells are active at a given position, they will not be able to activate a horizontally oriented bipole receptive field if sufficiently many vertically oriented on-cells are also active at positions within this receptive field. Each bipole receptive field can help to activate its bipole cell only if its *total* input is sufficiently positive. A bipole cell can only fire if *both* of its receptive fields receive positive total inputs. Sufficiently strong net positive activation of both receptive fields of a bipole cell enables the cell to generate feedback to like-oriented on-cells at the first competitive stage via an on-center off-surround interaction. Thus both bottom-up inputs and top-down cooperative feedback access the first competitive stage via an on-center off-surround interaction among like-oriented on-cells. On-cells which receive the most favorable combination of bottom-up inputs and top-down signals remain active within the emergent boundary segmentation, as in Fig. 15.

22. Formal Dynamics of the Monocular BC System

As of this writing, properties of the BC System and the FC System have been used to explain and predict a perceptual and neural data base of unprecedented size and variety. The remainder of the article briefly summarizes the dynamical equations which have been used to formally define and simulate monocular properties of the OC Filter and CC Loop of the BC System.

A. Oriented Masks

To define a mask, or oriented receptive field, centered at position (i,j) with orientation k, divide the elongated receptive field of the mask into a left-half L_{ijk} and a right-half R_{ijk}. Let all the masks sample a field of preprocessed inputs. If S_{pq} equals the preprocessed input to position (p,q) of this field, then the output J_{ijk} from the mask at position (i,j) with orientation k is

$$J_{ijk} = \frac{[U_{ijk} - \alpha V_{ijk}]^+ + [V_{ijk} - \alpha U_{ijk}]^+}{1 + \beta(U_{ijk} + V_{ijk})} \tag{17}$$

where

$$U_{ijk} = \sum_{(p,\,q)\in L_{ijk}} S_{pq}, \tag{18}$$

$$V_{ijk} = \sum_{(p,\,q)\in R_{ijk}} S_{pq}, \tag{19}$$

and the notation $[p]^+ = \max(p, 0)$. The sum of the two terms in the numerator of Eq. 17 says that J_{ijk} is sensitive to the orientation and amount-of-contrast, but not to the direction-of-contrast, received by L_{ijk} and R_{ijk}. The denominator term in Eq. 17 enables J_{ijk} to compute a ratio scale in the limit where $\beta(U_{ijk} + V_{ijk})$ is much greater than 1. In all of our simulations, we have chosen $\beta = 0$.

B. On-Center Off-Surround Interaction within Each Orientation

Inputs J_{ijk} with a fixed orientation k activate potentials w_{ijk} at the first competitive stage via on-center off-surround interations: each J_{ijk} excites w_{ijk} and inhibits w_{pqk} if

$|p - i|^2 + |q - j|^2$ is sufficiently small. All the potentials w_{ijk} are also excited by the same tonic input I, which supports disinhibitory activations at the next competitive stage. Thus

$$\frac{d}{dt} w_{ijk} = -w_{ijk} + I + f(J_{ijk}) - w_{ijk} \sum_{(p,\,q)} f(J_{pqk}) A_{pqij},\tag{20}$$

where A_{pqij} is the inhibitory interaction strength between positions (p,q) and (i,j) and $f(J_{ijk})$ is the input signal generated by J_{ijk}. In our runs, we chose

$$f(J_{ijk}) = BJ_{ijk}.\tag{21}$$

Sections (C) and (D) together define the on-cell subfield of the dipole field described in Section 21.

C. Push-Pull Opponent Processes between Orientations at Each Position

Perpendicular potentials w_{ijk} and w_{ijK} elicit output signals that compete at their target potentials x_{ijk} and x_{ijK}, respectively. For simplicity, we assume that these output signals equal the potentials w_{ijk} and w_{ijK}, which are always nonnegative. We also assume that x_{ijk} and x_{ijK} respond quickly and linearly to these signals. Thus

$$x_{ijk} = w_{ijk} - w_{ijK}\tag{22}$$

and

$$x_{ijK} = w_{ijK} - w_{ijk}.\tag{23}$$

D. Normalization at Each Position

We also assume that, as part of this push-pull opponent process, the outputs y_{ijk} of the second competitive stage become normalized. Several ways exist for achieving this property.[106] We have used the following approach.

The potentials x_{ijk} interact when they become positive. Thus we let the output $O_{ijk} = 0(x_{ijk})$ from x_{ijk} equal

$$O_{ijk} = C[w_{ijk} - w_{ijK}]^+\tag{24}$$

where C is a positive constant and $|p|^+ = \max(p, 0)$. All these outputs at each position interact via a shunting on-center off-surround network whose potentials y_{ijk} satisfy

$$\frac{d}{dt}y_{ijk} = -Dy_{ijk} + (E - y_{ijk})O_{ijk} - y_{ijk}\sum_{m\neq k}O_{ijm}. \tag{25}$$

Each potential y_{ijk} equilibrates rapidly to its input. Setting $\frac{d}{dt}y_{ijk} = 0$ in Eq. 25 implies that

$$y_{ijk} = \frac{EO_{ijk}}{D + O_{ij}} \tag{26}$$

where

$$O_{ij} = \sum_{m=1}^{n}O_{ijm}. \tag{27}$$

Thus if D is small compared to O_{ij}, then $\sum\limits_{m=1}^{n}y_{ijm} \cong E$.

E. Opponent Inputs to the Cooperative Stage

The next process refines the BCS model used in Ref. 85. It helps to realize the Postulate of Spatial Impenetrability that was described in Section 21. The w_{ijk}, x_{ijk}, and y_{ijk} potentials are all assumed to be part of the on-cell subfield of a dipole field. If y_{ijk} is excited, an excitatory signal $f(y_{ijk})$ is generated at the cooperative stage. When potential y_{ijk} is excited, the potential y_{ijK} corresponding to the perpendicular orientation is inhibited. Both of these potentials form part of the on-cell subfield of a dipole field. Inhibition of an on-cell potential y_{ijK} disinhibits the corresponding off-cell potential $\bar{y}_{ijK}$, which sends an inhibitory signal $-f(\bar{y}_{ijK})$ to the cooperative level. The signals $f(y_{ijk})$ and $-f(\bar{y}_{ijK})$ thus occur together. In order to instantiate these properties, we made the simplest hypothesis, namely that

$$\bar{y}_{ijK} = y_{ijk}. \tag{28}$$

F. Oriented Cooperation: Statistical Gates

The cooperative potential z_{ijk} can be supraliminally activated only if both of its cooperative input branches receive enough net positive excitation from similarly aligned competitive potentials (Fig. 9). Thus

$$\frac{d}{dt}z_{ijk} = -z_{ijk} + g\left(\sum_{(p,q,r)} [f\,(y_{pqr})]F_{pqij}^{(r,k)} \right)$$

$$+ g\left(\sum_{(p,q,r)} [fnot(y_{pqr})]\,G_{pqij}^{(r,k)} \right) \tag{29}$$

In Eq. 29, $g(s)$ is a signal function that becomes positive only when s is positive, and has a finite maximum value. A slower-than-linear function

$$g(s) = \frac{H[s]^+}{K + [s]^+} \tag{30}$$

was used in our simulations. A sum of two sufficiently positive $g(s)$ terms in Eq. 29 is needed to activate z_{ijk} above the firing threshold of its output signal $h(z_{ijk})$. A threshold-linear signal function

$$h(z) = L[z - M]^+ \tag{31}$$

was used. Each sum such as

$$\sum_{(p,q,r)} f\,(y_{pqr})\,F_{pqij}^{(r,k)} \tag{32}$$

and

$$\sum_{(p,q,r)} f\,(y_{pqr})\,G_{pqij}^{(r,k)} \tag{33}$$

is a spatial cross-correlation that adds up inputs from a strip with orientation (approximately equal to) k that lies to one side or the other of position (i,j), as in Figs. 31 and 32. The orientations r that contribute to the spatial kernels $F_{pqij}^{(r,k)}$ and $G_{pqij}^{(r,k)}$ also approximately equal k. The kernels $F_{pq}^{(r,k)}$ and $G_{pq}^{(r,k)}$ are defined by

$$F_{pqij}^{(r,k)} = \left[\exp\left[-2\left(\frac{N_{pqij}}{P} - 1\right)^2\right] [|\cos(Q_{pqij} - r)|]^R [\cos(Q_{pqij} - k)]^T \right]^+ \tag{34}$$

and

$$G_{pqij}^{(r,k)} = \left[- \exp\left[- 2\left(\frac{N_{pqij}}{P} - 1\right)^2\right] \left[\,|\cos(Q_{pqij} - r)|\,\right]^R \left[\cos(Q_{pqij} - k)\right]^T \right]^+,$$

$$(35)$$

where

$$N_{pqij} = \sqrt{(p - i)^2 + (q - j)^2}\,,\tag{36}$$

$$Q_{pqij} = \arctan\left(\frac{q - j}{p - i}\right),\tag{37}$$

and P, R, and T are positive constants. In particular, R and T are odd integers. Kernels F and G differ only by a minus sign under the $[\,\ldots\,]^+$ sign. This minus sign determines the polarity of the kernel; namely, whether it collects inputs for z_{ijk} from one side or the other of position (i,j). Term $\exp\left[-2\left(\frac{N_{pqij}}{P} - 1\right)^2\right]$ determines the optimal distance P from (i,j) at which each kernel collects its inputs. The kernel decays in a Gaussian fashion as a function of N_{pqij}/P, where N_{pqij} in Eq. 36 is the distance between (p,q) and (i,j). The cosine terms in Eqs. 34 and 35 determine the orientational tuning of the kernels. By Eq. 37, Q_{pqij} is the direction of position (p,q) with respect to the position of the cooperative cell (i,j) in Eq. 29. Term $|\cos(Q_{pqij} - r)|$ in Eqs. 34 and 35 computes how parallel Q_{pqij} is to the receptive field orientation r at position (p,q). By Eq. 37, term $|\cos(Q_{pqij} - r)|$ is maximal when the orientation r equals the orientation of (p,q) with respect to (i,j). The absolute value sign around this term prevents it from becoming negative. Term $\cos(Q_{pqij} - k)$ in Eqs. 34 and 35 computes how parallel Q_{pqij} is to the orientation k of the receptive field of the cooperative cell (i,j) in Eq. 29. By Eq. 37, term $\cos(Q_{pqij} - k)$ is maximal when the orientation k equals the orientation of (p,q) with respect to (i,j). Positions (p,q) such that $\cos(Q_{pqij} - k) < 0$ do not input to z_{ijk} via kernel F because the $[\,\ldots\,]^+$ of a negative number equals zero. On the other hand, such positions (p,q) may input to z_{ijk} via kernel G due to the extra minus sign in the definition of kernel G. The extra minus sign in Eq. 35 flips the preferred axis of orientation of kernel $G_{pqij}^{(r,k)}$ with respect to the kernel $F_{pqij}^{(r,k)}$ in order to define the two input-collecting branches of each cooperative cell, as in Figs. 8 and 9. The product terms $\pm\,|\cos(Q_{pqji} - r)|^R \cos(Q_{pqij} - k)^T$ in Eqs. 34 and 35 thus determine larger path weights from dipole field on-cells whose positions and orientations are nearly parallel to the preferred orientation k of the cooperative cell (i,j), and larger path weights from dipole field off-cells whose positions and orientations are nearly perpendicular to the preferred orientation k of the cooperative cell (i,j). The powers R and T determine the sharpness of orientational tuning: Higher powers enforce sharper tuning.

G. On-Center Off-Surround Feedback within Each Orientation

We assume that each z_{ijk} activates a shunting on-center off-surround interaction within each orientation k. The target potentials v_{ijk} therefore obey an equation of the form

$$\frac{d}{dt}v_{ijk} = -v_{ijk} + h(z_{ijk}) - v_{ijk}\sum_{(p,q)}h(z_{pqk})W_{pqij}. \tag{38}$$

The bottom-up transformation $J_{ijk} \to w_{ijk}$ in Eq. 20 is thus similar to the top-down transformation $z_{ijk} \to v_{ijk}$ in Eq. 36. Functionally, the $z_{ijk} \to v_{ijk}$ transformation enables the most favored cooperations to enhance their preferred positions and orientation as they suppress nearby positions with the same orientation. The signals v_{ijk} take effect by inputting to the w_{ijk} opponent process. Equation 20 is thus changed to

$$\frac{d}{dt}w_{ijk} = -w_{ijk} + I + f(J_{ijk}) + v_{ijk} - w_{ijk}\sum_{(p,q)}f(J_{pqk})A_{pqij}. \tag{39}$$

At equilibrium, the computational logic of the BCS is determined, up to parameter choices, by the equations

$$J_{ijk} = \frac{[U_{ijk} - \alpha V_{ijk}]^+ + [V_{ijk} - \alpha U_{ijk}]^+}{1 + \beta(U_{ijk} + V_{ijk})}, \tag{17}$$

$$w_{ijk} = \frac{I + BJ_{ijk} + v_{ijk}}{1 + B\sum_{(p,q)}J_{pqk}A_{pqij}}, \tag{40}$$

$$O_{ijk} = C[w_{ijk} - w_{ijK}]^+, \tag{24}$$

$$y_{ijk} = \frac{EO_{ijk}}{D + O_{ij}}, \tag{26}$$

$$z_{ijk} = g\left(\sum_{(p,q,r)}[f(v_{pqr}) - f(v_{pqR})]F^{(r,k)}_{pqij}\right) + g\left(\sum_{(p,q,r)}[f(v_{pqr}) - f(v_{pqR})]G^{(r,k)}_{pqij}\right), \tag{41}$$

and

$$v_{ijk} = \frac{h(z_{ijk})}{1 + \sum_{(p,q)} h(z_{pqk}) W_{pqij}}. \tag{42}$$

Wherever possible, simple spatial kernels were used. For example the kernels W_{pqij} in Eq. 38 and A_{pqij} in Eq. 39 were both chosen to be constant within a circular receptive field:

$$A_{pqij} = \begin{cases} A \text{ if } (p - i)^2 + (q - j)^2 \le A_0 \\ 0 \quad \text{otherwise} \end{cases} \tag{43}$$

and

$$W_{pqij} = \begin{cases} W \text{ if } (p - i)^2 + (q - j)^2 \le W_0 \\ 0 \text{ otherwise.} \end{cases} \tag{44}$$

Concluding Remarks: Collective Computation and the Gestalt Program

The textural segmentation process in preattentive vision is exquisitely context-sensitive. As illustrated above, a scenic element at a given location can be part of a variety of larger groupings, depending on what surrounds it. Indeed, the precise determination even of what acts as an *element* at a given location can depend on patterns at nearby locations.

One of the greatest sources of difficulty in understanding visual perception and in designing fast object recognition systems is such context-sensitivity of perceptual units. Since the work of the Gestaltists,[107] it has been widely recognized that local features of a scene, such as edge positions, disparities, lengths, orientations, and contrasts, are perceptually ambiguous, but that combinations of these features can be quickly grouped by a perceiver to generate a clear separation between figures, and between figure and ground. Indeed, a figure within a textured scene often seems to "pop out" from the ground.[108] The "emergent" features by which an observer perceptually groups the "local" features within a scene are sensitive to the global structuring of textural elements within the scene.

The fact that these emergent perceptual units, rather than local features, are used to group a scene carries with it the possibility of scientific chaos. If every scene can define its own context-sensitive units, then perhaps object perception could only be described in terms of an unwieldy taxonomy of scenes and their unique perceptual units. One of the great accomplishments of the Gestalt psychologists was to suggest a short list of rules for perceptual grouping that helped to organize many interesting examples. These rules characterized conditions for perceptual grouping into such intuitively specified perceptual categories as proximity, colinearity, or "good continuation." As is often the case in pioneering work, the rules were neither always obeyed nor exhaustive. No justification for the rules was given other than their evident plausibility. More seriously for practical applications, no effective computational algorithms were given to instantiate the rules.

Until recently, no theory had provided a *raison d'etre* for textural grouping or a computational framework for dynamically explaining how textural elements are grouped, in real-time, into easily separated figures and ground. One manifestation of this gap in contemporary understanding can be found in the image processing models that have been developed by workers in artificial intelligence. In this approach, curves are analyzed using different models from those that are used to analyze textures, and textures are analyzed using different models from the ones used to analyze surfaces.[109-110] All of these models are built up using geometrical ideas — such as surface normal, curvature, and Laplacian — that were used to study visual perception during the nineteenth century.[111] These geometrical ideas were originally developed to analyze *local* properties of physical processes. Instead, the visual system's context-sensitive mechanisms routinely synthesize figural percepts that are not reducible to local luminance differences within a scenic image. Such emergent properties are not just the effect of local geometrical transformations.

Our recent work suggests that nineteenth century geometrical ideas are fundamentally inadequate to characterize the designs that make biological visual systems so efficient. This claim has arisen from the discovery of new mechanisms that are not designed to compute local geometrical properties of a scenic image. These mechanisms are defined by parallel and hierarchical interactions within very large networks of interacting neurons. The visual properties that these equations compute emerge as the collective modes of network interactions, rather than from local transformations.

Notwithstanding their lack of an adequate formal mechanism, the Gestaltists' intuitions were insightful and fundamentally correct. Our results show that several apparently different Gestalt rules can be analyzed using the context-sensitive reactions of a single BC System. Taken together, these results suggest that a *universal* set of rules for perceptual grouping of scenic edges, textures, and smoothly shaded regions by preattentive vision is well on the way to being characterized.

Acknowledgments

We wish to thank Cynthia Suchta and Carol Yanakakis for their valuable assistance in the preparation of the manuscript and illustrations.

References

1.   E. Mach, The analysis of sensation and the relation of the physical to the psychical, C.M. Williams (Trans.), revised by S. Waterlow, Open Court Publishing Co., London (1914).

2.   J. Beck, Perceptual grouping produced by changes in orientation and shape, Science, **154**, 538 (1966).

3.   J. Beck, K. Prazdny and A. Rosenfeld, A theory of textural segmentation, in "Human and machine vision," J. Beck, B. Hope and A. Rosenfeld (eds.), Academic Press, New York (1983).

4.   B. Julesz, Binocular depth perception of computer-generated patterns, Bell System Technical Journal, **39**, 1125 (1960).

5.   G.A. Kaplan, Kinetic disruption of optical texture: The perception of depth at an edge, Perception and Psychophysics, **6**, 193 (1969).

6.   G. Kanizsa, Margini quasi-percettivi in campi con stimolazione omegenea, Revista di Psicologia, **49**, 7 (1955).

7.   S. Grossberg, Studies of mind and brain: Neural principles of learning, perception, development, cognition, and motor control, Reidel Press, Boston (1982).

8.   S. Amari and M.A. Arbib (eds.), Competition and cooperation in neural networks, Springer-Verlag, New York (1982).

9.   E. Basar, H. Flohr, H. Haken and A.J. Mandell (eds.), Synergetics of the brain, Springer-Verlag, New York (1983).

10.   G.A. Carpenter and S. Grossberg, Associative learning, adaptive pattern recognition, and cooperative-competitive decision making by neural networks, in "Hybrid and optical computing," H. Szu (ed.), SPIE, in press (1985).

11.   S. Grossberg, Adaptive resonance in development, perception, and cognition, in "Mathematical psychology and psychophysiology," S. Grossberg (ed.), American Mathematical Society, Providence, RI (1981).

12.   S. Grossberg (ed.), The adaptive brain, I: Cognition, learning, reinforcement, and rhythm, Elsevier/North-Holland, Amsterdam (1986).

13.   S. Grossberg, (ed.), The adaptive brain, II: Vision, speech, language, and motor control, Elsevier/North-Holland, Amsterdam (1986).

14.   S. Grossberg and M. Kuperstein, Neural dynamics of adaptive sensory-motor control: Ballistic eye movements, Elsevier/North-Holland, Amsterdam (1986).

15.   R. Hecht-Nielsen, Nearest matched filter classification of spatiotemporal patterns, in "Hybrid and optical computing," H. Szu (ed.), SPIE, in press (1986).

16.   D. Hestenes, Maximum entropy and Bayesian spectral analysis and estimation problems, C.R. Smith (ed.), Reidel Press, Boston (1986).

17.   J.P.E. Hodgson (ed.), Oscillations in mathematical biology, Springer-Verlag, New York (1983).

18.   T. Kohonen, Self-organization and associative memory, Springer-Verlag, New York (1984).

19.   D.S. Levine, Neural population modelling and psychology: A review, Mathematical Biosciences, **66**, 1 (1983).

20.   J.J. Hopfield, Neural networks and physical systems with emergent collective computational abilities, Proceedings of the Naitonal Academy of Sciences USA, **79**, 2554 (1982).

21. J.J. Hopfield, Neurons with graded response have collective computational properties like those of two-state neurons, Proceedings of the National Academy of Sciences USA, **81**, 3088 (1984).

22. J.J. Hopfield and D.W. Tank, Computing with neural circuits: A model, Science, **233**, 625 (1986).

23. S. Grossberg, Nonlinear difference-differential equations in prediciton and learning theory, Proceedings of the National Academy of Sciences USA, **58**, 1329 (1967).

24. S. Grossberg, Some nonlinear networks capable of learning a spatial pattern of arbitrary complexity, Proceedings of the National Academy of Sciences, **59**, 368 (1968).

25. S. Grossberg, On learning and energy-entropy dependence in recurrent and nonrecurrent signed networks, Journal of Statistical Physics, **1**, 319 (1969).

26. S. Grossberg, Some networks that can learn, remember, and reproduce any number of complicated space-time patterns II, Studies in Applied Mathematics, **49**, 135 (1970).

27. S. Grossberg, Pavlovian pattern learning by nonlinear neural networks, Proceedings of the National Academy of Sciences USA, **68**, 828 (1971).

28. S. Grossberg, Pattern learning by functional-differential neural networks with arbitrary path weights, in "Delay and functional-differential equations and their applications," K. Schmitt (ed.), Academic Press, New York (1972).

29. S. Grossberg, Classical and instrumental learning by neural networks, in "Progress in theoretical biology," Vol. 3, R. Rosen and F. Snell (eds.), Academic Press, New York (1974).

30. S. Grossberg and J. Pepe, Spiking threshold and overarousal effects in serial learning, Journal of Statistical Physics, **3**, 95 (1971).

31. M.A. Cohen and S. Grossberg, Absolute stability of global pattern formation and parallel memory storage by competitive neural networks, IEEE Transactions on Systems, Man, and Cybernetics, **SMC-13**, 815 (1983).

32. S.A. Ellias and S. Grossberg, Pattern formation, contrast control, and oscillations in the short term memory of shunting on-center off-surround networks, Biological Cybernetics, **20**, 69 (1975).

33. S. Grossberg, Contour enhancement, short-term memory, and constancies in reverberating neural networks, Studies in Applied Mathematics, **52**, 217 (1973).

34. S. Grossberg, Pattern formation by the global limits of a nonlinear competitive interaction in n dimensions, Journal of Mathematical biology, **4**, 237 (1977).

35. S. Grossberg, Competition, decision, and consensus, Journal of Mathematical Analysis and Applications, **66**, 470 (1978).

36. S. Grossberg, Decisions, patterns, and oscillations in the dynamics of competitive systems with applications to Volterra-Lotka systems, Journal of Theoretical Biology, **73**, 101 (1978).

37. S. Grossberg, Biological competition: Decision rules, pattern formation, and oscillations, Proceedings of the National Academy of Sciences USA, **77**, 2338 (1980).

38. S. Grossberg and D. Levine, Some developmental and attentional biases in the contrast enhancement and short term memory of recurrent neural networks, Journal of Theoretical Biology, **53**, 341 (1975).

39. S. Amari, Competitive and cooperative aspects in dynamics of neural excitation and self-organization, in "Competition and cooperation in neural networks," S.I. Amari and M. Arbib (eds.), Springer-Verlag, New York (1982).

40. S. Geman, The law of large numbers in neural modelling, in "Mathematical psychology and psychophysiology," S. Grossberg (ed.), American Mathematical Society, Providence, RI (1981).

41. T. Kohonen, Associative memory—A system-theoretical approach, Springer-Verlag, New York (1977).

42. T.R. Dixon and D.L. Horton, Verbal behavior and general behavior theory, Prentice-Hall, Englewood Cliffs, NJ (1968).

43. J. Jung, Verbal learning, Holt, Rinehart and Winston, New York (1968).

44. J.A. McGeogh and A.L. Irion, The psychology of human learning, second edition, Longmans and Green, New York (1952).

45. C.E. Osgood, Method and theory in experimental psychology, Oxford, New York (1953).

46. B.J. Underwood, Experimental psychology, second edition, Appleton-Century-Crofts, New York (1966).

47. S. Grossberg, On the serial learning of lists, Mathematical Biosciences, **4**, 201 (1969).

48. G.A. Carpenter and S. Grossberg, Neural dynamics of category learning and recognition: Attention, memory consolidation, and amnesia, in "Brain structure, learning, and memory," J. Davis, R. Newburgh and E. Wegman (eds.), AAAS Symposium Series, in press (1986).

49. G.A. Carpenter and S. Grossberg, A massively parallel architecture for a self-organizing neural pattern recognition machine, Computer Vision, Graphics, and Image Processing, in press (1986).

50. G.A. Carpenter, A geometric approach to singular perturbation problems with applications to nerve impulse equations, Journal of Differential Equations, **23**, 335 (1977).

51. G.A. Carpenter, Periodic solutions of nerve impulse equations, Journal of Mathematical Analysis and Applications, **58**, 152 (1977).

52. G.A. Carpenter, Bursting phenomena in excitable membranes, SIAM Journal on Applied Mathematics, **36**, 334 (1979).

53. G.A. Carpenter, Normal and abnormal signal patterns in nerve cells, in "Mathematical psychology and psychophysiology," S. Grossberg (ed.), American Mathematical Society, Providence, RI, pp. 49 – 90 (1981).

54. G.B. Ermentrout and J.D. Cowan, A mathematical theory of visual hallucination patterns, Biological Cybernetics, **34**, 137 (1979).

55. G.B. Ermentrout and J.D. Cowan, Temporal oscillations in neuronal nets, Journal of Mathematical Biology, **7**, 265 (1979).

56. W.J. Freeman, Nonlinear dynamics of paleocortex manifested in the olfactory EEG, Biological Cybernetics, **35**, 21 (1979).

57. H.R. Wilson and J.D. Cowan, Excitatory and inhibitory interactions in localized populations of model neurons, Biophysical Journal, **12(1)**, 1 (1972).

58. W.S. McCulloch and W. Pitts, A logical calculus of the ideas imminent in nervous activity, Bulletin of Mathematical Biophysics, **5**, 115 (1943).

59. E.R. Caianiello, Outline of a theory of thought and thinking machines, Journal of Theoretical Biology, **1**, 204 (1961).

60. F. Rosenblatt, Principles of neurodynamics, Spartan Books, Washington, DC (1962).

61. J. Beck, Effect of orientation and of shape similarity on perceptual grouping, Perception and Psychophysics, **1**, 300 (1966).

62. J. Beck, Textural segmentation, in "Organization and representation in perception," J. Beck (ed.), Erlbaum, Hillsdale, NJ (1982).

63. J. Beck, Textural segmentation, second order statistics, and textural elements, Biological Cybernetics, **48**, 125 (1983).

64. T. Caelli, On discriminating visual textures and images, Perception and Psychophysics, **31**, 149 (1982).

65. T. Caelli, Energy processing and coding factors in texture discrimination and image processing, Perception and Psychophysics, **34**, 349 (1983).

66. T. Caelli and B. Julesz, Psychophysical evidence for global feature processing in visual texture discrimination, Journal of the Optical Society of America, **69**, 675 (1979).

67. W.R. Garner, The processing of information and structure, Erlbaum, Hillsdale, NJ (1974).

68. J.R. Pomerantz, Perceptual organization in information processing, in "Perceptual organization," M. Kubovy and J.R. Pomerants (eds.), Erlbaum, Hillsdale, NJ (1981).

69. J.R. Pomerantz, Global and local precedence: Selective attention in form and motion perception, Journal of Experimental Psychology: General, **112**, 516 (1983).

70. J.R. Pomerantz and S.D. Schwaitzberg, Grouping by proximity: Selective attention measures, Perception and Psychophysics, **18**, 355 (1975).

71. D.L. Stefurak and R.M. Boynton, Independence of memory for categorically different colors and shapes, Perception and Psychophysics, **39**, 164 (1986).

72. A. Treisman, Perceptual grouping and attention in visual search for features and for objects, Journal of Experimental Psychology: Human Perception and Performance, **8**, 194 (1982).

73. A. Treisman and G. Gelade, A feature-integration theory of attention, Cognitive Psychology, **12**, 97 (1980).

74. A. Treisman and H. Schmidt, Illusory conjunctions in the perception of objects, Cognitive Psychology, **14**, 107 (1982).

75. A. Treisman, M. Sykes and G. Gelade, Selective attention and stimulus integration, in "Attention and performance VI," S. Dornic (ed.), Erlbaum, Hillsdale, NJ, pp. 333 – 361 (1977).

76. H.L.F. von Helmholtz, Treatise on physiological optics, J.P.C. Southall (Trans.), Dover, New York (1962).

77. E.H. Land, The retinex theory of color vision, Scientific American, **237**, 108 (1977).

78. M.A. Cohen and S. Grossberg, Neural dynamics of brightness perception: Features, boundaries, diffusion, and resonance, Perception and Psychophysics, **36**, 428 (1984).

79.   A.L. Yarbus, Eye movements and vision, Plenum Press, New York (1967).

80.   J. Krauskopf, Effect of retinal image stabilization on the appearance of heterochromatic targets, Journal of the Optical Society of America, **53**, 741 (1963).

81.   R.M. Pritchard, Stabilized images on the retina, Scientific American, **204**, 72 (1961).

82.   R.M. Pritchard, W. Heron and D.O. Hebb, Visual perception approached by the method of stabilized images, Canadian Journal of Psychology, **14**, 67 (1960).

83.   L.A. Riggs, F. Ratliff, J.C. Cornsweet and T.N. Cornsweet, The disappearance of steadily fixated visual test objects, Journal of the Optical Society of America, **43**, 495 (1953).

84.   S. Grossberg, Outline of a theory of brightness, color, and form perception, in "Trends in mathematical psychology," E. Degreef and J. van Buggenhaut (eds.), North-Holland, Amsterdam (1984).

85.   S. Grossberg and E. Mingolla, Neural dynamics of form perception: Boundary completion, illusory figures, and neon color spreading, Psychological Review, **92**, 173 (1985).

86.   H.J.M. Gerrits, B. de Haan and A.J.H. Vendrick, Experiments with retinal stabilized images: Relations between the observations and neural data, Vision Research, **6**, 427 (1966).

87.   H.J.M. Gerrits and J.G.M.E.N. Timmerman, The filling-in process in patients with retinal scotomata, Vision Research, **9**, 439 (1969).

88.   H.J.M. Gerrits and A.J.H. Vendrick, Simultaneous contrast, filling-in process and information processing in man's visual system, Experimental Brain Research, **11**, 411 (1970).

89.   N. Kawabata, Perception of the blind spot and similarity grouping, Perception and Psychophysics, **36**, 151 (1984).

90.   L.E. Arend, J.N. Buehler and G.R. Lockhead, Difference information in brightness perception, Perception and Psychophysics, **9**, 367 (1971).

91.   R.H. Day, Neon color spreading, partially delineated borders, and the formation of illusory contours, Perception and Psychophysics, **34**, 488 (1983).

92.   A.R.H. Gellatly, Perception of an illusory triangle with masked inducing figure, Perception, **9**, 599 (1980).

93.   G. Kanizsa, Contours without gradients or cognitive contours? Italian Journal of Psychology, **1**, 93 (1974).

94.   J.M. Kennedy, Illusory contours and the ends of lines, Perception, **7**, 605 (1978).

95.   J.M. Kennedy, Subjective contours, contrast, and assimilation, in "Perception and pictorial representation," C.F. Nodine and D.F. Fisher (eds.), Praeger, New York (1979).

96.   J.M. Kennedy, Illusory brightness and the ends of petals: Change in brightness without aid of stratification or assimilation effects, Perception, **10**, 583 (1981).

97.   T.E. Parks, Subjective figures: Some unusual concomitant brightness effects, Perception, **9**, 239 (1980).

98.   T.E. Parks and W. Marks, Sharp-edged versus diffuse illusory circles: The effects of varying luminance, Perception and Psychophysics, **33**, 172 (1983).

99. S. Petry, A. Harbeck, J. Conway and J. Levey, Stimulus determinants of brightness and distinctness of subjective contours, Perception and Psychophysics, **34**, 169 (1983).

100. C. Redies and L. Spillmann, The neon color effect in the Ehrenstein illusion, Perception, **10**, 667 (1981).

101. H.F.J.M. van Tuijl, A new visual illusion: Neonlike color spreading and complementary color induction between subjective contours, Acta Psychologica, **39**, 441 (1975).

102. H.F.J.M. van Tuijl and C.M.M. de Weert, Sensory conditions for the occurrence of the neon spreading illusion, Perception, **8**, 211 (1979).

103. D.H. Hubel and T.N. Wiesel, Functional architecture of macaque monkey visual cortex, Proceedings of the Royal Society of London (B), **198**, 1 (1977).

104. S. Grossberg and E. Mingolla, Neural dynamics of perceptual grouping: Textures, boundaries, and emergent segmentations, Perception and Psychophysics, **38**, 141 (1985).

105. S. Grossberg, On the development of feature detectors in the visual cortex with applications to learning and reaction-diffusion systems, Biological Cybernetics, **21**, 145 (1976).

106. S. Grossberg, Neural substrates of binocular form perception: Filtering, matching, diffusion, and resonance, in "Synergetics of the brain," E. Basar, H. Flohr, H. Haken and A.J. Mandell (eds.), Springer-Verlag, New York (1983).

107. M. Wertheimer, Untersuchungen zur Lehre von der Gestalt, II, Psychologische Forschung, **4**, 301 (1923).

108. U. Neisser, Cognitive psychology, Appleton-Century-Crofts, New York (1967).

109. B.K.P. Horn, Understanding image intensities, Artificial Intelligence, **8**, 201 (1977).

110. D. Marr and E. Hildreth, Theory of edge detection, Proceedings of the Royal Society of London (B), **207**, 187 (1980).

111. F. Ratliff, Mach bands: Quantitative studies on neural networks in the retina, Holden-Day, New York (1965).

112. L. Glass and E. Switkes, Pattern recognition in humans: Correlations which cannot be perceived, Perception, **5**, 67 (1976).

113. C.F. Nodine and D.F. Fisher (eds.), Perception and pictorial representation, Praeger, New York (1979).

114. S. Grossberg and E. Mingolla, The role of illusory contours in visual segmentation, in "Proceedings of the international conference on illusory contours," G. Meyer and S. Petry (eds.), Pergamon Press (1986).

Large-Scale Computations on a Scalar, Vector and Parallel "Supercomputer"

E. Clementi, J. Detrich, S. Chin, G. Corongiu,
D. Folsom and D. Logan
with
R. Caltabiano, A. Carnevali, J. Helin, M. Russo,
A. Gnudi and P. Palamidese
IBM Corporation
Data Systems Division, Dept. 48B MS 428
Neighborhood Road
Kingston, New York 12401

Abstract

We discuss two experimental parallel computer systems lCAP-1 and lCAP-2 which can be applied to the entire spectrum of scientific and engineering applications. These systems achieve "supercomputer" levels of performance by spreading large scale computations across multiple cooperating processors — several with vector capabilities. We outline system hardware and software, and discuss our programming strategy for migrating codes from a conventional sequential system to parallel. The performance of a variety of applications programs is analyzed to demonstrate the merits of this approach. Finally, we discuss lCAP-3 an extension to this computing system, which has been recently assembled.

I. Introduction

Mathematical models require computation to secure concrete predictions. Successes in relatively simple cases spurs interest in more complex situations. Failure brings forward more sophisticated, and usually more elaborate, models. In either case, computational demands increase, and we can see this tendency in effect throughout the physical sciences, most areas of engineering, and many other quantitative sciences. These demands constantly press the limits of currently available computer resources.

Somewhat specialized computer hardware and software has emerged in response to these demands. Examples are array processors such as the IBM 3838; these have become an important computational resource in the reduction of seismic data, image processing, and a number of other applications where computational demands fit the array processor architecture. Another group is the high-end processors with vector architecture, such as the CRAY series, the CDC CYBER

403

205, and the recently announced IBM 3090 with vector attachment. When a computation can effectively utilize vector architecture, such machines will outperform even the most powerful conventional general-purpose machine by a substantial margin. Such performance has given rise to the term "supercomputer".

An independent but complementary approach has been the development of parallel computing systems, where many processors can be concurrently applied in cooperation on a single calculation. This idea is certainly not new. It has already been the subject of numerous research projects and a very vast literature.[1] Interest in this type of computing has gained recent impetus as vector-oriented supercomputing matures and its limitations emerge. Increasing the speed of the already very fast cycle times of high-end vector processors requires improvements in basic technology that are difficult to achieve. Parallelism offers another road to faster computation. The two approaches, vector and parallel execution, are found to be in fact complementary; we see this for example where both the current CRAY series and the IBM 3090 series offer a combination of vector and parallel architectures.

Our laboratory's interest in parallel computing originated with the computational demands of our work in theoretical chemistry and biophysics. We have already discussed some of our researches in parallel computing elsewhere,[2] and we expand and update these discussions here. Our development of new computing resources has in fact given added impetus to our scientific researches, as demonstrated in more detail below. One result is expansion of our researches to include modelling of continuous media, such as fluid flow, and graphics development to assist in handling the data our models produce. In all these cases, and in fact for a very broad class of applications, parallel computing comes quite naturally, as the examples we present below demonstrate. As we shall see, parallel computing is actually inherent in the computational models, which makes it possible to realize results with somewhat less effort than might be expected.

The computing system we have assembled is intended to respond to the newly emerging viewpoint in science and engineering, the "global simulation" approach, exemplified in the companion paper submitted for this volume. In the "global simulation" one attempts to realistically simulate complex problems. In the past, theory and mathematical models have stripped reality of many "details", because of its unbearable complexity. Thus, often, many-body effects, non-linear terms and boundary conditions have been either over-simplified or ignored, or not even identified. But slowly we are coming to the understanding that "reality" is just a collection of interacting "details". With "global simulations" the demand on computational hardware, system softwares, physical and mathematical modelling is thus suddenly increased non-linearly. Whereas some "details" could be approached via vectors, some will require parallelism and some will request "super-scalar". The space within which the "global simulation" evolves is most definitely at least three-dimensional; namely, scalar, vector, and parallel.

But, a fourth dimension is already looming on the horizon to respond to the sudden manifold increase in the computational demand for storage, data flow speed, and execution accuracy and speed. This has spurred research in data flow machines and languages at one level and silicon compilations and systolic architectures at the other extreme. The new dimension requires computers to be more compact and to contain in a reasonable space the components it needs, to be built of new, presently unknown materials, to respond to drastically new cooling and packing requirements, to have homostatic capabilities built in as a primary and diffuse function at the hardware and system levels and at the maintenance and scheduling levels, up to the programming languages. However, this is not sufficient. The new dimension must·"compromise" with the users' physical and mental limitations, and therefore, it must attempt to mimic some of the users' characteristics, particularly at the level of languages where "colloquial speech" and "pattern recognition" must have a notable role and where today's precompilers will be more and more like today's expert systems. In time this new dimension will, as a necessary response to the new computer's complexity, be a very peculiar "application" of artificial intelligence with quite new languages. Very simply, the characterization "scalar, vector, and parallel" is much too "hardware oriented"; system softwares with intelligence and application programs must be present explicitly — an integral part — as coordinates of the space within which global simulation will occur. We shall refer to this still unrealized system as the "VI generation computer".

II. Strategy

Our approach to research in parallel processing has been, and remains, very pragmatic. This implies no criticism of other, more elaborate, approaches to parallel processing; it simply reflects our basic priorities, which are to secure a workable parallel processing system as quickly as possible, and begin using it on the large-scale scientific and engineering applications that are central to our research. This "consumer orientation" is perhaps the most distinctive feature of the initial phase of our project.

Many of the characteristics of our parallel strategy follow from these priorities. These characteristics are: 1) parallelism based on relatively few (less than 20), but quite powerful processors, with 64-bit hardware; 2) architecture as simple as possible, but extendable; 3) system software that varies as little as possible from that normally available for sequential processing; 4) initial applications programming entirely in FORTRAN, since this is the most widely used scientific application language; 5) migration from sequential execution to parallel execution with a minimal amount of recoding; 6) enhance, whenever feasible, each processor with vector features; 7) extend gradually and hierarchically, from loosely to tightly coupled, with a variety of approaches.

Our system is based on tens of processors, rather than hundreds or thousands of processors, in order to avoid questions concerning the binding together of large number of processors into a single system, since this would complicate and delay our attempts to apply parallel processing to our applications. It follows that the individual processors must be quite powerful in order to secure "supercomputer" performance, and hardware supporting 64-bit floating-point precision is essential for many of the applications of interest to us.

In line with our priorities, we have avoided the time-consuming effort of developing specialized hardware, preferring instead to acquire products which are available "off the shelf." Specifically, we have selected the Floating Point Systems FPS-164 (currently with the MAX vector boards) or FPS-264 attached processors (AP) for our parallel processors. This choice was dictated by the fact that the FPS-164 was the only 64-bit attached processor available when we began. In principle, we could just as well substitute something else, such as a group of IBM 3090 computers, for the attached processors in our configuration, but these machines are far more expensive, and we want a configuration that does not cost too much.

We have IBM hosts (IBM 4381, IBM 3081, IBM 3084, or IBM 3090, at present) for our AP's. One advantage of this choice for the host computer is the multiple-channel I/O architecture used for the AP attachments. These channels (3 Mbytes/sec.) were originally the only paths for data and synchronization between processors, giving rise to our term "loosely coupled array of processors" for our architecture. There are currently additional data paths, bringing us closer to a "tightly coupled array of processors" type of architecture. Our architecture is currently implemented in two distinct systems, called simply lCAP-1 and lCAP-2. They are distinguished by the operating system on the host, and also have different IBM host machines. A consequence is development of software to support parallel processing for two distinct operating systems, namely VM/SP and MVS/XA.

There have been many experiments in migrating software applications to parallel processing on our system. Ideally, one would like a completely automatic language processor to handle conversion from code for normal sequential execution to code supporting parallel processing. We are still a long way from this, but we have developed a number of aids to ease the labor of programming parallel processing, as described below. Prominent among these is a precompiler that recognizes experimental extensions to FORTRAN that we have introduced, and develops detailed code from this source code for compilation and parallel execution.

III. Brief History

The idea for the lCAP-type architecture was a rather natural evolution for Clementi and his group in April of 1983. At that time there was one FPS-164 attached to an IBM 4341. The experience on that AP and the perception of the "natural" parallel computations inherent in our codes for chemical research pushed us toward thinking in terms of parallel execution. In addition, the existence of multi-processor equipment at IBM supported the idea that this was reasonable and worth considering.

By November of that same year the original system had grown to include three FPS-164s. Our original software to support parallel execution, Virtual Machine Fortran-Accessible Communications Subroutines (VMFACS)[2] was already available, and experimentation with parallel execution had begun. In February of 1984 the three processors were expanded to six, and results from the first computations using our parallel configuration were presented shortly thereafter. In May 1984 the

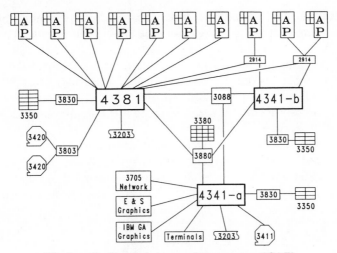

The loosely Coupled Array of Processors (ICAP)

Figure 1. Initial configuration for ICAP-1. Included are the 10 FPS-164's, the three IBM hosts (two 4341's and one 4381), printers, tape drives, and graphics stations.

full complement of ten FPS-164s were installed in IBM Kingston and became operational as the parallel computer now known as ICAP-1. A schematic diagram of the original ICAP-1 system is given in Fig. 1.

The initial success of ICAP-1 encouraged additional ventures with the ICAP architecture. Indeed, in the middle part of 1985 the IBM European Center for Scientific and Engineering Computing began operation with a copy of the ICAP-1 system. In a similar spirit, in May of 1985 an interim ICAP system for the Cornell Production Supercomputer Facility (PSF) was installed and operational, under joint sponsorship of IBM and the National Science Foundation. In October 1985 the Cornell PSF became fully operational with an IBM 3084 QX and four FPS-264s.

Our ICAP-1 system continued to evolve with acquisition of vector capability in the form of two MAX boards installed in each FPS-164 during June of 1985. Our precompiler software also emerged about this time from a number of experiments to evolve beyond the VMFACS software. In August of 1985 the first three FPS-264 machines were delivered to the ICAP-2 system at IBM Kingston, and three additional FPS-264 machines were added in October, 1985. At about the same time, the capabilities of ICAP-1 were further expanded by the acquisition of the first two SCA shared bulk memories. In January 1986 four more FPS-264s were added to the ICAP-2 system, bringing the total to ten. At this point, ICAP-1 had acquired the remainder of its five shared bulk memories. In May 1986 the IBM 3090-200 with vector facility was installed and the ICAP-1 configuration was upgraded by phasing out an IBM 4341 in favor of an IBM 3081. In June, 1986 we added five shared bulk memories to ICAP-2, and SCA delivered two large shared bulk memories of 512 Mbytes, one for ICAP-1 and one for ICAP-2. Additional transmission power was attempted by adding two FPS Busses, again one for ICAP-1 and one for

ICAP-2. In August, 1986 we started to plan for the first application on ICAP-3 as we call the merging of ICAP-1, ICAP-2 and the IBM 3090.

Additional description of our hardware configuration is provided in Section IV below. Aspects of our operating systems, particularly as they affect our implementations of parallel execution, are discussed in Section V. In Section VI we present the strategies and considerations we use to modify our applications programs for effective parallel execution on our system. Section VII describes the precompiler language we use to support our programming for parallel execution. A sampling of the results we have achieved using our parallel system for scientific computations is presented in Section VIII. Finally, in Section IX we wrap up our current experience with our system, and discuss some of the further developments we can foresee for our researches.

IV. Present Configuration for ICAP-1 and ICAP-2

As mentioned previously there are at present two parallel processing systems working in our laboratory. Both share the same fundamental architecture of a host computer and attached processors (AP) on the I/O channels of the host. Originally, these I/O channels were the only communications paths between the different processors; for this reason we have called our architecture a *loosely* coupled array of processors (ICAP). The first of these systems, called ICAP-1, is hosted either by an IBM 4381 or 3081 and attaches to 10 FPS-164 processors. The second and more powerful system, called ICAP-2, employs as host an IBM 3084 and has ten FPS-264 processors as slaves. In spite of these differences the two systems are very similar (aside from operating system considerations which will be discussed in the following section).

The ICAP-1 system is configured so that six FPS-164s are connected to an IBM 3081 host and the remaining four are attached to an IBM 3814 switching unit so they can be switched between the IBM 3081 host and a secondary host, which is at present an IBM 4381. The FPS-164 processors are attached to the IBM host through IBM 3 Mbyte/sec. channels available on these hosts. Connected to these two hosts is a "front end" IBM 4341, which is also connected to a graphics station. The graphics station includes an Evans and Sutherland PS300 and three IBM 5080 graphics terminals with a large set of graphics packages for such diverse uses as CAD/CAM applications or molecular modelling. The three IBM systems are interconnected, channel to channel, via an IBM 3088 connector. Tape drives, printers, and a communication network interface complete the ICAP-1 configuration.

One attractive feature of the above system is that up to four of AP's may be switched between the secondary host IBM 4381 and the primary host IBM 3081. The latter system can be dedicated to production jobs, while the secondary host supports experiments with the system or debugging of applications, without risk of disturbing production runs.

Each FPS-164 has an independent central processing unit and memory, and is attached to its own local disk drives. The CPU on the FPS-164 runs at 5.5 million instructions per second, and several concurrent operations can take place on each instruction cycle. In particular, one 64-bit floating-point addition and one 64-bit floating-point multiplication can be initiated each cycle, so that peak performance is about 11 million floating point operations per second (11 Mflops). Of course, one must make the distinction between peak performance (a characteristic of the machine hardware) and sustained performance (depending on the application and the code which implements it as well as the hardware).

Each of the FPS-164's has 8 Mbytes of real random access memory. The memory on the IBM 3081 is 48 Mbytes, while the IBM 4381/3 has 32 Mbytes, and the IBM 4341 has 16 Mbytes. Thus, taken as a whole, there is 176 Mbytes real storage available on lCAP-1.

Each FPS-164 also has four 135 Mbyte disks, for a total of 5.4 gigabytes. In addition there are banks of IBM 3350 and IBM 3380 disks accessible to the host computers, totalling about 25 gigabytes of disk storage.

Floating Point Systems also supplies the FPS-164/MAX; this is a special-purpose board that can be added to the FPS-164 to augment performance, particularly on matrix operations. Each MAX board contains two additional adders and two additional multipliers, and so adds 22 Mflops to peak attainable performance. Up to 15 boards can be placed in a single FPS-164, converting it to a machine with a peak performance of 341 Mflops. At present each of our AP's has been equipped with two MAX boards. This has upgraded our peak performance from 110 to 550 Mflops. Ultimately our system could grow to 3410 Mflops peak capability, but (recalling the distinction between peak performance and realized performance) it is clearly desirable to first explore the gains that one can realistically obtain with only a few 164MAX boards per AP, so we have settled at 550 Mflops.

The lCAP-2 system has an IBM 3084 QX host and ten FPS-264's for attached processors. The FPS-264 is compatible with the FPS-164, and codes developed for either machine run on the other machine without modification. The CPU of the FPS-264 runs at a peak performance of 38 Mflops, or 3.5 times faster than the FPS-164. We also note the improved memory interleaving and larger program cache on the FPS-264, which helps bring up the fraction of peak performance realized in tests of sustained performance. Thus, we have observed realized performance on the FPS-264 between 3.5-4.0 times what is observed on the FPS-164. With ten FPS-264's, the peak performance of lCAP-2 is 380 Mflops. However, again we stress the distinction between peak performance and observed performance. We expect the average performance of the lCAP-2 system to surpass that of the lCAP-1/MAX, as the MAX boards are special purpose boards with limited use. This has been verified by some preliminary tests on lCAP-2 that are discussed in a later section. Each of the FPS-264's has 8 Mbytes of real memory. There are also two of the FD64 disk drives, totalling 1.2 gigabytes of disk storage on *each* machine. The IBM 3084 has 128 Mbytes of real memory and IBM 3350 and 3380 disk packs totalling 50 gigabytes of disk storage. Again, tape drives, printers and a communication network complete the lCAP-2 configuration.

Data conversion and communication between the host computer and the attached FPS-X64s are handled by hardware and software that is provided and supported by Floating Point Systems as a standard feature. An optimizing FORTRAN compiler and supporting utilities (including disk I/O) are also standard products for the FPS-X64. The compiler is capable of producing machine code that is reasonably effective in exploiting the FPS processor architecture. An extensive library of machine-coded mathematical subroutines is also provided, and these can be combined with FORTRAN code for additional gains in performance. We have found it worthwhile to augment this library with our own machine-coded subroutines, in order to get the best performance for the applications codes in our laboratory.[3]

A library of mathematical routines will be available for use on the MAX boards. When properly employed they may achieve a gain in performance that is impressive. For large matrix multiplications the processing speed increases by approximately 22 Mflops, i.e. the rated peak performance of the supplemental vector board. The applicability of the MAX boards in general application programs is now under investigation.

It should be noted that upgrades such as the MAX boards have no effect on the parallel programming strategy to be discussed later. The strategy is equally effective for AP's of any architecture or computational speed. In principle, we could substitute 10 vector-oriented "supercomputers" for our 10 FPS-X64's. However given the notable differences in cost between these options the latter one is unrealistically high.

As in most parallel processors, data flow between distinct processors presents a potential bottleneck to effective use of the lCAP systems. Some data flow is obviously necessary to maintain a coherent calculation across the processors, but we attempt to minimize such transfers to the extent we can. It is also clearly desirable to acquire faster data paths between processors than is provided by the 3 Mbyte/sec. channel connections between host and APs. One vehicle to accomplish this is the shared memory systems and associated softwares which were designed and developed at our request by Scientific Computing Associates, Inc. (SCA).

The SCA Shared Bulk Memory System (SBMS) is composed of three elements: one or more FPS-X64 Processors, one or more bulk memory chassis, and the "SCA Bus," a set of cables and interface cards to connect the other elements of the system. The configuration of these building blocks is highly flexible; the SCA Bus is the key to this flexibility. (See Fig. 2, inset a.) A single SCA Bus can connect from one to four APs with a bulk memory capable of holding up to 512 Mbytes. Each unit is capable of sustained data transfer rates of 64 Mbyte/sec., but when attached to the FPS-164 will operate at a rate of 44 Mbyte/sec. (i.e. the maximum achievable I/O bandwidth of the FPS-164). When attached to the FPS-264 the data transfer rate will be equivalent to the I/O bandwidth, 38 Mbyte/sec. The addition of these shared memory systems provides the ability to perform quickly large asynchronous transfers of data between processors. This constitutes a step from ICAP towards a more tightly Coupled Array of Processors.

Currently there are five shared bulk memories installed and operational on ICAP-1 and ICAP-2. Some preliminary results using these resources are given in a later section. This configuration gives the flexibility of allowing one ring with 10 processors, or alternatively a number of smaller rings. The multiple connections between processors in the configuration indicated in Fig. 2, inset b, avoids the possibility of one inoperative AP or one inoperative memory unit breaking up the system data paths, and the 5 shared memories give a potential peak total data transfer rate of 220 Mbyte/sec.

The drawback of this shared bulk memory configuration is that a given pair of APs may not be connected to the same shared memory, so that transfer of data between two APs may require one or more intermediate transfers among APs and the shared memory units to which they are connected. This is obviously inconvenient in terms of both programming and performance. We have remedied this difficulty by adding a single large shared bulk memory which can attach to and be shared by up to 12 FPS-X64's (see Fig. 2, inset c). This shared memory system is be based on the same hardware and software as its predecessor. There are two of these units, one for ICAP-1 and one for ICAP-2, each with 512 Mbytes capacity.

The software that drives the SBMS is designed in three levels: the Message-Level Software, the Device-Level Software, and the Diagnostics. The Message-Level Software is the primary interface for user programs. Through these routines, all of the features of the SBMS can be utilized with a minimum of understanding of the hardware configuration or its characteristics. An alternative, lower-level interface is provided by the Device-Level Software. These routines provide the most direct means of writing to and reading from the SBMS, and provide the user with the means to build individualized user interfaces. The Diagnostics will verify that the system is operating properly, and are used to isolate system failures to individual components of the system. The interested reader is referred to standard literature available from S.C.A., Inc. in particular to the "Shared Bulk Memory System, Software Manual, Version 2.0 (1985, SCA).

A second but complementary approach to modify our "loosely" coupled array of processors towards a more tightly coupled system is the incorporation of two (one for each system) fast common busses that link all of the FPS-X64's on a particular system. In Fig. 2 (inset b), we show two such Busses in a 10 way ICAP-type linkage. This bus, designed and tested by FPS, has a data bandwidth of 32 Mbyte/sec. for each node, with 22 Mbyte/sec. available for any one connection between the bus and a single AP. A prototype model of this bus and its associated software was delivered to us in November 1985. Now, two FPS Busses have been delivered and we are in the process of investigating its possible uses. The ICAP configuration with bus and shared bulk memories is shown in Fig. 3.

We have also acquired an IBM 3090-200, which is being installed at this time. This processor incorporates the recently announced vector attachment, and thus repres-

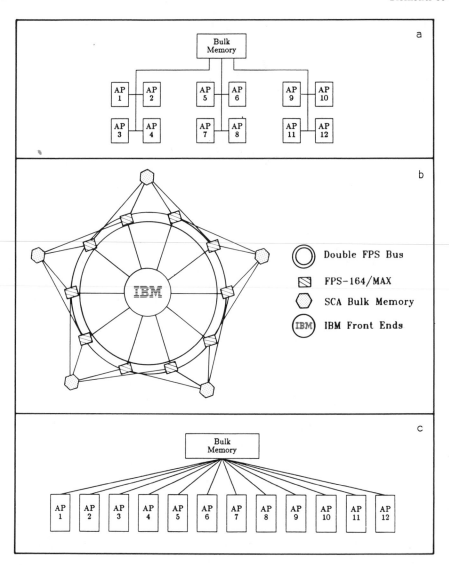

Figure 2. Inset a): schematic of the SCA bulk shared memories. Inset b): ICAP configuration with double ring of SCA bulk shared memories and Two FPS Busses with 10 IBM channels connecting to an IBM front end processor. Inset c): large SCA bulk shared memory connecting up to 12 array processors.

ents a very substantial addition to the computing resources at our disposal. Eventually this will grow to a model 3090-400, roughly twice as powerful. There are two tightly coupled processors on the 3090-200, each with its own vector facility, so it will be interesting to investigate the possibility of parallel execution with both inboard (on the 3090) and outboard (the attached processors) processors working together on the same computation; this would achieve a form of heterogeneous parallel processing. We are also acquiring a new ability to investigate how best to exploit both vector and parallel processing capability at the same time.

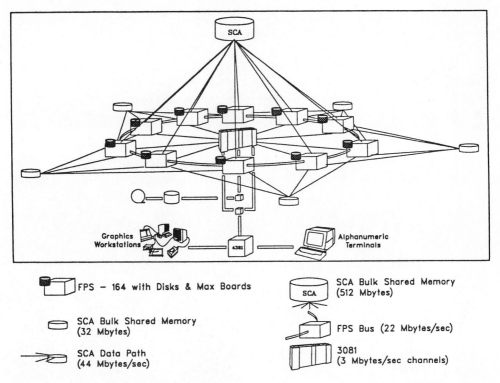

Figure 3. ICAP-1 configuration (ICAP-2 is virtually the same) including the five four-way 32-Mbyte SCA shared bulk memories, the 512-Mbyte SCA shared bulk memory, connected to 10 FPS-X64's, and the FPSBUS.

We can envision another step in this direction, where the 3090, the ICAP-1 system, and the ICAP-2 system are connected by a network with a capability to bring the three computing resources to bear on a single job. The extreme computational power of such a configuration is very attractive. One possibility would be to set up the 3090 to serve as supervisor and traffic director for the participation of the ICAP-1 and ICAP-2 systems, as well as having it participate in the calculation in its own right with its vector capability. There is a great deal of design work and experimentation before we can settle firmly on a configuration with such characteristics, but we have already begun some of these investigations. In the section on ICAP-3, we shall give additional details.

V. System Considerations

We have implemented our basic parallel approach on two configurations using different computer hardware; they also differ in terms of the operating systems employed. Here we take up some of the considerations the different operating systems raise, as they affect our approach to parallel execution. As we shall see, one can handle the details of implementation so that the broad structure is equivalent on both systems. This has the advantage that many aspects of our support for parallel execution can be readily applied to both systems, and in fact we have an avenue for easy migration of parallel codes from one system to the other.

The lCAP-1 system, hosted by either an IBM 4381 or IBM 3081, runs under the IBM Virtual Machines/ System Product (VM/SP) operating system.[4] For the AP's, we use the software provided for hosts running under this system by Floating Point Systems. We have not found it necessary to modify either set of software in order to run our applications in parallel.

VM/SP is a time-sharing system in which jobs run on virtual machines (VM) created by the system; these VM's simulate real computing systems. The standard software provided by Floating Point Systems to use the FPS-164's embodies the restriction that only one AP can be attached to a VM. Of course, for a task running in parallel, more than one AP is required. Our solution to this is to introduce extra "slave" VM's to handle the extra AP's we need. To make this work, one must have a way to communicate between different VM's; this is provided by the Virtual Machine Communication Facility (VMCF), which is a standard feature of VM/SP.[4]

A parallel task will consist of several FORTRAN programs, each running on a separate VM in the host system, and each controlling a particular AP on which additional FORTRAN code runs. On one of the VM's, the "master," is the part of the original FORTRAN code intended to be run on the host, combined with utility subroutines that handle communication with the "slave" VM's, and possibly also the utilities provided by FPS to handle any AP attached to the "master" VM. The logical structure of this system is illustrated in Fig. 4. The programs running on the "slave" VM's can be nothing more than transfer points for communication between the "master" program and the AP's attached to the "slaves". Since each VM is attached only to a single AP, the standard utilities provided by FPS[5] for communication between host and AP can be used without modification.

The lCAP-2 system is hosted by a 4 processor IBM 3084 running the Multiple Virtual Storage Extended Architecture (MVS/XA) operating system.[6] In contrast to VM/SP, this operating system has typically been employed as a batch oriented system, although it does incorporate time-sharing facilities. Here instead of slave VMs we deal in terms of subtasks that can be generated within a job by standard system facilities. In analogy to the case in the VM/SP system, we find we must deal in terms of multiple subtasks on the host to handle parallel execution, since the FPS utilities require for their integrity that only one AP be attached to a subtask. However, unlike the VM case, the subtasks, including the master, share the same area of virtual memory storage, so data transfer is merely a matter of passing addresses from the master to the slave subtasks. Some memory management capabilities must be provided to assure the integrity of this type of data transaction. The other function of VMCF in the VM system, synchronization between master and slaves, is assumed by the WAIT/POST logic in MVS, which is accessible as part of the standard system facilities.

The schemes described above for VM and MVS make rather light demands in terms of system programming, since they use services already embedded in the respective systems. In particular, we should note that no modification whatever of the utilities provided by Floating Point Systems is required, since the structures we use to support parallel execution are set up entirely within the IBM host system. Indeed, these structures will support parallel execution without APs, by running the slave FORTRAN code on the slave VMs or subtasks of the IBM host system. We actually run codes in this fashion, either for debugging, or to exploit the the possibilities for parallel execution inherent in a multiple-processor IBM host such as the IBM 3090, 3084, 3081 or 4381/3.

Use of the system services indicated above requires programming in assembler code. It is desirable to package access to these services once and for all in utility subroutines that can be invoked from normal FORTRAN code. Such utilities have been a feature of the lCAP system from its inception. The development of the first such set of utilities, called VMFACS,[2c] (Virtual Machine FORTRAN-Accessible Communications Subroutines), was one of the first steps in implementing our parallel system. Since that time we have experimented with utilities embodying alternative system facilities such as the Inter-User Communication Vehicle (IUCV),[4] and have also written other sets of communication softwares intended to make the implementation of the communication protocol more automatic and user friendly.[7a] Under MVS, the data transmission and subtask creation and synchronization functions have been packaged into a number of different FORTRAN-callable services, some from this laboratory, and others from other groups within IBM. Among these are DNL-5798, Paradigm, and the IBM Multi-Tasking Facility (MTF).[7b]

There remains a drawback to such approaches, which is that the user must select the appropriate utilities for his code and involve himself in the details of using them properly to achieve the effect he has in mind. We have found this is not a necessary burden. By introducing a language preprocessor, or precompiler, one can allow the user to move to a higher level where he defines the structure of the parallel program to the precompiler, and leaves the task of selecting and organizing the

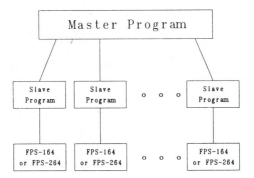

Figure 4. Basic structure for parallel execution of a program with the lCAP master/slave topology. This structure is true for both VM and MVS. In VM the slaves are actually secondary VM's, and in MVS the slaves are sub-tasks.

utilities to achieve the desired result to be handled automatically by the precompiler. In addition to sparing the user from unnecessary programming complications and the errors that can arise from them, we find at this level that the two operating systems look the same. Thus code developed for one system can immediately migrate to the other system, by reprocessing it through the precompiler. The structure of the precompiler language is described in a subsequent section.

The output of the precompiler is FORTRAN code, including calls to the appropriate utilities to support parallel execution. Thus the utilities developed for the precompiler are accessible for direct use by the programmer. These utilities are similar to the the utilities packages described above, but were developed specifically for the precompiler. Naturally they differ according to whether the VM or MVS system is being used.

Even if the user elects to work at the precompiler level, some differences between the two systems will be visible. The job control language to set up and run a parallel job differs in VM and MVS. Performance may also differ on the two systems, for all the reasons it can differ for sequential runs under VM and MVS. It is also possible to see differences in performance due to differences in the implementation of parallelism on the two systems, but this is rarely of any significance in practice.

VI. Parallel Programming Strategy and Considerations

Several aspects of our approach to programming for parallel execution have already been mentioned in the introduction. Over a period of time our group has developed a substantial collection of scientific applications codes, almost entirely written in FORTRAN, and one of the motivations for developing our parallel system was to run these applications for cases that would be too slow or unwieldy on a normal sequential system. Since that time, many additional applications codes have come into our laboratory, again typically in a form intended to run on a sequential system. Thus, much of our effort has been directed toward migrating existing applications to our parallel systems, and the parallel programming strategy we describe here has undoubtedly been influenced by this.

We begin with the observation that large-scale, typically CPU-bound, calculations almost invariably involve loops that are traversed many times. Most of the CPU time is consumed in such loops, so that if we adapt the tasks contained in these loops to parallel execution, we find that we actually have most of the code (as measured by execution time) running in parallel.

This is easy enough to accomplish. Let us suppose that our sequential FORTRAN code has a DO loop of the form

```
DO 500 I = 1,N
. . . . . . . . . . .
. . . . . . . . . . . . ,
```

with some computational kernel inside the loop (up to statement 500). Then, if we suppose that NCPU is the number of AP's available for parallel execution, we can keep the same computational kernel and modify the loop to read

DO 500 I = ICPU,N,NCPU

. . . .

. . . .

This portion of the program, with the computational kernel and modified loop, is dispatched to each of the NCPU AP's. Each AP must of course have a different value for the index ICPU, with $1 \leq ICPU \leq NCPU$.

This fundamental scheme has been applied to most of the application programs we have migrated to parallel execution, and has served us very well. Thus, after migration, a typical program flow would consist of an initial sequential part handling initial input, setup, etc., followed by a parallel part running simultaneously on several AP's. At the end of this portion, the results from the parallel execution must be gathered up and processed by another sequential portion. This may be a prelude to another period of parallel execution, or, ultimately, to development of final results and the end of the run.

There is an obvious limit on this scheme: the computational kernel for a particular value of I in the loop example above must not depend on results computed in earlier passes through the loop with a different value of I. Our experience so far indicates that this is not a severe restriction; indeed, we find many codes tend to fall naturally into such a form. There are some exceptions, of course, and, to start with, we have simply left the ones we have encountered in the sequential part of the code.

The reader may be surprised that such a simple approach to parallelism can yield good results so often, but this is what we have found. Of course there are cases where a more sophisticated strategy is required. Sometimes the loops we would like to find are hidden, because they are implicit, rather than put forth as explicit DO loops. We may also find that we need to rearrange the code somewhat to get loops that can be comfortably parallelized.

All this assumes that the large loop in question can be broken into parcels that can run in parallel with nearly equal run times for each parcel. The straightforward loop partitioning strategy sketched above breaks down when this is not true. The problem is that if one parallel parcel takes a significantly longer time than another, at least one of the parallel processors will be idle while it is waiting for the others to finish their portion of the parallel work. This cuts down on the efficiency of the program.

Parallel programs that avoid such situations are said to be load balanced. When a simple variation of the loop partitioning scheme fails to achieve good load balancing, it is usually because the execution time for a particular loop iteration is not predictable before the program is actually run for a specific case, and may vary considerably from iteration to iteration.

A less rigid scheme can be used in many cases to achieve load balancing. Instead of breaking up the parallel work into parcels equal to the number of processors, one can break it up into smaller parcels, where the number of parcels are much larger than the number of processors. At the beginning of the parallel work, each processor is given one parcel to do. The first processor to finish its assigned parcel is given the next parcel available to be done. After that, as each processor finishes, it is immediately assigned a new parcel of work, until all the parcels of parallel work are completed. Each processor is kept working all the time until the available parcels are exhausted. Hence the time that any one processor is idle can be no longer than the time to execute one parcel.

We have found a number of programs where this is the optimum approach to parallel execution. Typically we attempt this approach when the simple loop partitioning scheme is seen to give less than the hoped-for performance.

It would appear that one should make the parcels in this approach as small as possible, to assure good load balancing. In practice, the payoff of good load balancing has to be weighed against its cost in terms of additional communication overhead. Each time data passes between host and AP at the start or finish of a parallel task on an AP, the transmission takes time, and no useful computation is taking place during this time. This type of communication overhead does not occur in sequential programs, but is always a factor in programming for parallel execution.

The simplest way to hold down communication costs is to make the parcels for parallel execution as large as can be managed, so the number of instances of data communication are reduced. One should also carefully control the amount of data that must be transmitted to or from the AP for any one parcel. Sometimes one finds large intermediate arrays that are used in the parallel parts of the program, but need never be referenced in the sequential part. It is advantageous to arrange to create such arrays once and for all at the beginning of the run and maintain them on the APs from one parallel task to the next, throughout the run.

In order to understand the considerations that affect performance of programs for parallel execution more clearly, let us consider the (wall-clock) execution time for a program with some set of input data specific to that run. We denote this execution time by $T(N)$, where N is the number of APs participating in that particular parallel run. Part of this time will be taken up by the time to execute the sequential portion of the code, that is, the part of the code which for some reason was not modified to run in parallel; we use T_s to denote this time component. The remainder of the execution time can be divided into the time when the APs are engaged in parallel execution, which we denote by $T_p(N)$, and the time for data communication and other communications overhead, which we denote by $T_o(N)$. Accordingly, we have

$$T(N) = T_s + T_p(N) + T_o(N)$$

Note that T_s here is a constant which will not change as N, the number of APs in the run, changes.

We can compare this to the execution time for the corresponding sequential program for the same run; we denote this time by T_S. According to our discussion, we have

$$T_S = T_s + T_p(1)$$

We see that T_S is not identical to $T_p(1)$, since we are assuming, for the sake of simplicity, that there is no communications overhead in the sequential program.

A measure of the gain we achieve by parallel execution is the speedup ratio defined as

$$S(N) = T_S/T(N)$$

Ideally, we would like S(N) to be equal to N, but in reality it will always be somewhat less. The ratio given by

$$E(N) = S(N)/N$$

can be used as a measure of how efficient the code for parallel execution is in using all the parallel processors at run time.

The effectiveness of the load balancing the program achieves can be seen from the ratio $T_p(1)/T_p(N)$. Perfect load balancing would yield a value of N for this ratio, for each value of N. In practice we always see somewhat lower values of this ratio. Nevertheless, it serves as a useful measure of how well the load balancing strategies discussed above have actually been implemented in the particular code at hand.

Even with perfect load balancing for the parallel part, we see that the efficiency of parallel run will be degraded if the sum $T_s + T_o(N)$ is different from zero. There will always be some portion of the code that must run sequentially, although it is clearly worthwhile to minimize this to the extent possible. The same thing is true of communication overhead. However, as discussed above, modifying the program to hold down communication overhead may work against the strategies to achieve good load balancing.

Our discussion of communication overhead so far has been in terms of the way it can be affected by program organization, but it is obviously also affected by the system and hardware architecture the parallel configuration where the program runs. Thus, for example, one cannot assume that the communication overhead time $T_o(N)$ varies linearly as N, since it may be possible to handle some of the data communication to different APs at the same time. This capability actually exists to some extent on the lCAP systems. Furthermore, as indicated in Section V, part of the communications time is taken up by the use of the host operating system facilities to coordinate the calculation, and this will vary according to the system soft-wares used.

There are also some choices to be made in terms of hardware to be used for com-munication. Data to be passed between host and AP presently must pass over the host channel which run at 3 Mbyte/sec. This is a relatively slow data path for a parallel architecture, and in some cases this has serious effects on performance. The external shared bulk memory systems described in Section IV provide additional data paths, and, as discussed later, the use of these alternative data paths has some-times yielded a dramatic boost to parallel program efficiency.

In general, we can conclude that the efficiency $E(N)$ for a particular run must decrease as N increases. Good load balancing certainly becomes no easier to achieve as the number of processors increases, and communication overhead will increase as the number of processors increases. Even if we ignore these two effects, and assume that the sum $T_s + T_o(N)$ remains constant as N increases, we see that this constant term will eventually predominate in $T(N)$, so that adding an addi-tional processor to the calculation will yield only a very marginal decrease in the total execution time.

A particular program will typically be used for numerous runs of various sizes. As the size of a calculation increases, the times T_s and $T_o(N)$ do not necessarily increase at all, and they typically increase considerably less than the time $T_p(N)$ associated with the parallel part of the calculation. Furthermore, good load bal-ancing typically becomes easier to achieve as the size of the calculation gets larger. All these effects support the tendency of increased efficiency in parallel execution as the size of the calculation gets larger. This is just the trend we want, since parallel execution is needed most where the calculation is too large to be readily executed sequentially.

Combining these two tendencies yields another valuable conclusion: we can use an increased number of processors for a larger run and achieve an efficiency that a smaller run could only achieve with fewer processors. Hence the ability to vary the number of processors to be attached to a particular parallel application run is important. Our system allows this flexibility, and our programs incorporate it.

VII. ICAP Precompiler

In our scheme of parallel coding using the precompiler we have developed, the basic unit of parallel work is the subroutine. Thus the example of the modified loop given in the last section would be encapsulated in a FORTRAN subroutine. In place of an ordinary FORTRAN CALL statement (which one would find in a sequential program), a special precompiler directive statement is inserted. This can be viewed as essentially a fork and join process in which a master has a number of subroutines forking off to execute on the slaves, while the master continues exe-cuting its own program, which will eventually include gathering up the results from completion of the subroutines executing on the slaves and joining them together.

A comprehensive description of the precompiler language and its use is given else-where;[8] here we shall simply sketch the concepts and the way these concepts are realized in the precompiler language. Input to the precompiler are FORTRAN source files which contain precompiler directives. More than one input file is allowed, since the FORTRAN source code intended for use on the AP's will be compiled separately from the FORTRAN source code for programs to be run on the IBM host. Output from the precompiler is again FORTRAN source. There are two outputs, one consisting of code to be used by the master, and the other con-sisting of code intended to run on the slaves. A third source file contains the code to be compiled and run on the AP's subroutine. The difference between the input source code and the output source code is that the precompiler will implement the directives in the output code by means of calls to the utilities discussed in Section V.

All precompiler directives contain the characters 'C$' in the first two columns. This simplifies precompiler construction and has the additional advantage of making the directives transparent to the FORTRAN compiler, since they satisfy the FORTRAN convention for comments.

We begin with configuration of the parallel run. This consists of specifying the number of slaves in a parallel run (since this is a run time parameter, and not fixed by the form of the program), and also the AP's which are to attached to the slaves. The actual configuration information is specified in a FORTRAN COMMON block in the code to run on the master. This common block always has the name /LCAP$M/ and has the form

COMMON /LCAP$M/ NSL,NSLMIN,IDAP(10),SLPRG

where NSL is the number of slaves for the run, and IDAP(I) gives the AP number of the AP to be attached to the I-th slave, for I between 1 and NSL (the values of IDAP(I) for I > NSL are not used). We note that if IDAP(I) is zero, any available AP will be attached to that slave, while if IDAP(I) is negative, no AP will be attached to that slave. The latter option allows parallelism entirely on the IBM host, either for debugging purposes, or to take advantage of the capabilities of multiple-CPU processors like the IBM 3081, 3084, 3090, or 4381 model group III. NSLMIN is another integer to control attachment of the AP's: if some or all of the AP's are not available when requested, the run will wait for all of the AP's to become available if NSLMIN is negative or zero, and will abort if NSLMIN is positive and less than the NSLMIN slaves requested are currently available. The variable SLPRG is type CHARACTER*8, and has different significance under VM and MVS. In the case of VM, it denotes the name of the EXEC file used to con-figure the slave VM's, while in MVS, it is the name of the module for the slave program.

The user is responsible for making sure the /LCAP$M/ COMMON block contains the appropriate configuration specification prior to the point in the program where the configuration is actually requested. The configuration request takes the form of a precompiler directive of the form

C$ START

which can be viewed as a statement executing on the master that attempts to supply the requested slave configuration. In case the variable NSLMIN in the COMMON block /LCAP$M/ is positive, and fewer than NSL slaves are available, the parameters NSL and IDAP(I) in this COMMON block are modified to correspond to the actual number of slaves and the AP's actually attached to them.

In addition, the START directive initializes another special COMMON block on each of the slaves, with the special name /LCAP$S/. This COMMON block has the form

        COMMON /LCAP$S/ NSL,ISL,IDAP

where NSL is the number of slaves just as in the /LCAP$M/ COMMON block, and IDAP(ISL) is the ISL-th entry in the IDAP array in the /LCAP$M/ COMMON block. The variable ISL ranges between 1 and NSL, and is different for each slave. The variable ISL is the slave number used in some of the directives discussed below to denote which slave the directive controls.

The last in the group of configuration directives takes the form

C$ FINISH

which simply undoes the work of the START directive by detaching the AP's attached to the slaves and terminating the operation of the slaves configured for the run. It should be noted that the program flow for the master program must be such that the START directive is executed before any of the directives intended to control the slaves, and the FINISH directive must execute after any such directives.

There is also a group of directives to define parallel subroutines to the precompiler. These appear in the code intended to be executed on the slaves, and are regarded as non-executable. The first of these simply identifies a parallel subroutine, and is used as in the example

C$ SLROUTINE
        SUBROUTINE subr(arg,...)

where subr is used to denote the subroutine name and arg,... is used to denote the subroutine dummy arguments. That is, the second statement here is an ordinary SUBROUTINE statement as it normally occurs in FORTRAN. One must also identify data to be communicated between the master and the slave subroutine, and this is accomplished by means of directives of the form

```
C$    SLIN  < arg | / com / >,...
C$    SLOUT < arg | / com / >,...
C$    SLIO  < arg | / com / >,...
```

where arg denotes any of the arguments of the subroutine and com denotes any common block name appearing in the subroutine. That is, SLIN, SLOUT, and SLIO are followed by a list of names that can be arguments as in the SUBROU-TINE statement and/or common block names as in the COMMON statement, with the common block names having a slash '/' before and after. SLIN is used to indicate data that should be made available to the subroutine by the master at the beginning of execution, but is not required by the master as output at the end of subroutine execution. SLOUT is used to indicate data that is output to the master from the subroutine at the end of execution, but need not be provided by the master at the beginning of execution. SLIO combines SLIN and SLOUT to indicate data passed from the master, and then passed back to the master from the subroutine.

It is also possible for a subroutine intended for execution on the AP's to be a parallel subroutine, even though it runs on the the AP attached to the slave, instead of the slave itself. In this case the directives defining a parallel subroutine should be omitted, since the precompiler will recognize the extension to FORTRAN that FPS has introduced for use with the AP's.[5] Thus APROUTINE serves instead of the directive SLROUTINE, while APIN, APOUT, and APIO serve instead of, respectively, SLIN, SLOUT, and SLIO.

We come finally to the directives that the master program uses to control the slaves. To start subroutines executing on the slaves there are directives of the form

```
C$    EXECUTE ON SLAVE isl: subr (arg,...)
C$    EXECUTE ON ALL: subr (arg,...)
C$    EXECUTE ON ALL , USING isl: subr (arg,...)
```

where subr is the subroutine name and arg,... indicates the argument list for the subroutine. In the first variant of the EXECUTE directive, isl represents an integer expression that has a value designating the slave number of the slave where the subroutine is to run. The second variant allows one to start the same subroutine on all the slaves with a single directive, instead of repeating EXECUTE ON SLAVE for each slave individually. The third variant has the same purpose, but specifies the variable isl to be used for the slave number as each slave starts to execute the subroutine. This option is desirable because the specification of the variables in the argument list for the subroutine can depend on isl, as in the case of the EXECUTE ON SLAVE variant of the directive. This possibility turns out to be very useful in some parallel programs. In all cases the subroutine referenced by an EXECUTE directive must be defined to the precompiler as a parallel subroutine using the SLROUTINE directive or the APROUTINE statement as discussed above.

The EXECUTE directive starts execution of a subroutine on a slave, but after that, the master program continues executing regardless of whether the slave subroutines are finished or still executing. Thus it is necessary to stop execution of the master program explicitly if it should not continue without the output from one or more slaves. This is accomplished by directives of the form

C$ WAIT FOR SLAVE isl
C$ WAIT FOR ALL
C$ WAIT FOR ANY FREE isl

where isl is again an integer which gives the number of the slave being referenced. In the first variant, isl specifies a particular slave that the master needs data from. In the second variant, all slaves must complete. In the last variant, one is looking for any slave that is no longer executing a parallel subroutine, which could include a slave that never started executing. The variable isl in this case has a value supplied by the WAIT utility, and this value is the number of the "free" slave that was found.

There are several sub-directives associated with the EXECUTE directive, to help control the transfer of data between master and slave. All of these must immediately follow they EXECUTE directive that they modify. The simplest of these has the form

C$ PROTECT < arg | /com/ >,...

where arg is any of the arguments in the subroutine referenced in the EXECUTE directive, and com is any COMMON block referenced in the subroutine, provided that the argument or COMMON block is defined as SLIN or SLIO by a directive in the subroutine (SLROUTINE) or APIN or APIO by a statement in the subroutine (APROUTINE). Several arguments and/or common blocks can be referenced in the same PROTECT sub-directive. The function of the PROTECT sub-directive is to provide each slave with its own copy of the data referenced in the PROTECT which is being passed to the slave from the master. These copies are distinct from the data replica that remains on the master. This copy process will occur in any case if the EXECUTE applies to an APROUTINE, but a PROTECT can still be useful because it will insure that the copy is made at the time of the EXECUTE; otherwise data can be changed by program statements occurring after the EXECUTE even in cases where the variable cannot actually be shared between master and slaves.

Another sub-directive is of the form

C$ ADDING < arg | /com/ >,...

where arg is any of the arguments in the subroutine referenced in the EXECUTE directive, and com is any COMMON block referenced in the subroutine, provided that the argument or COMMON block is defined as SLOUT or SLIO by a directive in the subroutine (SLROUTINE) or APOUT or APIO by a statement in the subroutine (APROUTINE). Several arguments and/or common blocks can be referenced in the same ADDING sub-directive. Obviously, this sub-directive is intended to handle data passing from the parallel subroutine to the master, rather than the reverse case handled by the PROTECT sub-directive.

If an array is subject to an ADDING directive, there will always be separate array copies, one for the master, and one for each slave referenced in the EXECUTE directive. When the subroutine executing on a particular slave finishes and makes its data available to the master, the array copy from that slave for any array subject to the ADDING sub-directive will be added to the array copy on the master. There are actually several types of addition, depending on the type of the array (e.g., REAL*8, REAL*4, or INTEGER*4). The particular type of addition that is appropriate is automatically determined by the precompiler.

A programmer may wish to join array copies from the slaves with the arrays on the master in a manner other than what is provided by the ADDING sub-directive. In this case, one would define a merging subroutine on the master, and code this subroutine to join the arrays as required. The precompiler is informed of such a merging subroutine by a directive of the form

C$ MERGEROUTINE
      SUBROUTINE subr (arg,...)

where subr is used to denote the subroutine name and arg,... is used to denote the subroutine dummy arguments. This is similar to the SLROUTINE directive discussed above, but the reader should keep in mind that a SLROUTINE will execute on a slave, but a MERGEROUTINE always executes on the master.

The sub-directive that causes the MERGEROUTINE to be used with an EXECUTE directive is of the form

C$ MERGING BY subr (arg,...)

where subr is the name of the MERGEROUTINE and arg,... is the list of arguments for that subroutine. When a MERGING sub-directive is in effect, the output (SLOUT, SLIO, APOUT, or APIO) arrays or COMMON blocks from the subroutine executed on the slaves which are also referenced by the MERGEROUTINE subroutine will have separate replicas on each slave. When one of the slaves finishes and provides its output data to the master, that slave's version of the quantities referenced by arg,... will be passed to the MERGING subroutine, which will then execute. It will complete execution before the next slave data replica replaces the current replica, even if one or more slaves finishes execution while the MERGEROUTINE subroutine is still executing on the master.

We see here an interesting feature of the precompiler language, which is that arrays with the same FORTRAN name can contain different data on each slave and also on the master. We have found that this is a convenient feature for parallel programming, and the PROTECT, ADDING, and MERGING sub-directives allow this feature to be used in an orderly way, without serious confusion. We also note that use of the ADDING or MERGING sub-directives can improve efficiency by allowing the master to do some of the work it needs to do before all the slaves have completed.

It can be seen that the precompiler language is far more flexible and versatile than "fork and join" usually connotes. Although the identical program runs on each slave, there is no limit (apart from the normal constraints on memory for programs) on the parallel subroutines that can be introduced into that program. There is also no requirement that all slaves be executing the same parallel subroutine at the same time; any slave can execute any parallel subroutine, and the master can start their execution at any time that suits the needs of the program. For example, the master can use the WAIT FOR ANY FREE directive to find a free slave and immediately assign a new task to that slave. This possibility improves efficiency by allowing elimination of dead spots due to idle slaves.

Another feature of the precompiler language is that it embodies a model that avoids the possibility of deadlocks. A deadlock occurs when one processor comes to a point in its program where it cannot continue processing, and must wait until some event occurs as a result of the program running on another processor, but for some reason that event can never occur. In the model of parallel execution incorporated in the precompiler language, the slaves never wait for any event on another processor once they begin execution, so they cannot become deadlocked. The program on the "master" processor waits on the slaves, but they are guaranteed to finish (barring some condition such as an infinite loop which could occur just as readily in a sequential program). It is expected that the deadlock-free character of the precompiler language will be a considerable help in avoiding problems in debugging code for parallel execution.

As already noted, use of our precompiler is not confined to an lCAP-type configuration. It serves quite well to support parallel programming on multiple-CPU machines without AP's, and has actually been used this way in many instances on an IBM 4381 model group III, on an IBM 3081, and on an IBM 3084. We can put forward our precompiler as a general and versatile scheme to support parallel programming.

VIII. Applications and Practical Tests

The test of any parallel processing configuration is how it supports the applications programs it runs. As indicated in the introduction, the computational demands of our work in theoretical chemistry and biophysics yields several applications programs ready to serve as tests as soon as the system could support them. Subsequently, a number of additional codes have been migrated to our system, either through expansion of the researches in our laboratory, or our program where visitors are encouraged to try out our system on applications in their areas of expertise.

Here we describe the behavior of some the programs currently running on our system and examine their performance in parallel execution. Most of the programs we consider are well documented in the literature, so our discussion here will be confined to a brief description intended to give the reader an idea of the impact of parallel execution on these applications and the considerations that affect their performance. We first take up programs that are workhorses for our own researches, and later examine several programs that were acquired from our visitors.

We begin with our molecular quantum mechanics code, which is based on the Hartree-Fock model, and attempts to compute molecular structure and properties from first principles. This code is the product of much development in our laboratory over a considerable period of time.[9] It falls into two separate programs. The first is the integrals program, which computes integrals pertaining to motions and interactions of the electrons that determine the structure of the molecular system. The second uses these integrals as input to determine the behavior of the electrons by solution of the Roothaan-Hartree-Fock equations,[10] which are also called the self-consistent field (SCF) equations.

The integrals program repeatedly evaluates algebraic expressions with various sets of parameters. Evaluation of any one integral is fast, but any complete run will include huge numbers of these integrals. The computed integrals are stored on disk files which can be larger than a gigabyte in size. Computation of any one integral is independent of any other integral, which makes it easy to run the integrals program in parallel: one simply decides how the set of integrals is to be divided among the available APs. Once the run is set up, there is no communication between the tasks running on the different APs, and even disk I/O takes place entirely on the AP (parallel I/O). The time a particular task runs without interruption on the AP is the elapsed time for the entire run (a unique feature of this application), and this is typically on the order of hours. The only obstacle to ideal parallel execution is that we have not found a way to divide the set of integrals in the run that can guarantee perfectly even distribution among the APs. Thus parallel performance is limited only by load balancing, since there is no sequential portion and no communication overhead. The measured speed-up of the parallel integrals program on the lCAP system is shown is Fig. 5a. These results are for a 87-atom molecule; it is noted that this is not too large a calculation for our system. Indeed since the early days of lCAP-1 we have performed SCF computations on DNA fragments in excess of 90 atoms. As indicated above, one expects better efficiency for larger calculations, and this trend is in fact observed for the integrals program.

The SCF program iteratively improves on an initial guess for molecular electronic structure until convergence is achieved. Each iteration can be divided into several steps of which by far the most time-consuming is combining results from the last iteration with the integrals file generated by the integrals program to develop the current iteration. This step is the only one running in parallel at present. It involves heavy disk I/O (which again takes place in parallel) and also very substantial CPU time. The time for this step running in parallel is typically a fraction of an hour, and requires transmission of several hundred kilobytes of data between host and AP. The measured speed-up of this program is given in Fig. 5b, again for the case of our 87-atom molecule. Here we see performance degraded again by load balancing effects as in the integrals program, since the integrals files on various APs will differ somewhat in their size, and the amount of computation on a particular AP is proportional to the number of integrals it handles. There is also an effect from the sequential portion of the program (3% of the run time in this particular run). The effect of communication overhead is also present, but this is nearly negligible, since communication must take place only once during a period of several minutes of computation.

The next step in our hierarchy of models is modelling macroscopic properties of matter, using the data developed from our models at the molecular level. Two types of programs in this class are our Metropolis-Monte Carlo programs and our molecular dynamics programs. These are alternative approaches to description of the kinetic motion of molecules in bulk liquid or solution, and derivation of bulk properties such as density, heat capacity, etc. from this description.

Our Metropolis-Monte Carlo[11] programs deal with the liquid or solution in terms of its molecular constituents and the main task is evaluation of the change in the potential energy of the bulk each time a molecule is moved. It is this task that runs in parallel. All other tasks are so fast that they cannot benefit from parallel execution. There are several versions of our Monte Carlo code, depending on how elaborate (and hence realistic) an energy expression is being used. Thus the time a task runs in an AP without interruption varies from a fraction of a second up to several seconds. Data transmission between host and AP is on the order of a hundred bytes per task. The data, reported in the next section, are for one of our most elaborate Metropolis-Monte Carlo programs which is used to model water molecules interacting under the influence of four-body potentials.[12]

Our molecular dynamics[13] programs simulate the kinetic motion of molecules in bulk liquid or solution over a period of time divided into many time steps, with each time step involving the evaluation of many molecular energies and forces. Evaluation of these energies and forces is the bulk of the calculation and again is done in parallel for each time step. There are several versions of our molecular dynamics code, depending on the energy expression being used. Typical time for a task run in an AP without interruption is a fraction of a minute, and this involves data transmission between host and AP of as much as a few Mbytes. We again consider a version of the program simulating water, this time using up to two-body potentials.[14] The results for a simulation with 512 water molecules are given in Fig. 5c. Here two effects are an impediment to parallel performance. Foremost is the overhead associated with the transmission of data by all processors at every time step. This effect is more significant than in the SCF program because the time per iteration is much less. In the SCF each iteration is of the order of a few minutes, whereas for the MD each iteration is of the order of seconds. The second effect is the sequential portion of the program, which while only 2% of the program, still

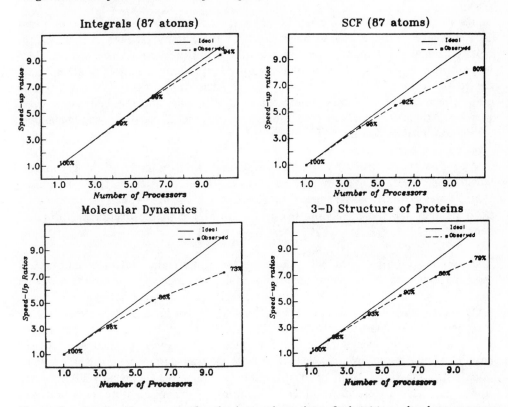

Figure 5. a): Speed-up curves for the integral portion of *ab initio* molecular program. The results are shown for a 27 atom system using a minimal basis set.

b): Speed-up curves for the SCF portion of the *ab initio* molecular program. The results are shown for a 27 atom system using a minimal basis set.

c): Speed-up curves for a 100-step run of a molecular dynamics simulation of 512 water molecules.

d): Speed-up curves for 1000 minimization steps in the 17 dimensional surface characterizing the glutamine-leucine conformational space.

yields a maximum speed-up of only 8.95 with 10 processors. Good load balancing is easy to achieve here, since we have identical molecules and each pair interaction takes equal amounts of time to compute.

Another program which we would like to discuss was originally written by H. Scheraga and co-workers at Cornell,[15] and later modified for parallel execution on the ICAP system. It is currently being used in a collaborative research effort between Prof. Scheraga's group and ours. In this program proteins are built up from segments of polypeptide chains; the arrangement of the constituents and their order on the chain are given, but spatial conformation as the chain attempts to fall into a preferred shape remains to be determined by the program. The program does this by computing energies of a particular conformation, or building up from conformational energies of parts of the chain; the preferred shape corresponds to the lowest conformational energy. Often thousands or ten of thousands of energy minimizations must be performed in order to ensure that the possible conformations have been adequately scanned. Each of these minimizations is independent from all remaining minimizations, and therefore this program naturally lends itself to parallel computation. It is notable that the simple loop partitioning scheme discussed in Section VI is not the best way to adapt this program to parallel execution, because the time each minimization takes is unpredictable before run time. However, there are sufficient minimizations so the alternative load balancing scheme outlined in Section VI is quite effective.

The results for the protein program on a rather small example of the glutamine-leucine dipeptide are given in Fig. 5d. This example consisted of 1,000 minimizations of the 17 dimensional space. In this example all three effects, the sequential portion, load balancing, and the overhead, contribute to the overall performance degradation.

The previous examples deal with quantum mechanical and statistical mechanical computations where there is an obvious granularity, namely, the interaction of one particle with the other particles of the system. But a very large number of applications deals with partial differential equations where the granularity — if limited to the operations at a grid point for one time step — might be as low as one or very few floating point operations. Fluid dynamics is, therefore, an ideal area to test parallel architectures. We recall that ILIAC IV was designed in the late 1960's exactly to solve this type of problems. Our program to handle the shallow water equations emerged from just such considerations. The shallow water equations are a means of studying the dynamics of a two-dimensional, incompressible, barotropic fluid under the hydrostatic approximation.[16] They are based on Newton's Law of Motion and the continuity equation (which ensures conservation of mass in fluids). They may be discretized by the finite difference representation over a grid of the physical system and then solved numerically. Parallelization can then be achieved by dividing the grid equally among the different processors. Each processor computes his own portion of the grid and the final result is obtained by merging the results of all processors to give the total grid.

Usually the evolution of the system over a period of time is studied; this evolution is modeled by dividing the total time into smaller time steps. For a particular time step, the results from the last time step are needed. In particular, each processor needs the values on the boundaries of its region of the grid from the adjacent regions on the other processors. As a result, while the numerical solution on the grid is highly parallelizable, this requires frequent data transfer among the processors. In particular, in the example studied here the computation time for each time step is of the order of 1.0 second (with 1 processor). Therefore, with 5 processors data transfer must occur on the average every 0.2 seconds.

The first results observed for a parallel version of this program modelling a small gridded region are given in Fig. 6a. The observed speed-up increases for both 2 and 3 processors, but thereafter increase only slightly, then actually begins to decreases. With 5 processors the maximum speed-up is obtained (2.03), but this is only 40%

efficient. The primary factor causing the degradation in this program is the overhead associated with communication between time steps.

In Fig. 6b we give results for the same program, modified to use the SCA shared bulk memory as a data path instead of the channel connections between host and AP. We see a marked increase in the performance with this modification. There is still degradation due to synchronization of the slaves between time steps. This now takes place between the APs through the shared memory. It is clear that we have gained a great deal in performance by using a faster data path than provided by the channels between host and AP.

Although our discussions here have centered on performance, the programs we have described so far are not test programs. All of them originated as programs to support research on questions of interest by production calculations. For instance, the parallel molecular quantum mechanics program has been used to throw new light on the nature of the bonding in DNA base pairs.[17] The parallel Metropolis-Monte Carlo program has been used to assess the effects of four-body interactions on the structure of water for the first time,[18] and the molecular dynamics program has also been used to model the structure of water.[19] The protein structure program has recently been used to predict the low energy conformations of all 400 naturally occurring dipeptides.[20] These conformations are currently being used as a starting point in building a much larger protein,[21] namely interferon, which has received attention for its possible role in combating viral or bacterial infection. Lastly, as already indicated, the program solving the shallow water equations is being used in oceanographic studies. The studies where we use molecular dynamics for problems generally analyzed with fluid dynamics are also of high scientific interest since they provide a firmer understanding of macrosystems based on continuity from micro-

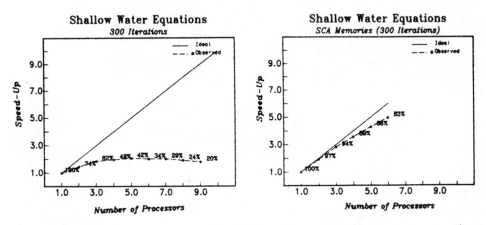

Figure 6. Inset a) Speed-up curves for 300 iterations of the shallow water equations, discretized by the finite difference method. These results are without the shared memory, and use the master on the IBM host as the transfer point for all data.

b): Speed-up curves for 300 iterations of the shallow water equations discretized by the finite difference method. The communication between the FPS-164's that is necessary each iteration is performed through the shared memory.

systems descriptions. For more details, we refer to the companion article in this volume.[22]

None of these calculations would be possible without a very powerful computing system. Our ability to produce results of significant scientific interest is perhaps the ultimate test of the power and usefulness of our parallel computing configurations.

We have encouraged visitors from other laboratories to try out our system on their own applications codes. Their experiences help us to better appreciate the general applicability of our system as an answer to computational demands. We present four such experiments as a sample of experiences. We give only a very brief sketch here, citing as references the visitor's own report for the reader who wishes more detail. Performance for all four codes is shown in Fig. 7. In Fig. 7a we have results from a high energy physics application from CERN.[23] This application is not vectorizable, but turns out to be highly parallelizable. Results in Fig. 7b come from a seismic application code developed at IBM Rome and IBM Palo Alto.[24] This code was originally thought to be very vectorizable but not parallelizable. Next, in

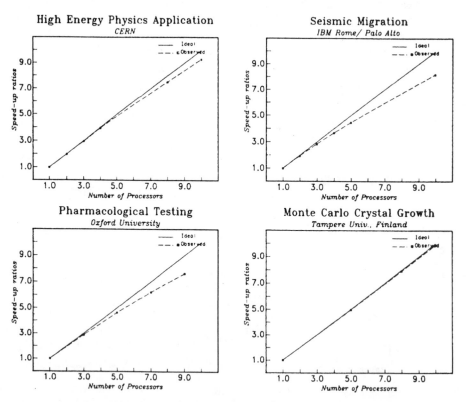

Figure 7. Speed-up curves for four application programs that have been migrated to parallel via the Visitor's Program. These applications include: a) a high energy physics application from CERN, b) a seismic data analysis code from IBM Rome/ Palo Alto, c) a pharmacological drug design code from Oxford University, and d) a Monte Carlo code simulating crystal growth from Tampere University, Finland.

Table I. Timing results for ICAP-1 with one, three, six and ten FPS-164's in parallel, with the IBM 4381 as the host processor. Times are given as total elapsed times for the job, in minutes. CRAY timings are for single-processor mode. None of the codes were optimized for the particular machine architecture, and standard FORTRAN was used as much as possible. On ICAP-1, this was VS FORTRAN 4.0 running under VM/SP 3.1 HPO 3.4 on the host and FPS System Release F01-100. On the CRAY-XMP this was CRAY FORTRAN 1.13 running under COS 1.12 Version 22.

JOB	Time (minutes)				
	1AP	3AP's	6AP's	10AP's	CRAY-XMP
Integrals (27 atoms)	71.7	24.0	12.3	7.8	7.6
SCF (27 atoms)	46.7	21.0	17.5	12.0	6.8
Integrals (42 atoms)	203.4	68.9	38.3	21.2	23.2
SCF (42 atoms)	108.5	44.9	34.1	22.0	13.5
Monte Carlo	162.1	57.8	32.0	22.0	20.4
Molecular Dynamics	99.6	34.6	19.3	13.7	17.0

Fig. 7c, we have results from a theoretical chemistry program from Oxford University used in pharmacological testing/drug design.[25] Last, in Fig. 7d, are results from a Monte Carlo program from Tampere University in Finland which is used to study crystal growth.[26]

These examples are only a few of many different applications from various scientific fields that have successfully been tried out on our parallel computing systems. Our experiences support the idea that these systems can be profitably applied across the entire spectrum of scientific and engineering applications.

IX. Performance

We have timed specific applications runs for four of the programs discussed in the previous section running sequentially (one FPS-X64), and running in parallel on three FPS-X64's, six FPS-X64's and ten FPS-X64's. One objective, of course, was to see how successful we are in exploiting our parallel system. We would also like to see how close we come to "supercomputer" performance, and so we have run these applications on a CRAY XMP as well. The applications code for the CRAY was developed under constraints analogous to the constraints for our parallel applications codes, that is, the minimum modifications required to run properly under that system. Efforts to modify the code to better exploit the vector architecture of the CRAY would certainly have resulted in faster timings for that machine, but, conversely, we could gain on our parallel system by adapting our code to the archi-

Table II. Timing results for lCAP-2 with one, three, six and ten FPS-264's in parallel, with the IBM 3084 as the host processor. Times are given as total elapsed times for the job, in minutes. CRAY timings are for single-processor mode. None of the codes were optimized for the particular machine architecture, and standard FORTRAN was used as much as possible. On lCAP-2, this was VS FORTRAN 4.0 running under MVS/XA SP 2.1.2 on the host and FPS System Release F02-100. On the CRAY-XMP this was CRAY FORTRAN 1.13 running under COS 1.12 Version 22.

JOB	Time (minutes)				
	1AP	3AP's	6AP's	10AP's	CRAY-XMP
Integrals (27 atoms)	19.1	6.5	3.3	2.3	7.6
SCF (27 atoms)	13.6	5.5	4.2	3.6	6.8
Integrals (42 atoms)	55.0	18.7	9.3	6.1	23.2
SCF (42 atoms)	30.0	11.1	7.0	5.3	13.5
Monte Carlo	60.0	20.9	11.4	7.7	20.4
Molecular Dynamics	29.9	10.6	5.9	4.2	17.0

tecture of the FPS-X64. As they stand, we believe our results are useful, even though they cannot be regarded as anything like a definitive comparison between the two systems. Our results for lCAP-1 and lCAP-2 are shown in Table I and Table II, respectively.

Comparing the two tables, we see that migration from lCAP-1 to lCAP-2 causes no appreciable degradation in relative parallel performance, but absolute performance is boosted by a factor of roughly 3 or 4, due mainly to the difference in perform-ance between the FPS-164 and the FPS-264. Comparison with the CRAY XMP timings indicates that parallel execution with 3 AP's on lCAP-2 is about as fast as using the CRAY. Again, we hasten to disclaim any definitive comparison. That is probably not possible. What is possible is a qualitative appreciation of the sense in which we apply the term "supercomputer" to our parallel configurations.

X. lCAP-3

In this final section we would like to sketch briefly lCAP-3, even if by doing so we contradict our "rules" not to discuss a configuration until we have not only the hardware and system software operational, but also tested it with at least one large application. We note, however, that for lCAP-3 there is an application under con-sideration, precisely the quantum-mechanical study of a Na^+ or a K^+ cation inter-acting with gramicidin A, a trans-membrane channel macromolecule most interesting for its selectivity toward very specific ions. The broader issue of the simulation is to understand transport of ions through membranes and in particular neuronic membranes[27] (and the next obvious step is to consider neuron networks). We plan to compute these ions at a few positions along the channel, the wave function of the latter being represented with a small basis set of gaussian functions; the wave function is considering *explicitly the entire macromolecule*, not only the fragment as previously reported. Computationally we are faced with solving the SCF-MO wave function for a few hundred atoms; but we are confident that lCAP-3 is well suited for this task.

As pointed out in the previous sections, lCAP-3 presently is the last evolutionary step from lCAP-1 and lCAP-2; however, we shall describe it below as if born ex-novo as "lCAP-3". In this viewpoint, the task calls for building a "supercomputing system" using existing components and system softwares as much as possible. The steps for solving the task are presented in the four insets a) , b) , c) , and d) of Fig. 8 and the completion is given in Fig. 9.

We start with thirty nodes, ten FPS-164, ten FPS-264 and ten IBM processors, precisely six of the dyadic IBM-4381, IBM-3081, IBM-3090 and four processors from the IBM-3084. The interconnections are standard IBM channels (see Fig 8 inset a) well suited to transmit both small or massive amounts of data (burst or stream). This configuration is already very powerful for a problem with large or intermediate granularity, and also contains vector capabilities. Indeed, even keeping in mind the relative informational value of the peak performance number, the above configuration is just about one Gigaflop system, with 424 Mbytes of fast random access memory and 1344 Mbytes of shared bulk memory distributed on the 30 nodes.

The next step for lCAP-3 is the 2nd-level of interconnections which is shown in inset b) of Fig. 8. The ten nodes on cluster 1 (lCAP-1) and the 10 nodes on cluster 2 (lCAP-2) can now directly inter-communicate at high speed. This communication is *ideal* for synchronization, message sending and receiving, and system communications in general, but less so for massive data transfer.

For the latter we move to the 3rd interconnection level; namely the two double rings of SCA memories (inset c) of Fig.8). This level is ideal for jobs that can be subpartitioned into tasks for four nodes (or less), but presents problems for situations where the 10 nodes need to exchange massive amounts of data. This request is satisfied at the next interaction level where we introduce into lCAP-3 the two massive SCA shared bulk memories with up to 12 connections. Thus we reach inset d) of Fig. 8, which shows two fully developed, independent and non-homogeneus clusters, channel to channel connected to the IBM-3090/200.

In Fig. 9, the two clusters are allowed to be interconnected by a more diversified network, which presently includes a) two connections from the massive memory on one cluster to two of the nodes on the other cluster, b) four direct connections from the IBM-3090 to four nodes on cluster 2 and c) two equivalent connections to two nodes on cluster 1. (Note that in Figs. 8 and 9 we have shown only one front

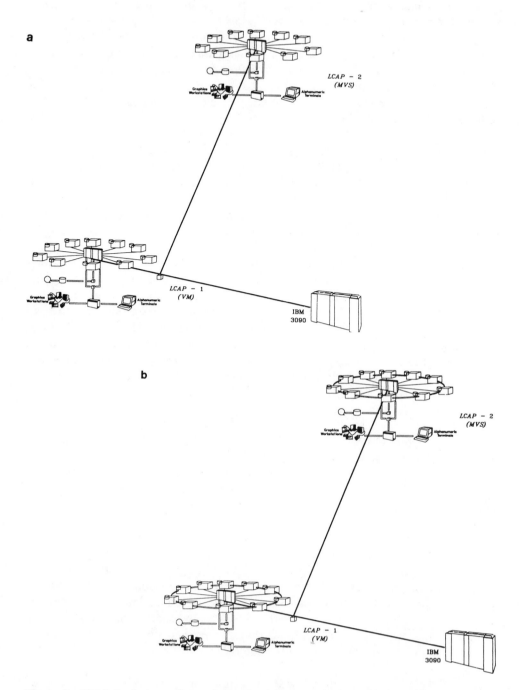

Figure 8. lCAP-3: a) the nodes and first-level interconnections; b) second-level interconnections.

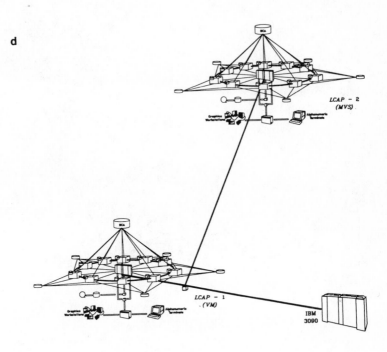

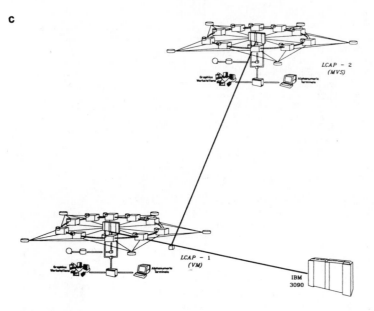

Figure 8 (continued). lCAP-3: c) third-level interconnections; d) fourth-level interconnections.

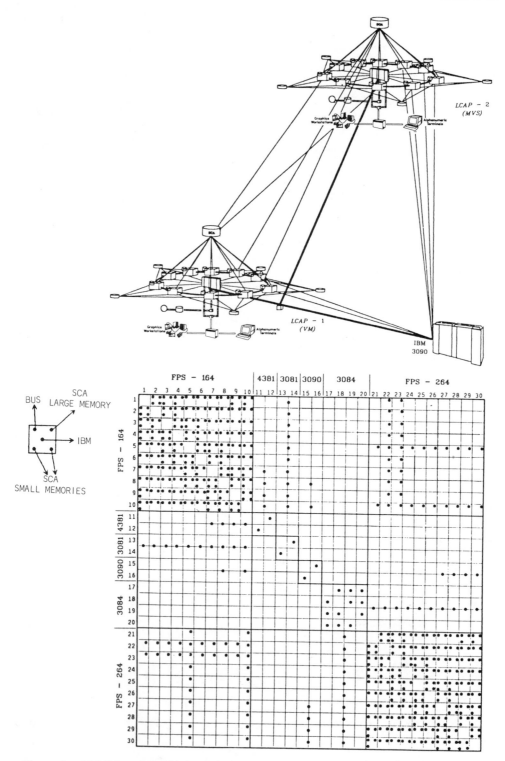

Figure 9. lCAP-3: the fifth-level interconnections (top) and the connectivity scheme (bottom).

end for lCAP-1 rather than two, an IBM-3081 and an IBM-4381, to simplify the drawings.)

At the bottom of Fig. 9 we have shown the connectivity scheme for lCAP-3. Nodes 1 to 10 refer to the APs on the first cluster, nodes 21 to 30 refer to the APs on the second cluster, and the nodes 11 to 20 refer to the pair of processors for the IBM-4381 (nodes 11, 12), IBM-3081 (nodes 13, 14), IBM-3090 (nodes 15, 16) and the four processors of the IBM-3084 (nodes 17, 18, 19, and 20). The addition of a third massive SCA shared bulk memory common to the two clusters and extension of the FPS busses to more than ten APs each, is under consideration. In addition we will, most likely, subtract two FPS from each cluster and build a mini cluster around the IBM-3090 in order to increase the connectivity between the two clusters. Our large scale test (computation of the wave function for gramicidin A) will help evaluate our needs.

XI. Discussion

We have seen that our systems can already effectively accommodate a large and diverse set of scientific and engineering applications. It is remarkable how many applications can benefit from even the simplest system capable of parallel execution. We claimed in the Introduction that parallel execution was "natural" to scientific and engineering calculations in the sense that parallel calculation was inherent in the models and the codes that implement these models. Our experiences, which were sampled in the last section, certainly support this idea.

If we look a little deeper we can see an explanation for this situation. In our Monte Carlo and molecular dynamics codes, we saw that the main task was to evaluate the potentials and forces that control the motions of each individual molecule in the simulation. If the potentials are two-body only, than the potential for a particular molecule is the sum of potentials of interaction with every other molecule in the simulation, and each of these potentials of interaction between pairs of molecules can be evaluated independently of any other pair potential. There is actually nothing special about two-body potentials; potentials involving three-body, four-body, etc. interactions will have the same property. Each distinct potential term can be evaluated independently from all the rest, so parallel computation is indeed quite natural.

Nothing in this argument contains the specific requirement that the particles in the simulation be molecules, so we find simulations with the same characteristic parallelism wherever classical mechanics can be applied to discrete particles. Generalization to the cases where quantum mechanics must be used instead is exemplified by the molecular modeling codes discussed above. Here the interaction between the electrons in the molecule are described in terms of integrals that can again be evaluated independent of all the others. There remain the cases where one models the behavior of a continuum, such as a body of water. But in fact such a continuum is broken up into discrete elements for modelling on the computer. There are many ways to do this, such as finite elements, finite differences or various spectral methods including Fourier transform. In every case, however, we may expect that the interaction of discrete elements will again be described in terms of computations that can be carried out in parallel. This is indeed what we saw in the example based on the shallow water equations presented above.

Graphics applications, are another obvious and natural example. The graphics application programs currently under development are primarily used as tools in analyzing, interpreting and presenting scientific data. Among these applications we note a parallel version of a molecular display program using ray-tracing techniques that are very compute intensive. Benchmark results of this program demonstrate almost linear speed-up with ten FPS-264's in parallel.[28]

Images from a computer simulation may be produced at various time intervals during the simulation. These images may then be transferred to either video tape or film to produce a motion picture. As examples, motion pictures of our molecular dynamics simulations of DNA, gramicidin, liquid water and ionic solutions are currently being produced. Our goal is to provide real-time graphics display of scientific results, namely, the graphics display of a scientific computation during the course of the simulation. In this way a scientist would be able to judge and correct theories and results without having to wait for the computation to complete.

Our systems are experimental, and continually evolving in terms of both hardware and software. In particular, we have seen the effect that use of shared bulk memory can have on performance in the case of the shallow water equations described in Section VIII above. We can expect to find many more applications where the shared memories and/or the fast bus will be an important component of our system.

All these planned and potential enhancements to our system involve new software as well as new hardware. A constant challenge is to develop and maintain a system that makes parallel programming and execution reasonably accessible to the programmers developing applications. We have made good progress in this area, but we have really only just begun our experimentations in such areas as scheduling of resources for parallel execution.

As already indicated, a notable feature of the ICAP systems is the flexibility to adjust the number of parallel processors to the demands of the applications run. A consequence is that several parallel runs can be on the system at the same time, using the same host. This raises new problems, in terms of allocating resources among the jobs requesting them. Conventional scheduling capabilities do not deal with such contingencies, since they assume that a job is a conventional sequential job, using only one processor at any given time.

We have made some progress toward acquiring a parallel job scheduling capability. Our current scheduler for the VM system[29] can maintain a queue of parallel jobs waiting for the requested processors to become available. Since it recognizes that several processors are dedicated to the same job, it also can be recognized that a program error on one processor affects the entire job; if the error is fatal, the entire job will abort, immediately freeing all the job's processors for the next job in queue. In addition to handling processor allocation, the scheduler has also recently acquired the capability to handle allocation of shared bulk memory resources among several parallel jobs. Most of these developments are equally applicable to the MVS system, and are being migrated there as they prove out on the VM system.

We also see that we have been taking our parallel applications programs on a case by case basis. Some general thumb rules have been found, as discussed in Section VI, but no deeper analysis has been put forward. Clearly, there is a need for some more sophisticated analysis, and eventually perhaps a preprocessor that could automate parts of our code migration to parallel mode.

In the meantime, the precompiler discussed in Section VII serves us well. We have emphasized its use in migrating already existing code to parallel execution, since much of our work has been concentrated in this area. However, we have all the tools we need to write new supercomputing tasks for our parallel system, and we expect to see more of this in the future. A fascinating and nearly inexhaustible area for future work is development of new methods and algorithms adapted to parallel execution.

In order to support our machine coding, especially for FPS processors, we are developing a novel package.[30] Using our graphics capabilities on the IBM-5080, the machine code is displayed in diagrammatic form as the coder develops it. The package of programs is interactive, and can check for some of the inconsistencies that can occur during the coding process. This approach has already demonstrated its usefulness in avoiding some of the normal obstacles to the development of high-performance machine-coded mathematical subroutines.[3]

In concluding, we can say our experiences give ample support for the belief that parallel execution will play an important role in the future of large-scale scientific and engineering computation, along with vector and sequential modes of execution. We might even go so far as to say we have had a peek at some of the features of future systems for parallel execution through our current experiments with our loosely coupled array of processors, although there is clearly much that remains to be learned.

Acknowledgments

We would like to thank Drs. V. Sonnad and A. Capotondi for permission to publish portions of their work prior to publication.

Appendix

Previously, we have briefly reported on the history of our parallel project. However, our interest predates the 1983 timeframe. It is worthwhile recalling that the late 1960s and early 1970s was the time for some most interesting experiments in parallel machines (like the ILIAC IV) and in neuron-neuron networks.

Let us also recall that in the late 1960s and early 1970s neuron modeling was carried out on two fairly distinct subjects. On one hand *single* neurons were accurately described with phenomenological membrane equations, analytic input/output functions, and even with sophisticated hardware models. These rather accurate descriptions stood in substantial contrast to the work of network modeling. With the argument that properties of a single neuron are unimportant when a large ensemble is regarded, neuronal models ranged from controlled statistical generators to binary and fuzzy logic. Probably the best known model of that time was the McCulloch and Pitts "neuron" model which did lead to the perceptron-theory. Undoubtedly, many of the early models contributed considerably to our understanding of neuronal system behavior. The question remains, however, whether a very rudimental model permits any explanation of the most important observation on living nervous systems.

To cope with this question, in 1973 and 1974 Clementi and Dr. Hayo Giebel — a postdoctoral visitor at the IBM San Jose Laboratory — undertook work on two problems. First, the modeling of a single neuron attempting to introduce as much realism as possible, then simulations on groups of such neurons interacting with each other.

The general idea was to mimic the firing of a *real* neuron, taking experimental data and *fitting it* to phenomenological equations, thus, in this way, producing a model. It is well known that most neurons have a great number of synaptic inputs which can affect their inner potential. If this potential exceeds a certain threshold, a specific mechanism starts producing a short impulse called a *spike*, which is of nearly constant size whenever it occurs. During the time of spike generation and immediately afterwards (absolute refractory period), the inner potential is almost unaffected by the inputs. The spike is transmitted by the neuron's axon to synapsis with other neurons (we hurry to say that this is not the only neuron-neuron interaction mechanism). Figure A1 shows our model. It consists of n inputs, a comparator and spike generator, C, with delay, τ, and two feedback blocks, F_v and F_x. If a single impulse arrives at one of the inputs, x_μ, the inner potential v responds with the function $r_\mu(t)$. The top right inset of Fig. A1 illustrates the qualitative shape of $r_\mu(t)$ for an input spike at $t=0$, as it is measured by impaling a nerve cell with a small electrode. After a delay, the function rises to a maximum and then decays back to zero again. Generally, the rise is much faster than the decay. The response function, $r_\mu(t)$, may be either positive (excitatory, depolarizing) or negative (inhibitory, hyperpolarizing). The exact shape of the input spike itself it not too important in this context as long as all spikes corresponding to a certain input x_μ are alike. If in a living neuron several spikes at different inputs occur together, their effect on v, in general, will not simply be the algebraic sum of the single effect.

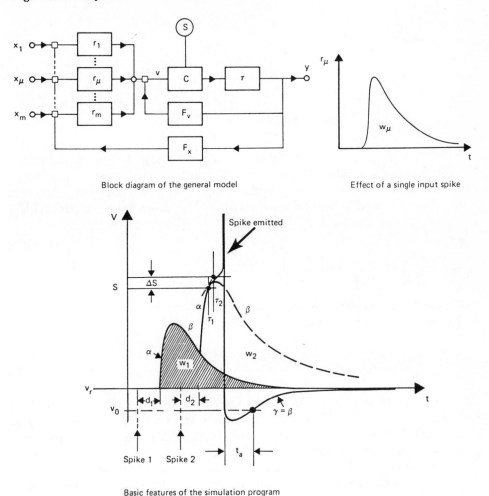

Block diagram of the general model

Effect of a single input spike

Basic features of the simulation program

Figure A1. Single neuron model. Top left: model generating spikes as output from incoming impulses. Top right: general form for response functions. Bottom: general features of the spike generation program and its parameters.

However, the linear superimposition is already a considerable part of the entire effect as shown in laboratory experiments. In addition, for a model which computes the time when a spike is emitted, the value of the inner potential is of interest only near the threshold S and need not be exactly described over its whole range. Therefore, the linear combination of inputs is a reasonable zeroth order approximation.

After the potential v exceeds the threshold S, a spike is generated with a certain delay τ (see Fig. A1, block c and τ). For the duration of the absolute refractory period, t_a, all inputs are disregarded (block F_x). The inner potential is reset to a certain value (starting potential v_0, block F_v) from which it decays to a resting value v_r if no further input spikes arrive. The function $r_\mu(t)$ can be approximated by a sum of two exponential functions, given below:

$$r_\mu(t) = w_\mu \frac{\alpha\beta}{\alpha - \beta} \left(e^{-\beta(t-d)} - e^{-\alpha(t-d)}\right) \text{ for } t > d \tag{1}$$

and

$$r_\mu(t) = 0 \text{ for } t \le d.$$

In Eq. 1 we select $\alpha > \beta$. The time d denotes the synaptic delay and w_μ is a certain positive or negative weight which is specific for each synaptic input. The time constant α is mainly responsible for the rise of r_μ while β describes the decay. The factor $\alpha\beta/\alpha - \beta$ is used to normalize the integral of the equation above, namely

$$\int_d^\infty x_\mu(t)dt = w_\mu \tag{2}$$

In the same way the decay of the inner potential v, after it is reset to v_0, can be approximated as an exponential function given below

$$v = v_r + v_0 \; e^{-\gamma(t - t_0 - t_a)} \tag{3}$$

with $t \ge t_0 + t_a$. Here t_0 is the time when the last spike was emitted, and t_a is the absolute refractory period. For additional details on our model we refer to a technical report.[1A] A computer program was written which was dimensioned for a maximum of 100 neurons, each one with, at most, 25 connections to other neurons. The basic features of the program are described in Fig. A1, bottom inset. In this figure the labels can be described as follows: w_μ = individual weights; d_μ = individual delay; $\tau = \tau_1 + \tau_2$ = delay of spike after exceeding the threshold; t_a = absolute refractory period; v_0 = starting potential; v_r = resting value of the potential; S = threshold; α, β and γ are the time constants in the exponential. Notice that τ_1 is a constant whereas $\tau_2 = \Delta S(dt/dv)$ and depends on the speed of v passing the threshold. These labels are the "parameters" which we shall fit using laboratory data on a single neuron.

Ten years ago this model was sufficiently complex; indeed, for a layer of 100 neurons with 25 connections and a maximum firing rate of 200 spikes per second, the simulation took about ten times the real time on an IBM 360/195, a "supercomputer" of the late 1960s.

As a first possibility for comparison of the model with experimental data, some fairly large neurons were chosen. Isolated ganglia of the Aplysia Californica were at the same time being investigated by Segundo, Perkel and others.[2A,3A] In Fig. A2 we compare the pattern of time obtained with our simulation program (once appropriate values for the parameters were fitted) and those obtained in the laboratory in the work by Segundo, Perkel and others. The actual experimental data are displayed in the upper right-hand corner of each inset. In one experiment we show the effect of spikes with different temporal distances; the next experiment simulates measurement where the threshold is reached and the spike is emitted; the bottom left experiment concerns an experiment with random inputs, and the last input shows an experiment with pairs of spikes. The overall comparison, simulation versus laboratory data, is reasonable and therefore we conclude our model is sufficiently realistic or, alternatively stated, that we can describe fairly well a single neuron and therefore we can pass this description into a network of such neurons. Of course we should add that we had ten parameters at our disposal in the fitting of the phenomenological model and this provides for much freedom and ensures a good fit (not necessarily a good model).

The second task we considered can be summarized in the following question: how many neurons must be assembled in order to manifest at least some of the very complex abilities of a neural system? To rephrase the problem, "what is the minimum number of neurons for which collective properties start to be manifested," or, "when does complexity start in a neuron-neuron network?" Because of the computational limitations of the time (1973 − 1976), we considered 96 neurons only with uniform time invariant parameters (the parameters were those optimized in the above study with the Aplysia). We have experimented with different types of neuron to neuron connections, some forming regular patterns and some randomly selected. In particular, we considered *small loops of neurons*. We define as a loop of size n, a system where each neuron v is connected only to neuron $v+1$ and where neuron n is connected to neuron 1. We have also considered *completely connected groups* where each neuron is connected to all the other neurons in the set. Obviously, many more different sequences of firing are possible compared with the case of simple loops. In order to get some estimate of the number of possibilities, we considered the case of uniform parameters and connections with high positive weights. In this context a neuron will fire if it is excited by at least one other neuron, provided that it is not in its absolute refractory period. Whenever a spike is emitted, it will activate every neuron which is not in its absolute refractory period. Due to their history they will fire at slightly different times, even if the delay d is uniform. The spikes emitted by those neurons will activate other neurons and so on until the same neurons will fire again. This time they have almost the same history and thus they will become more synchronized. *Partially connected groups* were also analyzed, either forming *layered structures* or with *random connections*. In the latter case, after an initial random stimulation, no further external input was applied. The questions arise: Which mode of activity is possible? Which mode is stable? We refer the interested reader to a second technical report which describes those experiments in detail. Here we report the main conclusions.

1A. H. Giebel, IBM Research Reports R.J. 1427 and R.J. 1428 (August 6, 1974).

2A. J.P. Segundo and D.H. Perkel, The Interneuron, Univ. of Calif. Press (1969).

3A. D.H. Perkel and J.P. Segundo, Science, **145**, 61 (1964).

By going from single neurons over small groups and completely connected groups to layers of neurons we were able to investigate some effects of the rhythmic activity, wave propagation, periodic and nonperiodic modes. Our main conclusion, however, was that a "collective mode" can be achieved with relatively few neurons, say no more than 20 – 30 neurons. On one hand it is clear that we must exceed this number by far if we are interested in setting up simulation of a complex system which might have some similarity with brain activity. On the other hand, *it is clear that we do not want to exceed such numbers if we wish to develop a parallel system simple enough not to encounter drastic communication problems.* Our decision to

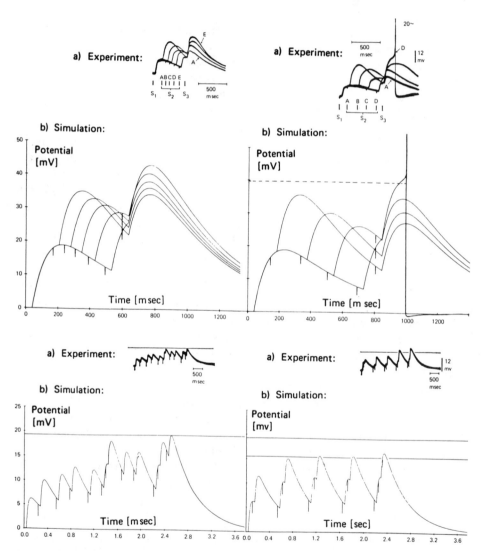

Figure A2. Comparison of experimental data from Aplysia Californica with our fitting, using the program of Fig. A1. The four sets of experiments reported are described in the text of this Appendix.

limit the parallelism of lCAP-1 and lCAP-2 to relatively few processors (less than 20) was based on the above conclusion. Finally, from our experiment it seemed reasonable to assume that "controlled complexity" could be achieved by connecting *small numbers* of *clusters of neurons* where the neurons within a given cluster are tightly and nearly uniformly connected; lCAP-1 and lCAP-2 are two such clusters.

References

1. As a sampling of this literature we give: a) "International Conference on Parallel Processing" August: (1981-84), Bellaire, MI.; (1985), St. Charles, ILL. b) R. W. Hockney, "Parallel Computers: Architecture, Programming and Algorithms," R. W. Hockney, C. R. Jesshope, Eds., Bristol, Adam Hilger, Ltd. (1981). c) "Parallel Processing Systems," J. Evans, Ed., Cambridge, NY, Cambridge Univ. Press (1982). d) Y. Wallach, "Alternating Sequential/Parallel Processing," Lecture Note in Computer Science, 124, Springer Verlag, Berlin, NY (1982).
2. a) E. Clementi, G. Corongiu, J. Detrich, S. Chin and L. Domingo, "Parallelism in Computational Chemistry". Int. J. Quantum Chem. Symp. **18**, 601 (1984); b) J. H. Detrich, G. Corongiu, and E. Clementi, "Monte Carlo Liquid Water Simulations With Four-Body Interactions", J. Chem. Phys. Lett. **112**, 426 (1984); c) G. Corongiu and J. H. Detrich, "Large Scale Scientific Applications in Chemistry and Physics on an Experimental Parallel Computer System," IBM Journal Res. and Dev., **29**, 422 (1985).
3. C.C.J. Roothaan "Improved Vector Functions and Linear Algebra for the FPS-164," Proceedings of the 1985 Array Conference, April 14-17, 1985, New Orleans, LA; R. Sonnenschein and J. Detrich, "Report on Vector Exponential Subroutines for FPS-164 Computers," IBM Technical Report **KGN-27** (November 15, 1985).
4. Virtual Machine/System Product System Programmer's Guide, Third Edition (Publication No. SC19-6203-2), International Business Machines Corp. (August, 1983).
5. FPS-164 Operating System Manual, **vols. 1-3** (Publication No. 860-7491-000B), Floating Point Systems, Inc. (January, 1983).
6. See the IBM Product description for MVS/XA - MVS/SP Version 2 JES2, IBM Product Number 5740-XC6.
7. a) S. Chin, and L. Domingo, "Parallel Computation on the loosely Coupled Array of Processors: Tools and Guidelines," IBM Technical Report **KGN-25**, September 15, 1985. b) D. L. Meck, "Parallelism in Executing FORTRAN Programs on the 308X: System Considerations and Application Examples," IBM Technical Report **POK-38** (April 2, 1984). For another set of FORTRAN-callable communications subroutines to support parallel execution on the IBM 308x under MVS, see IBM program offering DNL-5798, developed by P. R. Martin; the Program Description Operations Manual for this program offering is IBM publication number SB21-3124 (release date May 4, 1984). See also the IBM Product description of the VS FORTRAN Version 1 Release 4.1 Compiler which contains the description of the IBM Multi-Tasking Facility (MTF). IBM Product Number 5748-F03.

8. S. Chin, L. Domingo, A. Carnevali, R. Caltabiano, and J. Detrich, "Parallel Computation on the loosely Coupled Array of Processors: A Guide to the Pre-Compiler," IBM Technical Report KGN-42, November 25, 1985.

9. a). J. W. Mehl and E. Clementi, "IBM System/360 IBMOL-5 Program Quantum Mechanical Concepts and Algorithms," IBM Technical Report RJ #883 (June 22, 1971). b). R. Pavani and L. Gianolio, "IBMOL-6 Program - User's Guide," Ist. Ricerche G. Donegani, Novara-Italy, Technical Report DDC-771 (January 1977). c). E. Clementi, G. Corongiu, J. Detrich, S. Chin and L. Domingo, Int. J. Quantum Chem. Symp. **18**, 601 (1984).

10. a). R. S. Mulliken and C. C. J. Roothaan, Proc. Natl. Acad. Sci. USA **45**, 394 (1949). b). C. C. J. Roothaan, Rev. Mod. Phys. **23**, 69 (1951). c). C. C. J. Roothaan, Rev. Mod. Phys. **32**, 179 (1960).

11. N. Metropolis, A. W. Rosenbluth, M. N. Rosenbluth, A. H Teller and E. Teller, J. Chem Phys. **21**, 1078 (1953).

12. J. Detrich, G. Corongiu, and E. Clementi, Int. J. Quantum Chem. Symp. **18**, 701 (1984).

13. a) B. J. Alder, and T. E. Wainwright, J. Chem. Phys. **31**, 459 (1959). b) D. W. Wood in "Water-A Comprehensive Treatise," **vol.6**, pp. 279-409, Felix Franks, Ed., Plenun Press, NY, 1979.

14. H. L. Hguyen, H. Khanmohammadbaigi, and E. Clementi, J. Comp. Chem. **6**, 634 (1985).

15. a) F. A. Momany, R. F. McGuire, A. W. Burgess, and H. A. Scheraga, J. Phys. Chem. **79**, 2361 (1975). b G. Nemethy, and H. A. Scheraga, Q. Rev. Biophys. **10**, 239 (1977) c) G. Nemethy, M. S. Pottle, and H. A. Scheraga, J. Phys. Chem. **87**, 1833 (1983). d) H. A. Scheraga, Carlsberg Res. Commun. **49**, 1 (1984).

16. A. Capotondi, S. Chin, V. Sonnad, and E. Clementi, "Parallel Resolution of the Shallow Water Equations Using an Explicit Finite Difference Algorithm," to be published.

17. S. Chin, L. Domingo, G. Corongiu, and E. Clementi, "Hydrogen Bond Bridges in DNA Base Pairs," IBM Technical Report **KGN-21**, February 20, 1985.

18. J. Detrich, G. Corongiu, and E. Clementi, Chem. Phys. Lett. **112**, 426 (1984).

19. M. Wojcik and J. Detrich, "Dynamic Structure of Water," IBM Technical Report **KGN-29**, August 21, 1985.

20. K. Gibson, S. Chin, E. Clementi, and H. A. Scheraga, in preparation.

21. K. Gibson, S. Chin, M. Pincus, E. Clementi, and H. A. Scheraga, "Parallelism in Conformational Energy Calculations of Proteins," in Lecture Notes in Chemistry, M. Dupuis, Ed., (1986).

22. E. Clementi, G. C. Lie, L. Hannon, D. C. Rapaport, M. Wojcik, "Global *Ab Initio* Simulation: Study of a Liquid as an Example," this volume.

23. F. Carminati, R. Mount, H. Newman, and M. Pohl, CERN Technical Report L3-313/1984.

24. J. Gazdag and P. Squazzero, Geophysics **49**, 2 (1984); L. Domingo and E. Clementi, "Parallel Computation of Migration of Seismic Data on the LCAP," IBM Technical Report **KGN-17**, January 17, 1985.

25. P. E. Bowen-Jenkins and M. Dupuis, "Parallel Computation of Molecular Electronic Similarity," IBM Technical Report **KGN-46**, November 6, 1985.

26. K. Kaski, J. Nieminen, and J. D. Gunton, Phys. Rev. D **31**, 2998 (1984).

27. a) S. Fornili, D. P. Vercauteren and E. Clementi, Biochim. & Biophys. Acta **771**, 151 (1983); b) K. S. Kim and E. Clementi, J. Am. Chem. Soc. **107**, 5504 (1985); c) K. S. Kim, P. K. Swaminathan and E. Clementi, J. Phys. Chem. **83**, 2870 (1985).

28. W. L. Luken, N. Liang, R. Caltabiano, E. Clementi, R. Bacon, J. Warren, and W. Beausoleil, "Application of Parallel Processing to Molecular Modeling Graphics" (to be published).

29. J. Helin and J. Detrich, "Parallel Computation on the Loosely Coupled Array of Processors: Sheduler for the VM System", Technical Report **KGN-52** (in preparation, July 1986).

30. A. Gnudi, N. Liang, A. Carnevali, C.C.J. Roothaan, J. Detrich, and S. Chin "Interactive Pipelined Loop Constructor", (in preparation, August 1986).

Index

P

Q

R

S